Motor Innervation
of Muscle

THESLEFF

Motor Innervation of Muscle

Edited by

S. Thesleff

Department of Pharmacology,
University of Lund,
Sölvegatan 10, Lund, Sweden

1976

Academic Press

LONDON   NEW YORK   SAN FRANCISCO

A Subsidiary of Harcourt Brace Jovanovich, Publishers

ACADEMIC PRESS INC. (LONDON) LTD
24–28 Oval Road
London NW1

US edition published by
ACADEMIC PRESS INC.
111 Fifth Avenue,
New York, New York 10003

Library of Congress Catalog Card Number: 76 48387

ISBN: 012 6 85950 7

PRINTED IN GREAT BRITAIN BY
WILLIAM CLOWES & SONS, LIMITED
LONDON, BECCLES AND COLCHESTER

Contributors

S. G. Cull-Candy, *Department of Biophysics, University College London, London, England*

B. W. Fulpius, *Department of Biochemistry, Sciences II, University of Geneva, Switzerland*

E. Gutmann, *Institute of Physiology, Czechoslovak Academy of Sciences, Prague, Czechoslovakia*

Edith Heilbronn, *Department of Biochemistry, Research Institute of National Defence, Dept. 4., Sundbyberg, Sweden*

J. Heuser, *Department of Physiology, School of Medicine, University of California, San Francisco, U.S.A.*

B. Katz, *Department of Biophysics, University College London, London, England*

G. G. Lunt, *Department of Biochemistry, University of Bath, Bath, Somerset, England*

T. Lömo, *Institute of Neurophysiology, University of Oslo, Oslo, Norway*

L. G. Magazanik, *Sechenov Institute of Evolutionary Physiology and Biochemistry, Academy of Sciences of the U.S.S.R., Leningrad, U.S.S.R.*

R. Miledi, *Department of Biophysics, University College London, London, England*

R. Rahamimoff, *Department of Physiology, Hebrew University, Medical School, Jerusalem, Israel*

A. Takeuchi, *Department of Physiology, School of Medicine, Jutendo University, Hongo, Tokyo, Japan*

F. Vyskočil, *Institute of Physiology, Czechoslovak Academy of Sciences, Prague, Czechoslovakia*

Preface

Scientific advance usually reflects the introduction of new techniques or concepts. In the field of nerve-muscle interactions the appearance, a quarter of a century ago, of the intracellular recording technique and the formulation of the quantal release theory for the transmitter served as catalysers bringing the discipline a major step forward.

During the last few years I believe that we are again experiencing a similar sudden increase in knowledge. This time it is the introduction of techniques for the isolation, characterization and quantitation of end-plate receptors and the presence of sophisticated methods for studying the ultrastructure of the synapse and the molecular interactions between transmitter and receptor which are progressing. New approaches also seem fundamentally to change our outlook on long-term (trophic) interactions between nerve and muscle as well as on immune reactions at the end-plate. As a result of these developments an avalanche of scientific papers has appeared bringing an astounding and sometimes bewildering amount of new data.

My first reaction to this tremendous output of information was the feeling that this was the time to bring together distinguished scientists, active in the field, and let them, in their own words, describe and explain the research leading to present progress. Unfortunately such a conference is a difficult and highly costly task not easily accomplished in a world suffering from economic austerity. Fortunately, Academic Press came to my rescue by suggesting a more feasible alternative, *i.e.* that I be responsible for a *book* on "Motor Innervation of Muscle". This gave me a chance, as Editor, to invite some of the foremost scientists in the field and to ask them to present their work as separate chapters in such a treatise. The result you see in this volume. I have myself tremendously enjoyed the reading and have learned much from these contributions and my hope is that the reader will have the same experience.

I would like to express my sincere gratitude and special appreciation to the international group of distinguished scientists who made this treatise possible and for their willing collaboration, help, patience and understanding.

Stephen Thesleff
University of Lund, Sweden

August, 1976

Contents

3 *Morphology of Synaptic Vesicle Discharge and Reformation at the Frog Neuromuscular Junction.* J. HEUSER

1

Characterization, Isolation and Purification of Cholinergic Receptors

B. W. Fulpius

Dept. of Biochemistry, Sciences II, University of Geneva, Switzerland

I Introduction

Skeletal muscle fibres are electrically excitable, i.e. an impulse is generated when their cellular membrane is depolarized beyond a critical threshold. This impulse propagates along the whole muscle fibre and causes its contraction to occur. Under physiological conditions, the depolarization results from an indirect activation of the muscle fibre through the motor nerve. When an action potential reaches and depolarizes a motor nerve terminal, it triggers the secretion of roughly 500 000 molecules of acetylcholine into the synaptic cleft. These molecules diffuse through the cleft to the postsynaptic membrane where some are hydrolyzed at once by an enzyme, acetylcholinesterase, while others react in a reversible manner with recognition sites, the acetylcholine receptors (AChR). The interaction of acetylcholine molecules with AChRs is assumed to modify, in a still unknown manner, the properties of the postsynaptic membrane which thereby becomes transiently highly permeable to small cations.

In order to understand the sequence of events underlying the specific effect of acetylcholine molecules on the ionic conductances of the postsynaptic membrane, it is first necessary to characterize, at the molecular level, the individual membrane components of the so-called subsynaptic area, namely acetylcholinesterase, AChR and the cation pore. Among these three components, AChRs and the pore occupy a

1

key position because of their ability to transduce a chemical stimulus into an electrical signal.

The concept of AChR rests on the physiological response to acetylcholine. In this respect, AChR refers to an hypothetical macromolecule located in the postsynaptic membrane and having with acetylcholine molecules a transitory interaction connected, through many steps, to a measurable cellular response. If one wants to study directly this hypothetical molecule, the following two conditions are required.

1. A source of AChR quantitatively and qualitatively suitable for biochemical studies.

2. A specific reagent for the detection, *in vitro*, of AChR, because once the many different membrane components are dissociated, a molecule such as AChR can only be recognized through its specific interactions with certain ligands.

The first condition happens to be especially well satisfied by the electric organs of certain fishes. The embryological origin of these structures is similar to that of skeletal muscle. The amount of AChR present in these organs is estimated to be considerably higher than that present in innervated muscles or in the brain. In spite of numerous attempts, the limitations attached to the second condition have delayed the isolation and characterization of a molecule that exhibits the properties expected of AChR. In order to understand these limitations, one has first to consider the reversible interaction of a specific ligand A, with the receptor for that ligand R, according to the following scheme

$$R + A \rightleftharpoons RA, \qquad (1)$$

where RA is the ligand-receptor complex.

To be of valid use, a ligand should have a high affinity for the receptor. The affinity is characterized by the three following parameters.

1. The *forward rate constant*, k_1, which can be measured in a simple way by using the equation†

$$k_1 = \frac{0 \cdot 693}{(A)_i \cdot t_{\frac{1}{2}}}, \qquad (2)$$

† (2) is fully derived in the Appendix.

where $t\frac{1}{2}$ is the half-time of the association reaction between A and R and $(A)_i$ the concentration of A at $t = 0$.

2. The *reverse rate constant*, k_{-1}, that can be simply measured by using the equation

$$k_{-1} = \frac{0\cdot693}{t\frac{1}{2}},\tag{3}$$

where $t\frac{1}{2}$ is the half-time of the dissociation of the complex RA.

3. The *dissociation constant*, K_D, which can be calculated from the equation

$$K_D = \frac{k_{-1}}{k_1}.\tag{4}$$

According to equations (2) to (4), two ligands can have identical affinities for the same receptor as measured by their dissociation constants even if they differ in their respective forward and reverse rate constants. This is illustrated in the following theoretical numerical example.

Let us consider A_1 and A_2, two ligands present in solution at a concentration much greater than that of the receptor, for example $6\cdot93 \times 10^{-8}$M for a receptor concentration of 10^{-9}M. Their half-times of association with R are taken to be 1 and 100 s, respectively; their half-times of dissociation from ligand-receptor complexes 69·3 and 6930 s, respectively. According to equations (2) and (3), for A_1 and A_2 respectively, the k_1's are $1\cdot10^7$ M^{-1} s^{-1} and $1\cdot10^5$ M^{-1} s^{-1}, the k_{-1}'s 10^{-2} s^{-1} and 10^{-4} s^{-1}. From equation (4) it follows that the dissociation constants K_D's for A_1 and A_2 are identical, i.e. 10^{-9}M. This example shows that a complex RA having a k_1 of 10^7 M^{-1} s^{-1}, which is a low estimate, and a K_D of 10^{-9}M would exhibit a half-life of about one minute. In a binding assay based on the separation of RA from an excess of A by gel filtration, sedimentation on sucrose density gradient or extensive washing of a filter disk bearing RA, such a half-life is very short indeed and an extensive dissociation of RA would follow. This represents an absolute limitation to the use of a rapidly dissociating ligand such as dimethyltubocurarine in the kind of assay described above. It should be noticed that an increase in $[A]$ in order to prevent (RA) dissociation, according to equation (1), will be of no use because other molecules, R', having a low affinity for A will be detected as $R'A$ as shown in equation (5).

$$R + A \rightleftharpoons RA. \qquad (5)$$

with

$$R'$$
$$+$$
on the left, and
$$R'A$$

below, forming the scheme:

There is no way to distinguish between RA and $R'A$ with a simple assay. Actually the use of high concentrations of a rapidly dissociating ligand have brought erroneous results such as the identification of AChR as a mucopolysaccharide (for a review, see Hasson–Voloch, 1968). Obviously, a solution would consist in using equilibrium dialysis as a binding assay. For practical reasons, however, this method has not been used very often when a great number of fractions had to be assayed in the course of a purification procedure. From the preceding discussion, another solution would be to find ligands dissociating more slowly. Chang and Lee (1963) were the first to demonstrate that certain neurotoxins isolated from the venom of some snakes have a curare-like action and that these compounds bind almost irreversibly to AChR. The availability of these specific nicotinic compounds whose rate of dissociation is very favourable for assaying AChR activity has relieved the limitations attached to the use of rapidly dissociating ligands. Accordingly, these toxins are being used at the present time for detecting nicotinic cholinergic receptor molecules in almost every laboratory involved in this field.

II Methods used in the identification and purification of AChR

A ASSAY OF AChR ACTIVITY

The conditions required for the direct study of AChR molecules have been described in Section I. One of these is the ability to use an appropriate assay that rests on the detection of AChR by a slowly reversible ligand of high nicotinic specificity. As a consequence, most of the published assays are based on the binding, by AChR, of one of the snake venom neurotoxins, mainly cobra α-toxin and α-bungarotoxin. These assays differ, in other respects, only in the method used to separate the free toxin from toxin bound to AChR. They have been recently reviewed (Karlin, 1974). For assay purposes, the neurotoxins (small basic proteins of molecular weight about 8000 daltons) are radioactively labelled with either tritium or iodine.

We should consider in some detail the limitations associated with (a) the use of neurotoxins for AChR detection and (b) the labelling procedure. First, neurotoxins were selected as appropriate ligands though the best approach would be to measure the binding of acetylcholine. This choice was based on the kinetic properties and the nicotinic specificity of the neurotoxins. Consequently, whenever using toxins in the search for AChR, complementary acetylcholine binding studies should be undertaken to check the affinity of the putative AChR for the transmitter itself. Second, since the labelling procedure may be accompanied by a loss of specificity for the nicotinic AChR, the biological activity of the radioactive toxin derivative should always be checked. Actually, any reliable labelling procedure should satisfy the two following basic criteria. (a) The labelling reaction ought to be such that modified toxin molecules may be readily separated from unreacted ones. This is essential because of the small percentage of the molecules present which are labelled, only approximately 5% in the case of the iodination reaction (Berg *et al.*, 1972) and therefore the biological activity of the unfractionated reaction mix will not be representative of the labelled molecules; the only ones however to be taken into account in the AChR assay. (b) The modified toxin molecules, once purified, must still be biologically active. The following tests of biological activity have been used: ability to block neuromuscular transmission (Barnard *et al.*, 1971; Chang *et al.*, 1973; Almon *et al.*, 1974a), LD_{50} or LD_{100} doses on intact animals (Menez *et al.*, 1971; Cooper and Reich, 1972; Karlsson *et al.*, 1972a; Chang *et al.*, 1973; Eterovič and Bennett, 1974), serial dilutions with unlabelled toxins and measure of the binding of AChR (Berg *et al.*, 1972; Alper *et al.*, 1974; Ong and Brady, 1974; Devreotes and Fambrough, 1975; Fulpius *et al.*, 1975).

Different labelling procedures currently in use are summarized in Table I. Any method which does not permit separation of reacted toxin molecules from unreacted ones and the control of the biological activity should be considered with great caution.

1. In the method using the reduction with sodium (^{3}H)-borohydride of the pyridoxal phosphate derivatives of the main neurotoxin from the venom of the cobra *Naja naja siamensis*, mono-, di- and trisubstituted toxins were obtained (Cooper and Reich, 1972). Repeated chromatography of the reaction mixture on phosphocellulose gave a complete separation of the main and side-products (Fulpius *et al.*,

TABLE I

Radioactive labelling procedures used with various neurotoxines

Isotope	Method	Specific activity Ci/mmole	Further purification	Biological activity	References
^{125}I	Chloramin T	10–23			Potter (1974)
		225–450	+	unchanged	Alper et al. (1974)
		<1	+		Clark et al. (1972)
		26–57	+	unchanged	Devreotes and Fambrough (1975)
		68–76		60%	Berg et al. (1972)
		200–1500	+		Brockes and Hall (1975a)
	Lactoperoxidase	<1	+		Eldefrawi and Eldefrawi (1973a)
	KI nitrite	3	+	unchanged	Almon et al. (1974b)
	ICl	160–400	+	unchanged	Vogel et al. (1972)
^{3}H	ICl followed by catalytic reduction	14	+	60–70%	Menez et al. (1971)
		15	+	unchanged	Eterovic and Bennett (1974)
	Reduction of pyridoxal phosphate derivative	2	+	66%	Cooper and Reich (1972)
					Fulpius et al. (1975)
	Reductive formylation	2	+	75%	Ong and Brady (1974)
		6			Bosmann (1972)
	Acetylation	1	+	70%	Karlsson et al. (1972a)
		7		unchanged	Barnard et al. (1971)
				80%	Chang et al. (1973)

1975). Moreover, the measure of the dissociation constants for AChR of these different products showed that only one monosubstituted toxin displays a binding constant practically identical to that of the unreacted toxin. Thus only this species, once separated, can be validly used for detecting AChR activity. It has subsequently been used for competition studies on highly purified AChR (Maelicke *et al.*, 1976).

2. It has been shown by Chang *et al.* (1973) that acetylation of α-bungarotoxin with tritiated acetic anhydride leads to polyacetylated toxin molecules, some having a reduced specificity for nicotinic receptors. The use of a mixture of such molecules without further purification can lead to erroneous results, for example an overestimate of the binding capacity of a given tissue (Bosmann, 1972).

B CHOICE OF TISSUE

Tissues known to be rich either in acetylcholine or in enzymes playing a major role in acetylcholine metabolism (acetylcholinesterase and choline acetyltransferase) are likely to contain high concentrations of AChR. These are nervous tissues rich in cholinergic synapses but also non-neural tissues with a cholinergic innervation (skeletal, cardiac and smooth muscle) and even non-innervated tissues such as human placenta and red cells. Of special interest are the fish electric tissues whose concentration of acetylcholinesterase is very high. They may be regarded as modified skeletal muscles with a very rich innervation of cholinergic type. Among them, that of the electric eel, *Electrophorus electricus*, has relatively fewer synapses than that of the electric ray, *Torpedo*. These tissues are the richest sources known of AChR. Brain and skeletal muscle are also suitable, especially denervated mammalian muscle in which the total amount of AChR increases by a large factor within approximately two weeks following denervation. Estimates, by several authors, of the AChR content of different tissues are assembled in Table II. The essential parameter in the selection of the starting material is the binding capacity expressed in this Table in nmoles of ligand bound per kg of fresh tissue. The values reported show large variations and should be considered critically. For example, the use of an altered radioactive neurotoxin (Bosmann, 1972) or high concentrations of a reversible ligand (O'Brien and Gilmour, 1969) will favour the detection of non-specific material and thereby overestimate the specific binding

TABLE II

AChR content of different tissues

Tissue or source of electric organ	Binding capacity nmoles/kg	Ligand used
Electrophorus electricus		
Eldefrawi *et al.* (1971)	20–30	(^{3}H) muscarone
Raftery *et al.* (1971)	35–46	(^{125}I) α-bungarotoxin
Olsen *et al.* (1972)	53	(^{3}H) cobra toxin *Naja nigricollis*
Chang (1974)	83	(^{125}I) α-bungarotoxin
Potter (1974)	75	(^{125}I) α-bungarotoxin
Torpedo californica		
Salvaterra and Moore (1973)	850	(^{125}I) α-bungarotoxin
Schmidt and Raftery (1973)	750	(^{125}I) α-bungarotoxin
Michaelson *et al.* (1974)	1950	(^{125}I) α-bungarotoxin
Torpedo marmorata		
Miledi *et al.* (1971)	1090	(^{131}I) α-bungarotoxin
Cohen *et al.* (1972)	307	(^{3}H) cobra toxin *Naja nigricollis*
Eldefrawi and Eldefrawi (1973a)	1720	(^{3}H) acetylcholine
Narcine entemedor		
Schmidt and Raftery (1972)	900	(^{125}I) α-bungarotoxin
Mammalian skeletal muscle (a) Innervated		
Berg *et al.* (1972)	2·5[a]	(^{125}I) α-bungarotoxin
Almon *et al.* (1974b)	8·5	(^{125}I) α-bungarotoxin
Alper *et al.* (1974)	1	(^{125}I) α-bungarotoxin
Colquhoun *et al.* (1974)	3·8	(^{125}I) α-bungarotoxin
(b) Denervated		
Miledi and Potter (1971)	60	(^{131}I) α-bungarotoxin
Berg *et al.* (1972)	23	(^{125}I) α-bungarotoxin
Almon *et al.* (1974a)	128	(^{125}I) α-bungarotoxin
Alper *et al.* (1974)	53	(^{125}I) α-bungarotoxin
Colquhoun *et al.* (1974)	44	(^{125}I) α-bungarotoxin
Dolly and Barnard (1975)	80	(^{3}H) α-bungarotoxin
Nervous tissue (a) Mammalian brain		
O'Brien and Gilmour (1969)	18 000	(^{3}H) muscarone
Bosmann (1972)	17 500	(^{3}H) α-bungarotoxin
Salvaterra and Moore (1973)	1·5–3·4	(^{125}I) α-bungarotoxin
(b) Hog brain		
Moore and Loy (1972)	0·6	(^{125}I) α-bungarotoxin
Romine *et al.* (1974)	1[b]	—
(c) Housefly head		
Eldefrawi and O'Brien (1970)	924–3220	(^{3}H) muscarone
Clarke and Donnellan (1974)	310	(^{3}H) decamethonium

[a] 9·4 in neonatal muscle.　　　　[b] measured by protein content.

capacity. Values of AChR such as 1 μmole/kg *Torpedo* electric tissue, 50 nmoles/kg *Electrophorus* electric tissue, 1 nmole/kg brain or skeletal muscle, 50 nmoles/kg denervated muscle seem the most reliable.

C EXTRACTION AND SOLUBILIZATION OF AChR

According to Singer and Nicolson (1972), membrane proteins are divided into two kinds termed "peripheral" and "integral" depending on the means needed to separate the proteins from intact membranes. Whereas peripheral membrane proteins can be obtained in lipid free form by mild treatments which leave the lipid matrix intact, integral membrane proteins can only be detached by disrupting the membrane. Actually, some of them are extracted without loss of their biological activity by using mild detergents, which form soluble detergent-protein complexes. Many membrane proteins are known to have oligomeric structures. Depending on their function within the membrane, the non-covalent interactions between the monomeric peptides may be reversible or not. In the latter instance, the detergent used for solubilizing integral membrane proteins is unable to break these protein-protein inter-actions. In such cases, mild detergents such as Triton X-100 allow merely the extraction of complex structures whose polypeptide composition may be further studied by using stronger detergents such as sodium dodecylsulphate. It should be noticed however that the biological activity can be lost following this last treatment.

AChR is widely considered to be an integral membrane protein. This is based on the demonstration by several groups that toxin-AChR complex is extracted from membranes by mild detergent and that extracted AChR still binds toxin (Miledi and Potter, 1971; Miledi *et al.*, 1971; Berg *et al.*, 1972; Eldefrawi *et al.*, 1972; Meunier *et al.*, 1972a; Raftery *et al.*, 1972; Carroll *et al.*, 1973; Lindstrom and Patrick, 1974). In addition, pretreatment of membrane fragments with 1 M NaCl removes most of the acetylcholinesterase but none of the AChR (Karlin and Cowburn, 1973; Lindstrom and Patrick, 1974). Although there is not yet any information on the precise receptor conformation, one can visualize that one hydrophobic part of AChR penetrates deeply into the membrane whereas the portion which contains the acetylcholine binding site is located on the membrane surface and is exposed towards the synaptic cleft.

In solution, AChRs exist as AChR-detergent complexes. Obviously the presence of detergent can hinder further biochemical studies of this macromolecule but, for the time being, no one has convincingly purified AChR from usual sources with a different procedure. The presence of detergent, in the course of a purification procedure, gives rise to additional difficulties because of the following reasons. (a) Most of the non-ionic detergents such as Triton X-100 or Tween 80, absorb strongly at 280 nm where maximum absorption by proteins occurs. This makes it impossible to measure, in a solution, the protein content by a simple spectroscopic method. Modifications of the method described by Lowry *et al.* (1951) should be used (Dulley and Grieve, 1975). (b) Large amounts of detergent can be bound to AChR and consequently the molecular size of the purified AChR will depend on the kind of detergent used and its concentration. A typical feature of the AChR–Triton X-100 complex is a combination of a low sedimentation coefficient and a high Stoke's radius because the partial specific volume $\bar{v}$ is 0·73–0·74 for AChR and 0·99 for Triton X-100 (Meunier *et al.*, 1972b). (c) Only non-ionic detergents such as Triton X-100, Tween 80 or Brij 35 can be used during electrophoresis and electrofocusing. Anionic detergents such as sodium deoxycholate should be removed. (d) Removal of detergent is accompanied by a change in the physical state of AChR. Molecular species heavier than the one described by Meunier *et al.* (1972b) are observed. They correspond to higher aggregates of the basic functional unit (Carroll *et al.*, 1973; Edelstein *et al.*, 1975). Triton X-100 can be exchanged for sodium cholate by centrifugation of an AChR solution in Triton X-100 on sucrose gradients containing sodium cholate followed by gel filtration on Sephadex G75 (Meunier *et al.*, 1974). Tween 80 can be eliminated by extensive washing, with low ionic strength buffer containing no detergent, of an hydroxylapatite column (on which purified AChR was adsorbed) followed by elution at the appropriate ionic strength (Klett *et al.*, 1973). Triton X-100 can be eliminated by desorption from the affinity gel of AChR with a solution lacking any detergent, followed by extensive dialysis. But even when a medium completely free of detergent is used, there are still some molecules of Triton linked to the AChR protein (Eldefrawi *et al.*, 1975). In this last report the percentage of Triton remaining after purification of the AChR was calculated to be 0·0007% w/v. Two recent reports dealing with AChR solubilization without detergent deserve a special comment. In the first one, Aharonov *et al.* (1975) achieved the solubilization by extensive dialysis of an *Electrophorus electricus* membrane preparation against low ionic strength buffer followed by controlled tryptic digestion. Although the major component obtained differed in its physicochemical parameters from detergent-solubilized AChR preparations, this extract was successfully used to elicit anti AChR antibody formation. In the second one, Kato and Tattrie (1974) utilized an unusual source of AChR, namely the optic ganglion of the squid *Loligo opalescens*. Curiously, with this material, AChR was solubilized without detergent and exhibited properties very similar to those from fish electric organ AChR.

D PURIFICATION OF AChR

A logical approach to the extraction of AChR consists in using (a) intact cells, (b) subcellular fractions such as the microsacs formed from tissue homogenates and (c) proteins solubilized after treatment of the homogenate with detergents and comparing the physical and pharmacological properties of AChR in these three preparations. We will discuss only the methods used in the purification of AChR, once solubilized from any convenient source. These ones related to (a) and (b) have been extensively worked out by Changeux' group (see for a review Cohen and Changeux, 1975). We should however first comment on an additional method called affinity partitioning since it has been used recently with electroplax membranes from *Torpedo californica* partially purified by density gradient procedures (Flanagan *et al.*, 1975). This method combines the ease of aqueous polymer two-phase system methodology with the specificity reached in affinity chromatography, namely by coupling an appropriate cholinergic ligand to one of the polymers. It follows that a protein which binds the ligand will be recovered in the phase in which the polymer-ligand partitions predominate. In the system described for the purification of membrane fragments bearing AChR, the polymer-ligand used was a quaternary ammonium derivative coupled to polyethylene oxide. This ligand was used previously in affinity chromatrography for purifying AChR (Chang, 1974). In the conventional aqueous polymer two-phase system (dextran-polyethylene oxide), less than 1% of the membranes bearing AChR was found in the polyethylene-oxide rich top phase and about 65% in the dextran-rich bottom phase. When the polymer-ligand described was used, there was a striking increase in the concentration of

AChR containing membranes in the bottom phase. According to the authors, the purification achieved with a single partitioning step is comparable with that achieved by sucrose density gradient fractionation. It will be especially interesting to test this method in order to extract cells or cellular membrane fragments bearing surface receptors since other affinity methods have frequently been of only limited use because of the difficulties in eluting bound particulate substances from a solid matrix.

Most of the techniques used in protein purification have been tested for their usefulness in purifying AChR.

1. Gel permeation chromatography that separates macromolecules according to their molecular sizes has been widely used. Some purification was reported by filtration on Sepharose beads (either 4B or 6B) (Klett et al., 1973). The exclusion of large aggregates is one of the only advantages of this method together with the fact that it seems to improve the yield and purity of AChR obtained on subsequent affinity chromatography. Gel permeation chromatography, a classical tool in obtaining estimates of molecular weights by comparing elution rates of unknown and known macromolecules, is not very useful in this respect, however. As mentioned earlier, the presence of detergent in the eluant makes reference markers unreliable for estimating molecular weights because such markers are usually unrelated hydrophilic proteins whose interaction with detergent cannot be necessarily compared with that of AChR. Moreover this technique is slow and therefore may allow proteolysis to occur. In addition, it dilutes AChR very much which represents yet another disadvantage, since during concentration processes the concentration of detergent will increase too.

2. On the basis of the positive charge of cholinergic agonists and antagonists, AChR is considered to be an acidic macromolecule and consequently, anion-exchange chromatography (i.e. DEAE-cellulose) was one of the first techniques used in purification procedures. Actually its pI is close to 5·0 (see Section IIIA). The chromatographic behaviour of AChR on DEAE-cellulose depends on the stage of purity of the receptor. Crude material is eluted at quite low ionic strengths whereas highly purified AChR requires buffers of high ionic strength (Klett et al., 1973). Ion-exchange chromatography has been widely used, alone or in conjunction with affinity chromatography (Biesecker, 1973).

3. No one single method has proven as effective in a single step as affinity chromatography. This technique uses a specific cholinergic

ligand coupled covalently to a solid matrix in order to adsorb AChR selectively. Chromatography results in the retention on the matrix (i.e. Sepharose) of AChR and elution of the molecules devoid of affinity for the ligand. AChR is recovered by displacing it from the solid support with a solution containing high concentrations of a ligand of similar cholinergic specificity. This ligand is dialysed or washed out in a subsequent step. The selection of the ligand to be coupled to the insoluble matrix is as difficult as the selection of the ligand used for assaying AChR activity. On the one hand, ligands having a fast rate of dissociation from AChR should be avoided in order to prevent a loss of AChR during the washing step. On the other hand, ligands with a very slow rate of dissociation such as α-bungarotoxin should be avoided too, because AChR will then stick almost irreversibly to the Sepharose-bound α-bungarotoxin. In this respect cobra α-toxins have more favourable dissociating properties and have often been coupled to Sepharose. It does not help to elute AChR with high concentrations of cobra toxin since the AChR-toxin complex obtained in this manner is not easy to dissociate. Authors having selected cobra α-toxin as the ligand linked on Sepharose have eluted AChR with high concentrations of a rapidly dissociating cholinergic ligand, an agonist (carbamylcholine, decamethonium) or an antagonist (hexamethonium, flaxedil, benzoquinonium) (see Table III). All these ligands are easy to eliminate by dialyses in a subsequent step. On the basis of pharmacological experiments, agonists and antagonists do not seem to act in an identical manner. It is not clear that they are equally suitable as eluting agents. For example, Eldefrawi and Eldefrawi (1973a) have shown that desorption by carbamylcholine, a cholinergic agonist, yields an AChR that binds acetylcholine, decamethonium, nicotine, dimethyltubocurarine and α-bungarotoxin, whereas desorption from the same affinity ligand by benzoquinonium, a cholinergic antagonist, gives a protein that binds only α-bungarotoxin but not acetylcholine. It should therefore be kept in mind that proteins having the properties of an AChR molecule are not necessarily identical depending on the ligand used in the assay for AChR activity and in the affinity chromatography.

Ligands less specific towards AChR than neurotoxins have also been used for affinity chromatography (Table III). For example, NaCl displaces from insolubilized quaternary ligands a very heterogeneous population of proteins with some affinity for α-bungarotoxin (Schmidt and Raftery, 1972, 1973) whereas it is totally ineffective when working

TABLE III

Methods of affinity chromatography used on different tissues for AChR
purification

Tissue or source of electric organ	Ligand on the gel[a]	Eluting ligand
Electrophorus electricus		
Biesecker (1973)	$-NH\Phi\overset{+}{N}(CH_3)_3$	Decamethonium 1 mM
Karlin and Cowburn (1973)	$-CO\Phi\overset{+}{N}(CH_3)_3$	Carbamylcholine 50 mM
Klett *et al.* (1973)	α-toxin *Naja naja siamensis*	Hexamethonium 50 mM
Chang (1974)	$-NH\Phi CH_2\overset{+}{N}(CH_3)_3$	Bis Q[b] 3 μM
	$-NHC_5H_4\overset{+}{N}CH_3$	Bis Q 3 μM
Lindstrom and Patrick (1974)	α-toxin *Naja naja*	Benzoquinonium 1 mM
Meunier *et al.* (1974)	$-NH\Phi(-OCH_2CH_2\overset{+}{N}(C_2H_5)_3)_2$	Flaxedil 2·5 mM
Torpedo californica		
Schmidt and Raftery (1973)	$-NH(CH_2)_3\overset{+}{N}(CH_3)_3$	NaCl gradient up to 0·5 M
Torpedo marmorata		
Karlsson *et al.* (1972b)	α-toxin *Naja naja siamensis*	Carbamylcholine gradient up to 0·5 M
Eldefrawi and Eldefrawi (1973a)	α-toxin *Naja naja siamensis*	Carbamylcholine 1·0 M
Gordon *et al.* (1974)	α-toxin *Naja naja siamensis*	Carbamylcholine 0·5 M
Torpedo nobiliana		
Ong and Brady (1974)	α-toxin *Naja naja atra* *Naja naja siamensis* *Bungarus multicinctus*	Carbamylcholine 0·1 M
Narcine entemedor		
Schmidt and Raftery (1972)	$-NH(CH_2)_3\overset{+}{N}(CH_3)_3$	NaCl gradient up to 0·1 M
Mammalian skeletal muscle		
Brockes and Hall (1975a)	α-toxin *Naja naja siamensis*	Carbamylcholine 1·0 M
Dolly and Barnard (1975)	$-NH\Phi\overset{+}{N}(CH_3)_3$	Flaxedil 2·1 mM
Granato *et al.* (1976)	α-toxin *Naja naja siamensis*	Hexamethonium 0·1 M
Mouse and Hog brain		
Romine *et al.* (1974)	α-toxin *Naja naja*	Carbamylcholine 0·2 M or Hexamethonium 0·2 M

[a] The small ligands are linked to agarose by an arm; the toxin is linked directly.
[b] 3,3′-(Bis(α-trimethylammonium)methyl)azobenzene.

with Sepharose-bound toxin (Karlsson *et al.*, 1972b). Affinity chromatography, even repeated twice, does not yield a pure protein unless it is followed by centrifugation in 5–20% sucrose gradients (Meunier *et al.*, 1974), anion-exchange chromatography (Biesecker, 1973) or chromatography on hydroxylapatite (Klett *et al.*, 1973).

III   Characterization of AChR

A   CHEMICAL NATURE

The numerous data available on the solubilization and extensive purification from electric organ membrane fragments of a macromolecule having the properties of AChR indicate that AChR and acetylcholinesterase are indeed different macromolecules, for only extremely low levels of acetylcholinesterase activity have been detected in highly purified AChR preparations (Carroll *et al.*, 1973; Eldefrawi and Eldefrawi, 1973a; Klett *et al.*, 1973; Chang, 1974; Gordon *et al.*, 1974; Weill *et al.*, 1974). The various amounts reported (0·01–0·5‰ of AChR) may well be due to traces of acetylcholinesterase contaminating the AChR preparation, although it could also come from a weak esterase activity, inherent to the receptor macromolecule. A recent report by Dolly and Barnard (1975) favours the first hypothesis. According to these authors, there was no detectable acetylcholinesterase activity in highly purified preparations of AChR from denervated mammalian muscles, a tissue not as rich in acetylcholinesterase as fish electric organs.

Most of the recent data on AChR characterization confirm the proposal by Nachmansohn (1955) that AChR is a protein. In addition the possibility of the presence of carbohydrate in highly purified AChR preparations was approached by studying the interaction of the receptor with a variety of plant lectins known to bind sugars. (a) From preliminary studies obtained by Meunier *et al.* (1974), AChR purified from *Electrophorus electricus* seems to carry a carbohydrate moiety containing at least D-mannose and N-acetyl-D-galactosamine. (b) According to Raftery *et al.* (1975), AChR isolated from *Torpedo californica* contains approximately 5% neutral sugars, i.e. mannose, galactose and glucose, in the ratio 8:2:1. From amino acid analysis, N-acetyl-D-glucosamine was also identified as carbohydrate constituent. Moreover when the purified AChR was subjected to dodecylsulphate gel electrophoresis, the three bands observed (see Table IV) show a positive

TABLE IV

Purification of acetylcholine receptor from different sources

Tissue or source of electric organ	Specific activity μmole/g	Ligand used in the binding assay	Dodecylsulphate gel electrophoresis Number of bands	Subunit molecular weights
Electrophorus electricus				
Biesecker (1973)	4·5	(^{14}C) Cobra toxin *Naja naja atra*	2	(44 000; 50 000)
Karlin and Cowburn (1973)	3·0	(^{3}H) MBTA[a]	3	(41 000, 47 000, 53 000)
Klett *et al.* (1973)	6·6–11·0	(^{3}H) Cobra toxin *Naja naja siamensis*	1	(160 000) No reduction
Chang (1974)	5·0–6·5	(^{125}I) α-bungarotoxin	3	(37 000, 42 500, 49 000)
Lindstrom and Patrick (1974)	7·5	(^{125}I) Cobra toxin *Naja naja*	2	(42 000, 54 000)
Meunier *et al.* (1974)	5·5–6·5	(^{3}H) Cobra toxin *Naja nigricollis*	2	(43–45 000, 48–54 000)
Patrick *et al.* (1975)	3·6	(^{125}I) α-bungarotoxin	3	(48 000, 54 000, 60 000)
Torpedo californica				
Weill *et al.* (1974)	4·0	(^{3}H) MBTA[a]	4	(39 000, 48 000, 58 000, 64 000)
Raftery *et al.* (1975)	13·0	(^{125}I) α-bungarotoxin	3	(40 000, 50 000, 65 000)
Torpedo marmorata				
Karlsson *et al.* (1972b) and Heilbronn *et al.* (1974)	7·1	(^{3}H) Cobra toxin *Naja naja siamensis*	2	(38 000, 42 000)
Eldefrawi and Eldefrawi (1973a)	7·8–10·4	(^{3}H) acetylcholine	1	(46 000)
Gordon *et al.* (1974)	5	(^{125}I) α-bungarotoxin	2	(37 000, 49 000)
Torpedo nobiliana				
Ong and Brady (1974)	12·5	(^{3}H) Cobra toxin *Naja naja siamensis*	4	(33 500, 35 500, 38 500 43 500)
Narcine entemedor				
Schmidt and Raftery (1972)	2·7	(^{125}I) α-bungarotoxin	3	(28 000, 38 000, 45 000)
Mammalian skeletal muscle				
Dolly and Barnard (1975)	3·5–6·0	(^{3}H) α-bungarotoxin	—	

[a] 4-(N-maleimido)benzyltri(^{3}H)methylammonium iodide.

reaction for carbohydrates. There is no data available on the role of sugar, if any, involved in the toxin binding. (c) The total sugar content of AChR purified from *Torpedo marmorata* by Heilbronn *et al.* (1974) was estimated at 5·4%, with mannose as the main carbohydrate (48%) followed by glucose (30%) and *N*-acetyl-D-glucosamine (16%). According to these authors, we should be careful in the interpretation of these low quantities of sugars since cellulose and agarose are used in the purification procedure. (d) Eldefrawi and Eldefrawi (1973a) found one mole of carbohydrate per 10 000 daltons of AChR protein purified from *Torpedo marmorata*, but they found the same concentration of carbohydrates in a blank consisting of 1% Triton in Ringer as well as in a bovine serum albumin preparation, both subjected to the same affinity procedure as AChR. (e) Working with *Torpedo nobiliana* as the source of AChR, Moore *et al.* (1974) reported that in the highly purified material glucosamine represented 3·8% of the protein. (f) Finally, the glycoprotein nature of AChR was also demonstrated in a preparation purified from mammalian muscle. Data from Brockes and Hall (1975b) suggest that concanavalin A binds to α-D-glucopyranosyl or α-D-mannopyranosyl-like residues present on both junctional and extra-junctional AChR purified from innervated and denervated rat diaphragm muscle.

The protein nature of the AChR binding site was further demonstrated in experiments using proteolytic enzymes. Klett *et al.* (1973) reported that the toxin binding capacity of AChR purified from *Electrophorus electricus* is lost after incubation with trypsin. AChR u.v. absorption spectrum is that of a typical protein, with peak absorption at 280 nm. A shoulder at 290 nm and the fairly strong absorption at 295 nm and higher wavelengths suggest the presence of tryptophan residues (Klett *et al.*, 1973; Meunier *et al.*, 1974; Heilbronn *et al.*, 1974; Moore *et al.*, 1974). Several groups reported amino acid analyses of AChR (Eldefrawi and Eldefrawi 1973a; Klett *et al.*, 1973; Heilbronn and Mattson, 1974; Meunier *et al.*, 1974; Michaelson *et al.*, 1974; Moore *et al.*, 1974; Eldefrawi *et al.*, 1975). Most of them confirm the presence of tryptophan which was questioned in an early report (Klett *et al.*, 1973). All of them show a relatively high content of proline, of aromatic and hydrophobic amino acids, a basic amino acid content of 11–12% and an acidic amino acid content of 21–23%. According to Capaldi and Vanderkooi (1972), data on the percentage of the polar residue (Asp, Asn, Glu, Gln, Lys, Ser, Arg, Thr, His) allow an estimation to be made of the protein polarity and hydrophobicity. Such parameters are fairly interesting to know for a molecule supposed to be deeply anchored within the membrane such as AChR. These last authors showed that most of the soluble proteins have polarities of $47 \pm 6\%$ whereas membrane proteins values are inferior to 40%. Meunier *et al.* (1973) and Moore *et al.* (1974) reported values of 47 and 46% for AChR purified from *Electrophorus electricus* and *Torpedo nobiliana* respectively. These results well above 40% contrast with the known hydrophobic behaviour of AChR. They should, however, be considered with caution because the amino acid analysis represents the average composition of a mixture of polypeptide chains (at least two, see Table IV), one possibly involved in the binding of the toxin, the other(s) interacting perhaps with the hydrophobic environment. Moreover the physicochemical parameter of protein polarity has been recently challenged by Barrantes (1975) when applied to membrane proteins. This author questions the validity of methods in which an estimate of the protein polarity and hydrophobicity is given by emphasizing the hydrophilic and hydrophobic moieties without taking into account their relative proportion. Barrantes proposes instead a statistical technique, i.e. discriminant function analysis. This method uses simultaneously the optimum combination of values derived from the hydrophilic and hydrophobic moieties of membrane proteins in order to distinguish between the integral and peripheral ones. According to his results, AChR is a typical integral protein.

The homogeneity of purified AChR protein and AChR-toxin complex has been assayed by isoelectric focusing. ^{125}I-α-bungarotoxin-AChR complex from a crude soluble fraction of *Electrophorus electricus* was electro-focused in the presence of a non-ionic detergent (Raftery *et al.*, 1971). A major radioactive component of pI 5·15 was obtained. With AChR purified from the same source, Biesecker (1973) obtained an almost identical value (5·2) with the toxin-receptor complex and a slightly more acidic one (4·7) with the receptor alone. Similar results were obtained with AChR purified from *Torpedo marmorata* and *californica*: 4·8 and 4·9 respectively (Eldefrawi and Eldefrawi, 1973a; Raftery *et al.*, 1973; Heilbronn *et al.*, 1974). Isoelectric focusing allows the separation of proteins differing in their isoelectric points by only 0·01 of a pH unit. This characteristic was exploited by Brockes *et al.* (1975b) in order to test a possible difference in the chemical nature of junctional and extrajunctional AChR purified from normal and denervated rat diaphragm muscle. These authors reported pI values of

5·3 and 5·1 for the electrofocalization of toxin-receptor complex from junctional and extrajunctional origin respectively. According to them, such results are consistent with a structural difference between the major species present in junctional and extrajunctional receptor preparations. AChR, because of its low pI value, is exposed during electrofocusing to rather extreme pH conditions. As its toxin binding properties are lost after exposure to pH values below pH 6 (Klett *et al.*, 1973), it follows that electrofocusing cannot be used as a preparative tool to purify AChR.

Klett *et al.* (1973) reported that the most highly purified AChR fraction from *Electrophorus electricus* is free of detectable lipid phosphorus, i.e. less than one atom of phosphorus per mole of (^{3}H)-neurotoxin binding activity. In this context, one should discuss the results obtained by De Robertis and his coworkers (see for a review De Robertis, 1971, 1973). Using organic solvents, these authors reported the extraction from nervous tissue and fish electric organs of hydrophobic lipoproteins having the characteristics of AChR. According to them, the proteolipid isolated from the electric organ of *Electrophorus electricus* and further chromatographed on a lipophilic Sephadex gel (LH 20) binds acetylcholine, hexamethonium and ^{131}I-α-bungarotoxin with K_i values in the range 10^{-6}–10^{-7}M. The value, which is much higher for α-bungarotoxin than the one reported by most authors (10^{-9}–10^{-10}M) using cellular fragments or aqueous extracts, expresses probably a non-specific binding. This can explain why De Robertis (1973) found many more receptor molecules per muscle fibre ($8\cdot4 \times 10^9$) than Barnard *et al.* (1971) per endplate of the same material (3×10^7). In addition, it has been shown by Levinson and Keynes (1972) that peak fractions from Sephadex LH 20 exhibiting acetylcholine binding activity could also be obtained when protein-free or receptor-free extracts were used. These last results together with the fact that organic solvents are expected to denature proteins have brought strong criticism towards De Robertis' studies, in particular the utilization of organic solvents in the extraction procedure (Cuatracasas, 1974; Karlin, 1974). Actually chloroformmethanol does not extract affinity labelled AChR (Karlin, 1973) nor toxin-AChR complexes (Potter, 1973).

B MOLECULAR SIZE

The specific activities for highly purified AChR (Table IV) range from 3 to 13 μmoles ligand bound per g protein, corresponding to weights of

77 000–333 000 daltons per site. More direct measurements of AChR molecular weight were also attempted by Sepharose chromatography in presence of detergents. Sharp and symmetrical peaks for AChR and AChR–toxin complex were reported with an apparent Stoke's radius of 7 nm corresponding to that of a globular protein having a molecular weight of about 54 000 daltons (Raftery et al., 1971; Meunier et al., 1972a; Fulpius et al., 1972; Biesecker, 1973). These findings are in disagreement with values obtained from density gradient centrifugation ($s = 9$ for AChR and 9·5 for AChR-toxin complex) and corresponding to globular proteins having a molecular weight of about 250 000 (Schmidt and Raftery, 1973; Meunier et al., 1974; Lindstrom and Patrick, 1974). As discussed earlier, this discrepancy is probably due to the binding of large amounts of detergent to the AChR protein resulting in an apparent decrease in AChR density. Of course asymmetry (Raftery et al., 1971) and abnormal hydration of the complex might also contribute to this anomaly. According to Meunier et al. (1972b), Triton X-100 (average molecular weight of 636) contributes up to 21% (160–170 molecules) of the total mass of AChR–Triton complex purified from Electrophorus electricus. Consequently, these authors corrected the apparent molecular weight (calculated from gel filtration experiments) of AChR without Triton from 470 000 to 360 000 daltons. Along these lines, Martinez-Carrion et al. (1975) reported that as much as 45% of the dry weight of AChR from Torpedo californica may be contributed by Triton X-100. They postulated that the detergent is bound to the hydrophobic domain of monomeric AChR as several smaller weight micellar aggregates (c. 70 000). This value corresponds to a Triton micelle aggregation number of 120. In addition, these authors determined AChR molecular weight by membrane osmometry in detergent solutions. According to the data obtained with this technique, the receptor is a globular protein of 270 000 ± 30 000 daltons. Sedimentation equilibrium experiments were performed by Carroll et al. (1973). Their measurements gave minimum molecular weights of 500 000 to 1 000 000 daltons. Such high values were obtained in solutions containing low concentrations of detergent. Subsequent addition of Triton (0·05% w/v) resulted in a marked reduction of the sedimentation coefficient. Other sedimentation velocity measurements gave somewhat similar results (Edelstein et al., 1975). With 0·113 mg residual Triton X-100 bound per mg AChR protein purified from Torpedo electric organ, these authors reported an apparent molecular

weight of 510 000 for AChR from *Torpedo californica* and 665 000 for *Torpedo marmorata*. The sedimentation pattern showed a unit with an apparent molecular weight of 330 000 as well as higher aggregates. Upon addition of Triton X-100 beyond 0·1%, the 330 000 unit became the principal component. From this last result it follows that the purified AChR might exist in multiples of a 330 000 unit. The question of knowing whether *in situ* AChR is the 330 000 unit or a higher aggregate is still unanswered. Similar values, although somewhat lower, were obtained by an entirely different technique: dodecylsulphate polyacrylamide gel electrophoresis performed after cross-linking of AChR or AChR-toxin complex. When suberimidate was used and an extensive crosslinking achieved (Hucho and Changeux, 1973), a single band was detected on the electrophoresis. Its molecular weight was estimated to be 230 000 ± 15 000. Biesecker (1973), using glutaraldehyde as crosslinking agent of the toxin–AChR complex obtained also a single component, at a slightly higher molecular weight (260 000 ± 25 000).

Various arguments have been used to explain why these estimates are lower than those obtained with the other methods described earlier. According to Hucho and Changeux (1973), such a discrepancy could be due to the fact that dodecylsulphate polyacrylamide gel electrophoresis of integral membrane proteins is known to give molecular weights consistently too low by a factor of 20% when the standards used are hydrosoluble proteins. By crosslinking toxin-receptor complex in a different manner, Gordon *et al.* (1974) obtained five bands corresponding to molecular weights of 191 000, 167 000, 130 000, 82 000 and 53 000. These authors compared these bands with those corresponding to molecular weights of 39 000, 49 000, 74 000, 93 000 and 148 000 obtained from purified AChR run on a dodecylsulphate polyacrylamide gel electrophoresis under denaturing and reductive conditions. They concluded that only one of the two low molecular subunits of AChR binds toxin. This confirms previous data by Biesecker (1973) who reported that only the subunit of molecular weight 44 000 binds toxin. Affinity labelling studies using (^{3}H)-MBTA† as a ligand provides similar evidence, i.e. a single polypeptide subunit of approximately 40 000 daltons was specifically labelled (Weill *et al.*, 1974). As seen in Table IV, most authors report one band with a molecular weight near 42 000 daltons, though one to three higher molecular weight compo-

† 4-(N-maleimido) benzyltri (^{3}H) methylammonium iodide.

nents (47 000–65 000) have been described. It is not yet clear whether or not there are two different polypeptide chains with different functions, one of them bearing the toxin binding site. The low molecular weight unit reported could well represent a degradative product caused by proteolysis. Thus, Patrick *et al.* (1975) report the complete absence of any component with an apparent molecular weight lower than 48 000 daltons when AChR was purified from *Electrophorus electricus* in the presence of the serine protease inhibitor phenyl-methyl-sulphonyl-fluoride. The 42 000 daltons component began to appear with limited digestion of the preparation with trypsin. In any case, the molecular weight of all the subunits (about 40 000–60 000) is smaller than that calculated from specific activities (77 000–100 000). Based on such non definitive numbers, models have been proposed by Raftery *et al.* (1975) and Edelstein *et al.* (1975). According to these latter authors, AChR consists of a protomer of about 100 000 daltons made of two polypeptide chains (for example 46 000 and 54 000), one of which binds the toxin. Within the membrane, the receptor is a hexamer (molecular weight 660 000) that can be split in half by Triton X-100. Such a model is favoured by electron micrographs (Cartaud *et al.*, 1973; Nickel *et al.*, 1973; Meunier *et al.*, 1974) showing doughnut-shaped particles, each made up of three to six subunits.

C BINDING OF CHOLINERGIC LIGANDS

We have noticed previously that AChR was not always assayed with the same nicotinic cholinergic ligand (Table IV) and discussed the reasons for specific choices such as α-bungarotoxin and acetylcholine. In addition, we have pointed out that in the affinity step of the purification procedure, there was no uniformity in the selection of the ligand, neither in the gel, nor in the eluting buffer (Table III). One should remember that the classification of the nicotinic cholinergic ligands has been made according to anatomical and physiological criteria. There is therefore no way, at least for the time being, of deciding whether or not these different nicotinic cholinergic ligands act similarly at the molecular level and whether or not the selection of one ligand or another for identifying AChR is of critical importance. In any case binding studies of a variety of ligands must be performed for two obvious reasons: (a) to see if the solubilization and the purification of AChR results in any major change of the binding properties of the receptor; (b) to compare, at the molecular level, the properties of several cholinergic ligands in order to provide, for example, a molecular basis for the classical agonist–antagonist distinction.

There is not enough space, within this chapter, to describe and analyse the large amount of data published on this subject. The reader should refer to the articles published by Eldefrawi and Eldefrawi (1973a), Meunier et al. (1974), Raftery et al. (1975) and Maelicke et al. (1976). We shall only discuss here whether α-bungarotoxin and acetylcholine binding sites are identical or not. This question is of importance in order to know whether the material purified by several authors using ^{125}I-α-bungarotoxin is an α-bungarotoxin-receptor, an acetylcholine-receptor or both. We will discuss data coming from three different laboratories.

1. According to the work performed by Raftery and his coworkers on AChR purified from *Torpedo californica*, there is clearly a nonequivalence of α-bungarotoxin and acetylcholine (as well as other cholinergic ligands) binding sites (Martinez-Carrion and Raftery, 1973; Moody et al., 1973; Michaelson et al., 1974; Raftery et al., 1975). The number of acetylcholine binding sites represents 50% of the number of α-bungarotoxin binding sites. Acetylcholine binds to one single site with a K_D of $2·3 \times 10^{-6}$M. This binding is non-cooperative in nature. It is in strict competition with the binding of another agonist, carbamylcholine.

2. Some of the preceding results appear in contradiction with those published by Eldefrawi's group who reported two types of acetylcholine binding sites on AChR purified from *Torpedo marmorata* with K_D's of $1·4 \times 10^{-9}$M and $2·2 \times 10^{-7}$M as well as negative cooperativity and autoinhibition of acetylcholine binding at concentrations above 10^{-6}M (Eldefrawi et al., 1972; Eldefrawi and Eldefrawi, 1973a). In another report, these authors described also positive cooperativity in the binding of acetylcholine to AChR at very low concentrations of this agonist (Eldefrawi and Eldefrawi, 1973b). Finally, using *Torpedo californica* as a source of AChR, they showed that AChR binds as much ^{3}H-acetylcholine as ^{3}H-α-bungarotoxin, i.e. 10 nmole/mg protein (Eldefrawi et al., 1975).

3. Still different conclusions were reached by Reich and his coworkers using an AChR preparation purified from *Electrophorus electricus*. From competition studies between (^{3}H)-cobra-toxin (*Naja naja siamensis*) and acetylcholine, they reported binding patterns for acetylcholine similar to those observed for other agonists with the

exception that the number of acetylcholine and toxin sites was the same whereas most other cholinergic ligands displaced two toxin molecules for each ligand bound (Maelicke *et al.*, 1976). The characteristics of these binding patterns were of an anticooperative type and there was mutual exclusive competition between acetylcholine and toxin, although this need not imply that acetylcholine and toxin sites are fully overlapping.

Obviously it is difficult to compare these three sets of results since they were obtained by different methods and with AChR preparations purified from different sources. Our purpose in reporting them in detail was not to explain the discrepancies but to underline that toxin and acetylcholine sites are most probably not completely overlapping. The differences mentioned above suggest "caution in the accepting of the specific activities of purified AChR based on toxin binding without protection controls" (Karlin, 1974).

IV   Concluding remarks

The nicotinic AChR is the first pharmacological receptor to have been isolated and purified. This has been made possible by using detergent and affinity chromatography. AChR is most probably a glyco-protein whose precise molecular weight and exact subunit composition are still under intensive study.

These recent progresses should allow, in the near future, a better understanding of the sequence of events underlying the specific effect of acetylcholine molecules on the ionic conductance of the postsynaptic membrane. As the presently used purification means are selective for the AChR recognition site, there is still no way of knowing whether the ionophore is a separated entity or represents an intrinsic function of AChR.

One way of solving this question will be to reconstitute with purified AChR and other well characterized membrane components, a membrane in which the specificity for cations and cholinergic ligands can be monitored electrically.

V   Acknowledgements

The preparation of this chapter was supported by a grant number 3. 072-73 of the Swiss National Fund for Scientific Research. I would like to thank

Drs. J. J. Dreifuss and A. Tartakoff for their most useful comments in reading the manuscript.

VI Appendix

Equation (2) is derived in the following way:

$$\frac{d(AR)}{dt} = k_1(A)\,(R) - k_{-1}\,(AR), \tag{a}$$

$$(R) = (R)_i - (AR)\Big] \tag{b}$$

$$(A) = (A)_i - (AR)\Big\} \text{where i means initial.} \tag{c}$$

By substituting (b) and (c) into (a)

$$\frac{d(AR)}{dt} = k_1[(A)_i - (AR)]\,[(R)_i - (AR)] - k_{-1}\,(AR).$$

(A) being present in excess, $(A)_i > (AR)$,

$$\frac{d(AR)}{dt} = k_1\,[(A)_i]\,[(R)_i - (AR)] - k_{-1}(AR)$$

$$= k_1(A)_i(R)_i - k_1(A)_i(AR) - k_{-1}(AR)$$

$$= k_1(A)_i(R)_i - [k_1(A)_i + k_{-1}]\cdot(AR).$$

$$\frac{d(AR)}{k_1(A)_i(R)_i - [k_1(A)_i + k_{-1}]\cdot(AR)} = dt. \tag{d}$$

Equation (d) is of the form $\dfrac{dx}{a + bx}$ where $b = -\,[k_1(A)_i + k_{-1}]$.

By integration, we have $\displaystyle\int \frac{dx}{a + bx} = \frac{1}{b}\ln\,(a + bx)$.

Similarly, from (d)

$$-\frac{1}{[k_1(A)_i + k_{-1}]}\,\ln\{k_1(A)_i(R)_i - [k_1(A)_i + k_{-1}]\cdot(AR)\}_0^t = t,$$

$$\ln\{k_1(A)_i(R)_i - [k_1(A)_i + k_{-1}]\cdot(AR)\} - \ln\,k_1(A)_i(R)_i = \\ -\,[k_1(A)_i + k_{-1}]t,$$

$$\frac{k_1(A)_i(R)_i - [k_1(A)_i + k_{-1}]\cdot(AR)}{k_1(A)_i(R)_i} = e^{[k_1(A)_i + k_{-1}]t},$$

$$1 - \frac{[k_1(A)_i + k_{-1}].(AR)}{k_1(A)_i(R)_i} = e^{-[k_1(A)_i + k_{-1}]t},$$

$$AR = \frac{k_1(A)_i(R)_i}{k_1(A)_i + k_{-1}} (1 - e^{-[k_1(A)_i + k_{-1}]t}). \tag{e}$$

From (e)

for $t = 0$
$$(AR)_0 = 0,$$

for $t = \infty$
$$(AR)_\infty = \frac{k_1(A)_i(R)_i}{k_1(A)_i + k_{-1}}. \tag{f}$$

From equations (f) and (e)

$$(AR)_\infty - (AR) = \frac{k_1(A)_i(R)_i}{k_1(A)_i + k_{-1}} - \frac{k_1(A)_i(R)_i}{k_1(A)_i + k_{-1}} (1 - e^{-[k_1(A)_i + k_{-1}]t}),$$

$$= \frac{k_1(A)_i}{k_1(A)_i + k_{-1}} [1 - (1 - e^{-[k_1(A)_i + k_{-1}]t})],$$

$$(AR)_\infty - (AR) = \frac{k_1(A)_i(R)_i}{k_1(A)_i + k_{-1}} \cdot e^{-[k_1(A)_i + k_{-1}]t}, \tag{9}$$

$$\ln[(AR)_\infty - (AR)] = -[k_1(A)_i + k_{-1}]t + \ln \frac{k_1(A)_i(R)_i}{k_1(A)_i + k_{-1}}.$$

From equation (f)

$$\ln[(AR)_\infty - (AR)] - \ln(AR)_\infty = -[k_1(A)_i + k_{-1}]t,$$

$$\ln\left[1 - \frac{(AR)}{(AR)_\infty}\right] = -[k_1(A)_i + k_{-1}]t. \tag{h}$$

At time $t = \frac{1}{2}$
$$\frac{(AR)}{(AR)_\infty} = \frac{1}{2},$$

$$\ln \frac{1}{2} = -[k_1(A)_i + k_{-1}] t\frac{1}{2} = 2.303 . \log \frac{1}{2} = -0.693.$$

As $k_1(A)_i > k_{-1}$

$$\therefore k_1 = \frac{0.693}{(A)_i . t\frac{1}{2}}. \tag{2}$$

VII References

Aharonov, A., Kalderon, N., Silman, I. and Fuchs, S. (1975). *Immunochemistry* 12, 765–771.

Almon, R. R., Andrew, C. G. and Appel, S. H. (1974a). *Science* 186, 55–57.

Almon, R. R., Andrew, C. G. and Appel, S. H. (1974b). *Biochemistry* **13**, 5522–5528.
Alper, R., Lowy, J. and Schmidt, J. (1974). *FEBS Lett.* **48**, 130–132.
Barnard, E. A., Wieckowski, J. and Chiu, T. H. (1971). *Nature Lond.* **234**, 207–209.
Barrantes, F. J. (1975). *Biochem. Biophys. Res. Commun.* **62**, 407–414.
Berg, D. K., Kelly, R. B., Sargent, P. B., Williamson, P. and Hall, Z. W. (1972). *Proc. Nat. Acad. Sci. U.S.A.* **69**, 147–151.
Biesecker, G. (1973). *Biochemistry* **12**, 4403–4409.
Bosmann, H. B. (1972). *J. Biol. Chem.* **247**, 130–145.
Brockes, J. P. and Hall, Z. W. (1975a). *Biochemistry* **14**, 2092–2099.
Brockes, J. P. and Hall, Z. W. (1975b). *Biochemistry* **14**, 2100–2106.
Capaldi, R. A. and Vanderkooi, G. (1972). *Proc. Nat. Acad. Sci. U.S.A.* **69**, 930–932.
Carroll, R. C., Eldefrawi, M. E. and Edelstein, S. J. (1973). *Biochem. Biophys. Res. Commun.* **55**, 864–872.
Cartaud, J. Benedetti, E. L., Cohen, J. B., Meunier, J.-C. and Changeux, J. P. (1973). *FEBS Lett.* **33**, 109–113.
Chang, C. C. and Lee, C. Y. (1963). *Arch. Int. Pharmacodyn.* **144**, 241–257.
Chang, C. C., Chen, T. F. and Chuang, S. T. (1973). *Brit. J. Pharmacol.* **47**, 147–160.
Chang, H. W. (1974). *Proc. Nat. Acad. Sci. U.S.A.* **71**, 2113–2117.
Clark, D. G., Macmurchie, D. D., Elliott, E., Wolcott, R. G., Landel, A. M. and Raftery, M. A. (1972). *Biochemistry* **11**, 1663–1668.
Clarke, B. S. and Donnellan, J. F. (1974). *Biochem. Soc. Trans.* **2**, 1373–1374.
Cohen, J. B., Weber, M., Huchet, M. and Changeux, J.-P. (1972). *FEBS Lett.* **26**, 43–47.
Cohen, J. B. and Changeux, J.-P. (1975). *Ann. Rev. Pharmacol.* **15**, 83–103.
Colquhoun, D., Rang, H. P. and Ritchie, J. M. (1974). *J. Physiol.* **240**, 199–226.
Cooper, D. and Reich, E. (1972). *J. Biol. Chem.* **247**, 3008–3013.
Cuatrecasas, P. (1974). *Ann. Rev. Biochem.* **43**, 169–214.
De Robertis, E. (1971). *Science* **171**, 963–971.
De Robertis, E. (1973). *In* "Drug Receptors" (H. P. Rang, ed.), pp. 257–272. MacMillan, London.
Devreotes, P. N. and Fambrough, D. M. (1975). *J. Cell Biol.* **65**, 335–358.
Dolly, J. O. and Barnard, E. A. (1975). *FEBS Lett.* **57**, 267–271.
Dulley, J. R. and Grieve, P. A. (1975). *Anal. Biochem.* **64**, 136–141.
Edelstein, S. J., Beyer, W. B., Eldefrawi, A. T. and Eldefrawi, M. E. (1975). *J. Biol. Chem.* **250**, 6101–6106.
Eldefrawi, A. T. and O'Brien, R. D. (1970). *J. Neurochem.* **17**, 1287–1293.
Eldefrawi, M. E., Eldefrawi, A. T. and O'Brien, R. D. (1971). *Proc. Nat. Acad. Sci. U.S.A.* **68**, 1047–1050.
Eldefrawi, M. E., Eldefrawi, A. T., Seifert, S. and O'Brien, R. D. (1972). *Arch. Biochem. Biophys.* **150**, 210–218.
Eldefrawi, M. E. and Eldefrawi, A. T. (1973a). *Arch. Biochem. Biophys.* **159**, 362–373.
Eldefrawi, M. E. and Eldefrawi, A. T. (1973b). *Biochem. Pharmacol.* **22**, 3145–3150.
Eldefrawi, M. E., Eldefrawi, A. T. and Wilson, D. B. (1975). *Biochemistry* **14**, 4304–4310.
Eterović, V. A. and Bennett, E. L. (1974). *Biochim. Biophys. Acta*, **362**, 346–355.
Flanagan, S. D., Taylor, P. and Barondes, S. H. (1975). *Nature Lond.* **254**, 441–443.

Fulpius, B. W., Cha, S., Klett, R. and Reich, E. (1972). *FEBS Lett.* **24**, 323–326.

Fulpius, B. W., Maelicke, A., Klett, R. and Reich, E. (1975). *In* "Cholinergic Mechanisms" (P. G. Waser, ed.), pp. 375–380. Raven Press, New York.

Gordon, A., Bandini, G. and Hucho, F. (1974). *FEBS Lett.* **47**, 204–208.

Granato, D., Fulpius, B. and Moody, J. F. (1976). *Experientia* **32**, 768–769.

Hasson-Voloch, A. (1968). *Nature Lond.* **218**, 330–333.

Heilbronn, E. and Mattson, Ch. (1974). *J. Neurochem.* **22**, 315–317.

Heilbronn, E., Mattsson, C. and Elfman, L. (1974). *Proc. 9th FEBS Meeting* **37**, 29–37.

Hucho, F. and Changeux, J.-P. (1973). *FEBS Lett.* **38**, 11–15.

Karlin, A. (1973). *Proc. 5th Int. Congr. Pharmacology, San Francisco* **5**, 86–97.

Karlin, A. and Cowburn, D. (1973). *Proc. Nat. Acad. Sci. U.S.A.* **70**, 3636–3640.

Karlin, A. (1974). *Life Sci.* **14**, 1385–1415.

Karlsson, E., Eaker, D. and Ponterius, G. (1972a). *Biochim. Biophys. Acta* **257**, 235–248.

Karlsson, E., Heilbronn, E. and Widlund, L. (1972b). *FEBS Lett.* **28**, 107–111.

Kato, G. and Tattrie, B. (1974). *FEBS Lett* **48**, 26–31.

Klett, R. P., Fulpius, B. W., Cooper, D., Smith, M., Reich, E. and Possani, L. D. (1973). *J. Biol. Chem.* **248**, 6841–6853.

Levinson, S. R. and Keynes, R. D. (1972). *Biochim. Biophys. Acta* **288**, 241–247.

Lindstrom, J. and Patrick, J. (1974). *In* "Synaptic Transmission and Neuronal Interaction" (M. V. L. Bennett, ed.) pp. 191–216. Raven Press, New York.

Lowry, O. H., Rosebrough, N. J., Farr, A. L. and Randall, R. J. (1951). *J. Biol. Chem.* **193**, 265–275.

Maelicke, A., Fulpius, B. W., Klett, R. P. and Reich, E. (1976). *J. Biol. Chem.* (in press).

Martinez-Carrion, M. and Raftery, M. A. (1973). *Biochem. Biophys. Res. Commun.* **55**, 1156–1164.

Martinez-Carrion, M., Sator, V. and Raftery, M. A. (1975). *Biochem. Biophys. Res. Commun.* **65**, 129–137.

Menez, A., Morgat, J.-L., Fromageot, P., Ronseray, A.-M. and Boquet, P. (1971). *FEBS Lett.* **17**, 333–335.

Meunier, J.-C., Olsen, R. W., Menez, A., Fromageot, P., Boquet, P. and Changeux, J.-P. (1972a). *Biochemistry* **11**, 1200–1210.

Meunier, J.-C., Olsen, R. W. and Changeux, J.-P. (1972b). *FEBS Lett.* **24**, 63–68.

Meunier, J.-C., Sugiyama, H., Cartaud, J., Sealock, R. and Changeux, J.-P. (1973). *Brain Research* **62**, 307–315.

Meunier, J.-C., Sealock, R., Olsen, R. and Changeux, J.-P. (1974). *Eur. J. Biochem.* **45**, 371–394.

Michaelson, D., Vandlen, R., Bode, J., Moody, T., Schmidt, J. and Raftery, M. A. (1974). *Arch. Biochem. Biophys.* **165**, 796–804.

Miledi, R., Molinoff, P. and Potter, L. T. (1971). *Nature Lond.* **229**, 554–557.

Miledi, R. and Potter, L. T. (1971). *Nature Lond.* **233**, 599–603.

Moody, T., Schmidt, J. and Raftery, M. A. (1973). *Biochem. Biophys. Res. Commun.* **53**, 761–772.

Moore, W. J. and Loy, N. J. (1972). *Biochem. Biophys. Res. Commun.* **46**, 2093–2099.

Moore, W. M., Holladay, L. A., Puett, D. and Brady, R. N. (1974). *FEBS Lett.* **45**, 145–149.

Nachmansohn, D. (1955). *Harvey Lectures* **49**, 57–99.

Nickel, E. and Potter, L. T. (1973). *Brain Res.* **57**, 508–517.

O'Brien, R. D. and Gilmour, L. P. (1969). *Proc. Nat. Acad. Sci.* **63**, 496–503.

Olsen, R. W., Meunier, J.-C. and Changeux, J.-P. (1972). *FEBS Lett.* **28**, 96–100.

Ong, D. E. and Brady, R. N. (1974). *Biochemistry* **13**, 2822–2827.

Patrick, J., Boulter, J. and O'Brien, J. C. (1975). *Biochem. Biophys. Res. Commun.* **64**, 219–225.

Potter, L. T. (1973). *In* "Drug Receptors" (H. P. Rang, ed.). pp. 295–312. MacMillan, London.

Potter, L. T. (1974). *Methods in Enzymology* **32**, part B, 309–323.

Raftery, M. A., Schmidt, J., Clark, D. G. and Wolcott, R. G. (1971). *Biochem. Biophys. Res. Commun.* **45**, 1622–1629.

Raftery, M. A., Schmidt, J. and Clark, D. G. (1972). *Arch. Biochem. Biophys.* **152**, 882–886.

Raftery, M. A., Schmidt, J., Martinez-Carrion, M., Moody, T., Vandlen, R. and Duguid, J. (1973). *J. Supramol. Struct.* **1**, 360–367.

Raftery, M. A., Bode, J., Vandlen, R., Michaelson, D., Deutsch, J., Moody, T., Ross, M. J. and Stroud, R. M. (1975). *In* "Protein-Ligand Interactions" (Horst Sund and Gideon Blauer, eds). pp. 328–355. Walter de Gruyter, Berlin and New York.

Romine, W. O., Goodall, M. C., Peterson, J. and Bradley, R. J. (1974). *Biochim. Biophys. Acta* **367**, 316–325.

Salvaterra, P. M. and Moore, W. J. (1973). *Biochem. Biophys. Res. Commun.* **55**, 1311–1318.

Schmidt, J. and Raftery, M. A. (1972). *Biochem. Biophys. Res. Commun.* **49**, 572–578.

Schmidt, J. and Raftery, M. A. (1973). *Biochemistry* **12**, 852–856.

Singer, S. J. and Nicolson, G. L. (1972). *Science* **175**, 720–731.

Vogel, Z., Sytkowski, A. J. and Nirenberg, M. W. (1972). *Proc. Nat. Acad. Sci. U.S.A.* **69**, 3180–3184.

Weill, C. L., McNamee, M. G. and Karlin, A. (1974). *Biochem. Biophys. Res. Commun.* **61**, 997–1003.

2

The Analysis of End-plate Noise—A New Approach to the Study of Acetylcholine/Receptor Interaction

B. Katz and R. Miledi

Dept. of Biophysics, University College, London, England

I Introduction

The transmitter substance acetylcholine acts on the postsynaptic end-plate membrane by increasing its permeability to small cations, enabling an inward current of sodium ions to pass through the membrane and to depolarize the muscle fibre. The effect is rapidly reversible and may be pictured as a brief transient opening of ionic gates or channels at localized membrane sites where specific acetylcholine receptors are present and interact with the transmitter molecules (see the equivalent circuit in Fig. 1). Normally, neuromuscular transmission proceeds in discrete steps, involving the release from the nerve terminal of packets, each of which contains thousands of acetylcholine molecules. The postsynaptic effect of an individual packet is registered as a miniature end-plate potential, a local depolarization of the muscle fibre usually amounting to an appreciable fraction of a millivolt. These *quantal* components of transmission have been recorded at many chemical synapses, but the elementary interactions between transmitter and receptor *molecules* have not so far been accessible to direct recording techniques. Indeed, the fact that end-plate depolarizations due to application of graded small doses of acetylcholine (Ach) appear to be continuously graded, had led to the conclusion that the molecular components of the ACh-potential are too small to be resolved individually,

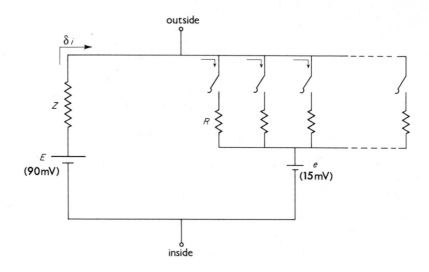

Fig. 1. Diagram illustrating ACh-induced ion channels in the end-plate membrane. The resting potential of the muscle fibre is represented by E, and the characteristic "input impedance" by Z (of the order of 10^5 ohm). The reaction between ACh molecules and end-plate receptors leads to the transient opening of a statistically varying number of channels in the membrane, represented by the closure of individual keys each in series with a channel resistance R (about 5×10^{10} ohm) and a voltage e (about 15 mV, negative inside). The result of the temporary closure of one key is the flow of a brief depolarizing current δi and a minute brief reduction of the p.d. between inside and outside. During steady ACh action many keys are closed at any given time, producing a maintained depolarization, but their numbers fluctuate from moment to moment leading to characteristic "ACh noise".

and that discrete spontaneous discharges like miniature end-plate potentials (m.e.p.p.) must arise from a synchronous impact of a very large number of transmitter molecules.

In recent years, however, an indirect approach has been introduced which enables one to obtain an approximate measure of the size and time course of the "molecular" depolarization due to the ACh/receptor interaction, and of the underlying current and conductance change. It is based on the finding that the depolarizing effect of a steady dose of ACh is invariably accompanied by a significant increase in "membrane noise", i.e. in voltage fluctuations which can be ascribed to the statistical variations in collision rate between ACh and receptor molecules and the consequent fluctuations in the number of "ion gates" which are open from moment to moment. Suppose, for example, that a given steady dose of ACh produces a maintained depolarization of 10mV, and that this is associated with an average number N of 50 000 conducting ion gates or channels through which depolarizing current passes. Furthermore, suppose that the current through each gate varies in all-or-none fashion between zero and a fixed intensity. On ordinary Poisson statistics, the number N would fluctuate from moment to moment with a standard deviation of $\pm \sqrt{N}$, i.e. ± 224 channels, or with a coefficient of variation of $1/\sqrt{N}$, i.e. 0.45%. Conversely, knowing the coefficient of variation, we can determine N: thus, if we were to measure the *average depolarizing current I* and its *standard deviation* or root mean square value σ_I, the ratio of the two measurements I/σ_I would give us, at least approximately, the standard deviation of the number of open channels ($\sqrt{N}$), and one could calculate the current flowing through an individual ion channel from $\sigma_I^2/I = I/N$. Knowing also the "driving voltage" V_0 (i.e. the difference between recorded membrane potential and the reversal potential of the e.p.p.), one can calculate the conductance of the ion channel as $\sigma_I^2/(IV_0)$.

The principal advantage of this approach is that we record amplitude fluctuations which involve summed effects of hundreds of molecular actions rather than unresolvably small single actions. The disadvantage is that one does not deal directly with discrete elementary discharges, but attempts to derive their characteristics from an analysis of continuous noise fluctuations. The basic assumption which underlies the whole procedure is that the relatively large average effect (depolarization, or membrane current) and the relatively small fluctuations are both built up by statistical superposition of the same elementary components or "shot effects". This is a very plausible assumption, but it cannot be proved at present, because such proof would require *direct* information on the size of the elementary event. A further assumption which has been made in the noise analysis is that the experimental doses of ACh and other agonists are confined to a range in which only a small proportion of the available receptors becomes occupied. This is reasonable, and to some extent justified by the fact that the characteristic properties of the ionic channel derived from the analysis were found to be independent of the applied dose (Anderson and Stevens, 1973; Katz and Miledi, 1972).

II Acetylcholine voltage noise

The initial studies were made by using the conventional intracellular recording technique with a single microelectrode inserted into the end-plate region of the muscle fibre, and ionophoretic application of a steady dose of ACh (Fig. 2). The mean depolarization and the additional

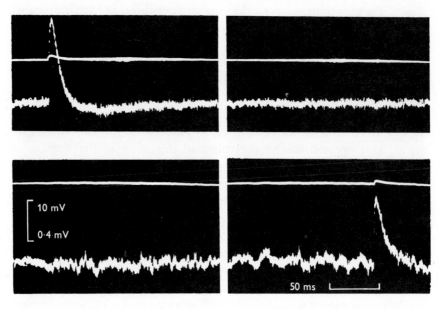

Fig. 2. Acetylcholine noise recorded from an end plate in frog sartorius (Katz and Miledi, 1972). In each block, the upper trace was recorded on a low-gain d.c. channel (scale 10 mV); the lower was simultaneously recorded on a high-gain condenser coupled channel (scale 0·4 mV). The top row shows controls (no ACh); the bottom row shows membrane noise during ACh application, by diffusion from a micropipette. In the bottom row of records, the increased distance between the low and high gain traces is due to upward displacement of the d.c. trace because of ACh-induced depolarization. Two spontaneous min. e.p.p.'s are also seen.

variance (mean square deviation) during the ACh application were measured and used to determine the size and time course of the elementary depolarization due to the transient opening of single ion channels by ACh/receptor interactions.

A AMPLITUDE OF ELEMENTARY DEPOLARIZATION

To give a typical example, a 5 mV ACh-induced depolarization at a frog end-plate was accompanied by excess noise of 25 μV r.m.s. value, in a fibre which had a resting potential of 90 mV and m.e.p.p.'s of 0·5 mV average amplitude.

If we make the simplifying assumption that both mean and variance of the depolarization arise from a linear superposition of randomly occurring, exponentially decaying voltage blips (similar to m.e.p.p.s but much smaller in size), of a shape $f(t) = ae^{-t/\tau}$, where a is amplitude and τ time constant of decay, then according to Campbell's theorem, the mean value V is

$$V = n \int f(t)\, \mathrm{dt} = na\tau \qquad (1)$$

and the variance

$$\overline{E^2} = n \int f^2(t)\, \mathrm{dt} = \frac{na^2\tau}{2} = \frac{Va}{2}, \qquad (2)$$

where n is the average frequency of the voltage blips. Hence, the elementary voltage amplitude, a, is

$$a = 2\overline{E^2}/V. \qquad (3)$$

For the above example, $a = 0·25$ μV, some three orders of magnitude smaller than the m.e.p.p., which suggests that an m.e.p.p. is produced by the synchronous opening of one or a few thousand ion gates. Assuming each ion gate is opened as a result of an effective interaction between one or a small number of receptor and of ACh molecules, the m.e.p.p. requires the simultaneous action of a few thousand ACh molecules (though the number released in one quantal packet may be substantially larger).

Katz and Miledi (1972) have pointed out several shortcomings in this calculation. (a) If the depolarization exceeds a few mV, departure from linear superposition of "shot effects" becomes appreciable and a correction for non-linearity has to be applied. (b) There is some uncertainty about the shape and in particular about the amplitude dispersion of the individual elementary blips. The main effect would be to reduce the factor 2 in equation (3) towards unity, in other words the value of a may have been overestimated by a factor of 2, and it may be more appropriate to use the variance-to-mean ratio E^2/V as a convenient approximate index of the amplitude of the molecular depolarization.

If it is true, as suggested by Katz and Miledi (1972) and by Anderson and Stevens (1973), that the elementary ion gates have uniform conductance, but life-times of random duration — all very brief compared to the membrane time constant, but exponentially dispersed—then the *amplitudes* of the individual *voltage blips* will also become widely dispersed with an approximately exponential distribution. The variation among individual blips will then contribute appreciably to the observed ACh noise, and equation (3) should be rewritten as

$$\overline{E^2} = \frac{V\bar{a}}{2}[1 + (c.v._a)^2], \qquad (4)$$

where $\bar{a}$ is the mean amplitude and $c.v._a$ the coefficient of variation of the elementary depolarization. If the individual values of a have a random exponential distribution, $c.v._a$ becomes unity, and the effect would be to make $\bar{a} = \overline{E^2}/V$.

B TIME COURSE OF ELEMENTARY DEPOLARIZATION

To obtain an indication of the time course of the elementary "blip" from noise records, the auto-correlation function or the so-called "power spectrum" (i.e. the spectral frequency distribution of the total variance) of the fluctuations are convenient guides (Katz and Miledi, 1972; Verween and de Felice, 1974). Power spectra of intracellularly recorded ACh-voltage noise (Fig. 3) have a characteristic shape, with a plateau at very low frequencies and a progressive decline at frequencies above several Hz (frog, 15°C), falling to half at about 15 Hz and reaching very low values above 100 Hz. A spectrum of this *general shape* would be obtained if the elementary components of the noise fluctuations had the shape $f(t) = ae^{-t/\tau}$, with τ approximately 10 mseconds. The power spectrum would then be given by

$$S = S_0/(1 + 4\pi^2 f^2\tau^2), \qquad (5)$$

where f is frequency (Hz), $S = d\overline{E^2}/df$ and S_0 the maximum (plateau value) of S at zero frequency (which can be shown to be $S_0 = 2V a\tau$). The value of τ can be conveniently determined from the "half-power" or "cut-off" frequency $f_{1/2}$, at which $S = \frac{1}{2}S_0$ and $f_{1/2} = 1/(2\pi\tau)$. The spectra actually obtained depart from equation (5) in certain respects, especially at the high frequency end and require a more complicated explanation than has been offered here. These matters are discussed in detail by Katz and Miledi (1972) and by Verween and

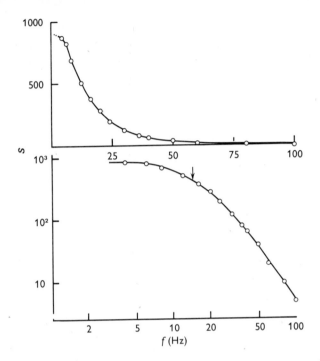

Fig. 3. "Variance spectrum" of intracellularly recorded ACh voltage noise (Katz and Miledi, 1972). Temperature 5·5°C. Upper part: linear plot (ordinates: $S = d\overline{E^2}/df$ in relative units; abscissae: frequency in Hz). Lower part: same results in double-logarithmic coordinates. Arrow indicates "half-power frequency" ($f_{1/2} = 1/2\pi\tau$).

de Felice (1974). What is clear, however, is that the intracellularly recorded ACh voltage noise is confined to a very low-frequency spectrum, because the membrane capacity of the muscle fibre with its large time constant acts as an effective shunt at all low frequencies. Intracellular voltage recording, therefore, provides little information on the time constant acts as an effective shunt at all but low frequencies. Intra-knowledge of these current pulses that one can draw firm conclusions about the conductance and particularly the "life-span" of the ionic gates.

III   Acetylcholine current noise

Two methods have been used to examine fluctuations of ionic current during the ACh response: (a) focal extracellular recording (e.g. Figs 4 and 5) and (b) voltage-clamping with intracellular feedback current. The former is relatively simple, requiring only a single microelectrode for registering the external potential drop due to the converging current which enters the active end-plate surface. This provides a faithful record of the time course of the membrane current and its fluctuations, but not of its absolute magnitude. The second method has the important advantage of determining the current intensity directly, while keeping the membrane potential at a predetermined level, but it introduces further technical difficulties because it requires the use of two intracellular electrodes plus high-gain feedback and monitoring systems. The voltage-clamp technique was first applied by Anderson and Stevens (1973) and has since been widely used for ACh noise measurements (Ben-Haim et al., 1975; Neher and Sakmann, 1975).

The power spectra obtained in this way differ from the intracellular results in showing a much higher "cut-off frequency" $f_{1/2}$, which confirms that the elementary current blip is much briefer than the resulting depolarization. But apart from the increased frequency scale, there is little change in the general shape of the curve, which suggests that the *average* current pulse can be fitted by a similar function, viz. $f(t) = be^{-t/\tau_1}$, where b is the amplitude and τ_1 the time constant of decay. One plausible interpretation, put forward by Katz and Miledi (1972) and by Anderson and Stevens (1973), is that individual current blips are all brief square pulses of intensity b, and of random durations which are exponentially distributed around a statistical mean value of τ_1.

On the basis of this assumption, Anderson and Stevens (1973) calculated that, at 20°C and with a resting potential of 70 mV, the value of b is of the order of 1·5 to 2 pA, and τ_1 around 1 ms, which agrees with the duration estimated from extracellular power spectra (Katz and Miledi, 1971, 1972). This corresponds to ion channels with a conductance of approximately 2 to 3 $\times$ 10^{-11} ohm^{-1} and an average life-span of 1 ms. Similar values have been obtained by Ben-Haim et al. (1975); somewhat larger values were given by Sachs and Lecar (1973). From our original measurements of voltage noise (Katz and Miledi, 1972) we had estimated a conductance of the order of 10^{-10} ohm^{-1}, but this was a more indirect and less accurate estimate; moreover, it was based on the assumption of a uniform pulse shape (equation 3) which accounts for a substantial part of the difference.

The noise measurements do not tell us how many molecules are involved in the membrane reaction leading to the opening of one elementary ion channel. Various estimates have been made based on the shape of the dose/response curve and on the "Hill coefficient" (Hill, 1910) derived from its initial curvature. The estimates vary between one and four ACh molecules, reacting with one to four receptor molecules or "monomeres" (Rang, 1975; Dreyer and Peper, 1975; Magleby and Terrar, 1975; Kuffler and Yoshikami, 1975; Colqhoun, 1975; Dionne and Stevens, 1975). There is agreement in any case that the interaction of quite a small number of molecules opens an ion gate which allows the influx of at least 10^4 Na ions across the end-plate membrane at 20°C (and even more at lower temperature when the gate remains open for a longer time).

IV   Effects of temperature and of membrane potential

According to Anderson and Stevens (1973), the size of the channel conductance is relatively constant and remains unaffected by changes in temperature or the level of the membrane potential, whereas the mean life-time of the open channel changes considerably. Lowering the temperature by 10°C lengthens the average channel duration 2·5–3 times (and even more if the membrane is hyperpolarized; Anderson and Stevens, 1973), so that a correspondingly larger total charge is transferred across the end-plate membrane and the size of the elementary depolarization increases appreciably (Katz and Miledi, 1972). It is of interest that the amplitude of the miniature e.p.p. barely increases when the temperature is reduced; its rise time is lengthened, but the rate of rise of the m.e.p.p. and the intensity of the underlying current are diminished. Evidently, fewer ionic gates are opened by the quantal packet of ACh as the temperature is lowered. Whether this is due to a change in ACh affinity for the receptor, or in the forward rate constant of the "gating reaction" remains to be seen.

By obtaining spectra of noise current at different levels of membrane potential, Anderson and Stevens (1973) showed that the channel duration lengthens when the end plate is hyperpolarized and shortens as a result of depolarization. The value of τ_1 approximately doubles for a 100 mV potential change. This explains the parallel changes in the time course of end-plate currents which had been found previously (Kordas, 1969; Magleby and Stevens, 1972a, b). It also conforms closely to the transient changes in ACh (or carbachol) current which have recently been observed to follow a step-change in membrane potential (Adams, 1974; Neher and Sakmann, 1975).

The view has been put forward that the life-time of the ion gate depends directly on the stability or the time constant of dissociation of an activated ACh/receptor complex (see Katz and Miledi, 1972; Anderson and Stevens, 1973), and that the reaction kinetics of this compound are influenced by the size and polarity of the membrane field, indicating that the ACh/receptor complex itself has a net charge or dipole moment.

The nature of the relation between the life-time of the ion channel and of the drug/receptor complex remains conjectural. The formal scheme proposed by several authors (del Castillo and Katz, 1957; Magleby and Stevens, 1972b; Katz and Miledi, 1972, 1973 a, b) is that an agonist like ACh proceeds in a two-step reaction forming first an intermediate, inactive compound with the receptor R, ACh + R $\rightleftharpoons$ AChR, which then changes to an activated depolarizing conformation AChR* whose life-time *may* be identical with that of the open channel conductance. Competitive inhibitors form an inactive compound without proceeding to the second reaction step. In the case of ACh and other agonists, both steps have been supposed to be of a reversible kind (ACh + R $\rightleftharpoons$ AChR $\rightleftharpoons$ AChR*), but for the second step this assumption is unnecessary, and there are various alternative ways in which the life of the "activated" complex may be terminated, e.g. by dissociating directly into ACh and either a normal or refractory form of the free receptor. As already mentioned, there are reasons for supposing—from the shape of the dose/response curve—that more than one ACh and receptor molecule, or monomere, may cooperate in this reaction.

Apart from temperature and membrane potential, there are other influences which have been shown to modify the gating period, among them a number of chemical agents such as alcohols and local anaesthetics which act on the receptor or, more probably, on membrane constituents surrounding it (see below). Interestingly, chronic denervation was found to lead, not only to the spread and formation of additional extrajunctional ACh receptors (Axelsson and Thesleff, 1959; Miledi, 1960), but to a marked prolongation of the ionic gating period and a consequent slowing and increase of ACh voltage noise (Katz and Miledi, 1972; Neher and Sakmann, 1975).

V   The effects of other end-plate agonists

When the effects of ACh and carbachol were compared (Katz and

Miledi, 1972), it became obvious that equipotent doses—in terms of the resulting depolarization—produce very different intensities of *voltage* noise, carbachol being a far less "noisy" agent than acetylcholine. On analysing their *current* spectra, it became clear that the main difference between the actions of the two drugs was in the duration, or stability, of the open ion gate, the elementary carbachol current lasting for only about 0·35 ms at 20°C compared with 1 ms for ACh. It follows that, at the molecular level, carbachol is less potent in transferring electric charge across and depolarizing the membrane than ACh. In terms of the hypothesis outlined above, this would be because the active conformation of the carbachol/receptor complex is less stable and disintegrates more rapidly than the ACh/receptor compound. At first sight, this seemed a somewhat surprising conclusion, because for practical purposes of drug application, carbachol is of course known to be a more stable agonist than ACh. But this is a different kind of stability: in noise measurements we are not concerned with the life-span of the free agonist molecule (which is much shorter for ACh than for carbachol because the latter is not hydrolysed by the specific esterase), but with the survival of the drug/receptor complex and its associated gating action, and this appears to be longer for ACh than for a number of similarly acting, but less potent substances.

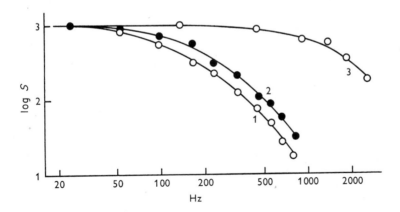

Fig. 4. Spectral distribution of end-plate *current* noise produced by different depolarizing agents. Coordinates as in the lower part of Fig. 3. Curve 1, suberyldicholine; curve 2, acetylcholine; curve 3, acetylthiocholine. Temperature approx. 22°C. The noise fluctuations were recorded with focal micro-electrodes applied externally to the end-plate surface. (Katz and Miledi, 1973*a*.)

Several other agonists have been studied (Katz and Miledi, 1973a; Colqhoun *et al.*, 1975; Rang, 1975): decamethonium and acetylthiocholine resemble carbachol in producing ion channels of relatively short duration and a small elementary depolarization; by contrast, suberyldicholine causes a somewhat longer molecular action than ACh itself (Fig. 4). It is not yet clear to what extent the conductance of the channel, and the ionic current intensity, depends on the nature of the agonist. There are indications that the channel conductance is also subject to modification, and that a short life-time is associated with a reduced conductance and peak current (Ben-Haim *et al.*, 1975; Colqhoun *et al.*, 1975). Furthermore, some agonists produce power spectra which can no longer be interpreted in terms of a single exponential time constant, or of an exponentially distributed duration of square pulses (Colqhoun *et al.*, 1975); it appears, therefore, that the simple assumption of a pre-formed channel of fixed "all-or-none" conductance may not be valid, and that the opening and particularly the closing of the ionic path may be a much more variable process.

VI   The action of antagonists

Among the postsynaptic end-plate blocking agents, it is customary to distinguish between competitive and non-competitive ACh inhibitors. The former are envisaged to occupy binding sites, so that for a given concentration fewer ACh molecules have access to free receptors. Non-competitive antagonists are visualized as interfering with adjacent sites, on the receptor or on its immediate environment so that ACh can still react with the receptors, but forms a less stable and less effective compound. In practice, the distinction has usually been made by examining the characteristic changes in the (log dose)/response curve brought about by the antagonist, and seeing whether the curve is simply displaced along the log-dose axis, or also reduced in maximum amplitude.

The noise analysis provides a new way of differentiating between these two types of ACh antagonists. A "competitive" inhibitor would be one which reduces the *molecular* effect of ACh in "all-or-none" fashion. The number of free receptors is lowered, hence the probability of individual ACh molecules reacting with them is reduced. But those receptors which are not occupied by the inhibitor will react in the normal way and give rise to the same elementary depolarization. Hence, the ratio $\overline{E^2}/V$ should remain unchanged. This is the result obtained with curare, a reversible blocking agent, and with α-bungarotoxin, a practically irreversible inhibition (Katz and Miledi, 1972, 1973d). The latter, because of its irreversible action, could not be classified as "competitive" in the usual sense, but presumably attaches itself to the same primary binding site as the curare molecule and the transmitter itself.

In contrast to curare and α-toxin, various other inhibitors were found which reduce the value of $\overline{E^2}/V$, that is, of the elementary depolarization, and indeed owe their postsynaptic blocking action entirely to this effect. Among such substances are atropine, dithiothreitol, procaine and several other local and general anaesthetic agents (Katz and Miledi, 1973c, 1975a; Landau and Ben-Haim, 1974; Ben-Haim et al., 1975). Spectral analyses show that the elementary ACh current pulses are shortened in duration (cf. Fig. 5) so that the molecular depolarizing potency of ACh is now reduced to that normally obtained with substances like carbachol or acetylthiocholine (see above). Procaine has a more complicated effect: the current spectrum separates into two components, a major fast and a minor very slow component, indicating a complex relaxation behaviour of the ion channels. This finding, at the molecular level, presumably accounts for the well-known transformation of the end-plate potential by procaine and various other local anaesthetics (Furukawa, 1957; Steinbach, 1968): the e.p.p. changes from the simple exponentially decaying shape to a complicated wave consisting of a brief initial spike followed by a slowly subsiding plateau.

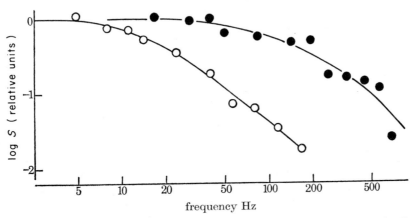

Fig. 5. Effect of a "non-competitive" ACh-antagonist on ACh noise (Katz and Miledi 1973c). ACh current noise, recorded extracellularly (as in Fig. 4). Hollow circles before, filled circles after atropine. Temperature 7°C.

There are indications that some of these "non-competitive" blocking agents not only modify the time course of the ACh/receptor reaction and shorten the life-time of the ion channel, but also reduce its peak conductance (Ben Haim *et al.*, 1975), thus detracting further from the simple concept of a fixed channel conductance operated by an all-or-none switching mechanism.

VII   The effect of aliphatic alcohols

Gage *et al.* (1975) have shown that a series of alcohols, ethanol to hexanol, increase the size and duration of m.e.p.p.s. This is due to a slowing of the decay of the underlying ionic current. In the case of ethanol, the effect on ACh noise was examined, and it was found that the mean life of the ion channel was lengthened, sufficiently so to account for the potentiating action on the m.e.p.p. As the aliphatic chain length is increased, more complicated changes are observed, resembling somewhat the two-time course action of procaine, and with octanol the main effect is a shortening rather than lengthening of the ionic current. The authors attribute their findings to possible small changes in dielectric constant and in the viscosity of the membrane phase which may have a strong influence on the kinetics of the supposedly polar ACh/receptor complex.

VIII   The effect of anti-cholinesterases

In the experiments of Anderson and Stevens (1973), the relaxation times derived from the power spectra of the current fluctuations were found to be identical with the exponential time constant of decay of the miniature end-plate current (m.e.p.c.) and also of the larger impulse-evoked end-plate current (e.p.c.). It was concluded, therefore, that the duration of the normal transmitter action is determined by the life-time of the single ion channel, and that the ACh molecules released spontaneously or by the nerve impulse do not survive long enough in the synaptic cleft to repeat their action. However, this conclusion probably does not apply at all temperatures, and it certainly does not hold when ACh hydrolysis in the cleft has been stopped by the use of cholinesterase inhibitors. At higher temperatures, even in the absence of anti-esterases, the duration of the m.e.p.c. slightly exceeds the average life-time of the single channel: e.g. at 20°C the latter is about 1 ms, while the decay constant of the m.e.p.c. is more nearly 1·5 mseconds. Nevertheless, it is clear that by the time the channels are closed, most of the ACh molecules have been destroyed or dispersed away from their site of action. After a suitable dose of an anti-esterase such as prostigmine, the situation is quite different: the m.e.p.c. is lengthened severalfold, while there is little or no prolongation of the elementary, molecular ACh action (Katz and Miledi, 1973b). The same result was obtained during enzymatic removal of cholinesterase, using a proteolytic enzyme to solubilize cholinesterase and clear it from the junction (Neher and Sakmann, unpublished observations). Evidently, under these conditions transmitter molecules are able to repeat their action and open several ion gates in succession before escaping from the synapse.

It seems clear from these experiments that the rapid termination of the nerve-evoked transmitter action which is normally observed is brought about by local hydrolysis rather than outward diffusion of ACh. Incidentally, hydrolysis of ACh serves an important second function, namely to provide choline molecules for the specific re-uptake process which exists in the membrane of the nerve terminal (Birks and MacIntosh, 1961; Collier and MacIntosh, 1969). A large part of the choline combines with *pre*synaptic receptors and is recaptured by the nerve ending. It is subsequently resynthesized in the terminal cytoplasm, thus providing a specific and effective recycling process for the released transmitter substance. ACh itself is taken up much less effectively (Collier and MacIntosh, 1969; Potter, 1968), so that in the anti-esterase treated preparation most of the released transmitter is lost into the circulation (which, in turn, provided the experimental means for the original detection of ACh and its role as the neuromuscular transmitter substance).

In the absence of hydrolysis, the action of the transmitter is terminated by its outward diffusion from the synaptic cleft. If it were a process of free diffusion, unretarded by local barriers or by binding to reactive sites, then this should still be a very rapid event, for the distance from the centre of the cleft to the outside is no more than a micron (see Eccles and Jaeger, 1958). Moreover, the process would have a small temperature coefficient, so that at sufficiently low temperature, one might expect the cleft to be effectively cleared of ACh molecules at the end of the gating period, even without hydrolysis. In practice, however, the end-plate current long outlasts that period, appreciably longer than would be expected on the basis of free diffusion (Eccles and Jaeger,

1958; Katz and Miledi, 1973b). It is very probable that an important factor in retarding diffusion is the presence of a large surplus of receptor binding sites on the postsynaptic membrane; in fact, Katz and Miledi (1973b, c) have calculated that the normal binding capacity of the postjunctional surface is sufficient to absorb two-thirds of the amount of ACh remaining in the cleft, and therefore lengthens local diffusion times by a factor of three. This view is supported by the finding that, after treatment with anti-esterases, substances like curare and α-bungarotoxin not only *reduce* the amplitude of the end-plate current, but *shorten* the duration towards the limit set by unretarded diffusion (Katz and Miledi, 1973b; see also Magleby and Terrar, 1975; cf. Fig. 6). Curare and α-

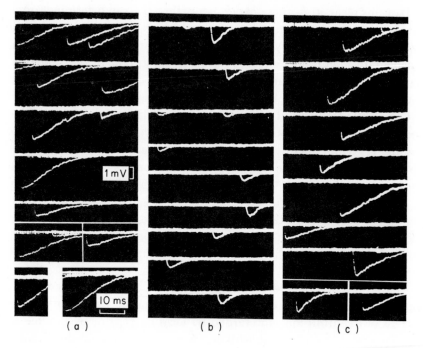

(a)   (b)   (c)

Fig. 6. The effect of curare on the size and time course of miniature end-plate currents in anti-esterase treated muscle (Katz and Miledi, 1973b). M.e.p.p.s. were recorded with a focal external electrode. Temperature 22·5°C. Low-Na Ringer. The time course of decay has been slowed as a result of prostigmine treatment. Between (a) and (b), curare was applied ionophoretically, which greatly reduced the size of the m.e.p.p.s. After curare efflux was stopped, records (b) and later (c) were taken, during the gradual recovery of amplitudes. Note reduction in size as well as duration during the curare action.

toxin have no direct influence on the life-span of the ionic gate, nor do they reactivate the inhibited esterase (cf. Eccles *et al.*, 1942); the shortening of the time course is most probably explained by the blockage of ACh-binding sites, and hence the removal of their retarding influence on ACh diffusion from the cleft.

When a muscle, after esterase inhibition, is subjected to a rapidly repeated stimulation of the motor nerve, an additional phenomenon is observed at the end plate, viz. a cumulative "slow wave" of depolarization which builds up to considerable height during successive impulses and takes a few seconds to subside after stimulation has ceased (Eccles *et al.*, 1942; Eccles and MacFarlane, 1949). Various explanations have been offered for this prolonged effect. Although it is clearly associated with drugs that have an anti-cholinesterase action (Eccles and Mac-Farlane, 1949), the suggestion has been made (Kuba *et al.*, 1974) that the very slow decay is due to quite a different side effect, namely an enormous prolongation of the "open life-time" of the ion gate, while the transmitter itself has been dispersed initially within several milliseconds. It would be technically impracticable to demonstrate such a very slow component by spectral analysis of ACh noise, for the expected fluctuations would occur in a frequency range so low that they could scarcely be distinguished from artifacts such as fluctuations in ACh concentration (see Katz and Miledi, 1975a). It is possible, however, to test the alternative interpretation, namely that the prolonged depolarization arises from a residue of ACh which accumulates during repetitive stimulation in the synaptic cleft and in the secondary folds and crevices attached to it. If this is the explanation, then—regardless of whether the very slow disappearance of the transmitter is due to physical diffusion barriers or to chemical binding—the whole period of the slow wave should be accompanied by rapid fluctuations whose size and time course should match normal ACh noise. In recent experiments (Katz and Miledi, 1975), we obtained an unequivocal answer to this question: the "slow wave" in the anti-esterase treated muscle in indeed accompanied by "transmitter noise" which, quantitatively, was found to differ little, if at all, from that obtained during ACh dosage. This is an interesting novel application of the method, in an experiment in which noise is evoked by nerve impulses rather than by depolarizing drugs, and the continued fluctuations provide evidence for continued presence of nerve-released acetylcholine molecules.

The purpose of this article was to review some of the published findings with the recently developed method of end-plate noise analysis, and to give an idea of the kind of problems that may be tackled with this approach. We may expect that there are numerous applications still to be exploited; the method will presumably be useful much more generally in studying kinetics of drug/receptor interactions at the molecular level. However, it is important to realize that the statistical interpretation of noise fluctuations in terms of single shot effects remains somewhat indirect and depends on assumptions whose validity cannot be checked until ways have been found of recording the single molecular event itself. At present, at least at the motor end plate, there still appears to be two orders of magnitude below the resolving power of our recording instruments, but perhaps by an improved recording technique or an improved cellular preparation, this difficulty could be overcome.

IX   References

Adams, P. R. (1974). Kinetics of agonist conductance changes during hyperpolarization at frog endplate. *Br. J. Pharmac.* **53**, 308–310.

Anderson, C. R. and Stevens, C. F. (1973). Voltage clamp analysis of acetylcholine produced end-plate current fluctuations at frog neuromuscular junction. *J. Physiol.* **235**, 655–692.

Axelsson, J. and Thesleff, S. (1959). A study of supersensitivity in denervated mammalian skeletal muscle. *J. Physiol.* **147**, 178–193.

Ben-Haim, D., Dreyer, F. and Peper, K. (1975). Acetylcholine receptor: modification of synaptic gating mechanism after treatment with a disulfide bond reducing agent. *Pflügers Arch.* **355**, 19–26.

Birks, R. I. and MacIntosh, F. C. (1961). Acetylcholine metabolism of a sympathetic ganglion. *Can. J. Biochem.* **39**, 787–827.

Collier, B. and MacIntosh, F. C. (1969). The source of choline for acetylcholine synthesis in a sympathetic ganglion. *Can. J. Physiol. Pharm.* **47**, 127–135.

Colquhoun, D. (1975). Mechanisms of drug action at the voluntary muscle endplate. *Ann. Rev. Pharmacol.* **15**, 307–325.

Colquhoun, D., Dionne, V. E., Steinbach, J. H. and Stevens, C. F. (1975). Conductance of channels opened by acetylcholine-like drugs in muscle end-plate. *Nature, Lond.* **253**, 204–206.

del Castillo, J. and Katz, B. (1957). Interaction at end-plate receptors between different choline derivatives. *Proc. R. Soc. B.* **146**, 369–381.

Dionne, V. E. and Stevens, C. F. (1975). Voltage dependence of agonist effectiveness at the frog neuromuscular junction: resolution of a paradox. *J. Physiol.* **251**, 245–270.

Dreyer, F. and Peper, K. (1975). Density and dose-response curves of acetylcholine receptors in frog neuromuscular junction. *Nature, Lond.* **253**, 641–643.

Eccles, J. C. and Jaeger, J. C. (1958). The relationship between the mode of operation and the dimensions of the junctional regions at synapses and motor end-organs. *Proc. R. Soc. B.* **148**, 38–56.

Eccles, J. C., Katz, B. and Kuffler, S. W. (1942). Effects of eserine on neuromuscular transmission. *J. Neurophysiol.* **5**, 211–230.

Eccles, J. C. and MacFarlane, W. V. (1949). Actions of anti-cholinesterases on end-plate potential of frog muscle. *J. Neurophysiol.* **12**, 59–80.

Furukawa, T. (1957). Properties of the procaine end-plate potential. *Jap. J. Physiol.* **7**, 199–212.

Gage, P. W., McBurney, R. N. and Schneider, G. T. (1975). Effects of some aliphatic alcohols on the conductance change caused by a quantum of acetylcholine at the toad end-plate. *J. Physiol.* **244**, 409–430.

Hill, A. V. (1910). The possible effects of the aggregation of the molecules of haemoglobin on its dissociation curves. *J. Physiol.* **40**, iv–vii.

Katz, B. and Miledi, R. (1971). Further observations on acetylcholine noise. *Nature New Biol.* **232**, 124–126.

Katz, B. and Miledi, R. (1972). The statistical nature of the acetylcholine potential and its molecular components. *J. Physiol.* **224**, 665–699.

Katz, B. and Miledi, R. (1973a). The characteristics of "end-plate noise" produced by different depolarizing drugs. *J. Physiol.* **230**, 707–717.

Katz, B. and Miledi, R. (1973b). The binding of acetylcholine to receptors and its removal from the synaptic cleft. *J. Physiol.* **231**, 549–574.

Katz, B. and Miledi, R. (1973c). The effect of atropine on acetylcholine action at the neuromuscular junction. *Proc. R. Soc. B.* **184**, 221–226.

Katz, B. and Miledi, R. (1973d). The effect of α-bungarotoxin on acetylcholine receptors. *Br. J. Pharmac.* **49**, 138–139.

Katz, B. and Miledi, R. (1975a). The effect of procaine on the action of acetylcholine at the neuromuscular junction. *J. Physiol.* **249**, 269–284.

Katz, B. and Miledi, R. (1975b). The nature of the prolonged end plate depolarization in anti-esterase treated muscle. *Proc. R. Soc. B.* **192**, 27–38.

Kordas, M. (1969). The effect of membrane polarization on the time course of the end-plate current in frog sartorius muscle. *J. Physiol.* **204**, 493–502.

Kuba, K., Albuquerque, E. X., Daly, J. and Barnard, E. A. (1974). A study of the irreversible cholinesterase inhibitor, diisopropylfluorophosphate, on time course of end-plate currents in frog sartorius muscle. *J. Pharmac. Exp. Ther.* **189**, 499–512.

Kuffler, S. W. and Yoshikami, D. (1975). The number of transmitter molecules in a quantum: an estimate from iontophoretic application of acetylcholine at the neuromuscular synapse. *J. Physiol.* **251**, 465–482.

Landau, E. M. and Ben-Haim, D. (1974). Acetylcholine noise: analysis after chemical modification of receptor. *Science* **185**, 944–946.

Magleby, K. L. and Stevens, C. F. (1972a). The effect of voltage on the time course of end-plate currents. *J. Physiol.* **223**, 151–171.

Magleby, K. L. and Stevens, C. F. (1972b). A quantitative description of end-plate currents. *J. Physiol.* **223**, 173–197.

Magleby, K. L. and Terrar, D. A. (1975). Factors affecting the time course of decay of end-plate currents: a possible co-operative action of acetylcholine on receptors at the frog neuromuscular junction. *J. Physiol.* **244**, 467–496.

Miledi, R. (1960). The acetylcholine sensitivity of frog muscle fibres after complete or partial denervation. *J. Physiol.* **151**, 1–23.

Neher, E. and Sakmann, B. (1975). Voltage-dependence of drug-induced conductance in frog neuromuscular junction. *Proc. Nat. Acad. Sci.* **72**, 2140–2144.

Potter, L. T. (1968). Uptake of choline by nerve endings isolated from the rat cerebral cortex. *In* "The Interaction of Drugs and Subcellular Components of Animal Cells" (P. N. Campbell, ed.), pp. 293–304. Churchill, London.

Rang, H. P. (1975). Acetylcholine receptors. *Quart. Rev. Biophys.* **7**, 283–399.

Sachs, F. and Lecar, H. (1973). Acetylcholine noise in tissue culture muscle cells. *Nature New Biol.* **246**, 214–216.

Steinbach, A. B. (1968). Alteration by xylocaine (lidocaine) and its derivatives of the time course of the end-plate potential. *J. Gen. Physiol.* **52**, 144–161.

Verween, A. A. and DeFelice, L. J. (1974). Membrane noise. *Progr. Biophys. Molec. Biol.* **28**, 189–265.

3

Morphology of Synaptic Vesicle Discharge and Reformation at the Frog Neuromuscular Junction

J. Heuser

Dept. of Physiology, School of Medicine, University of California, San Francisco, U.S.A.

I Introduction

As efforts to determine the structural basis of synaptic transmission have progressed beyond the early pure descriptions of synaptic fine structure in the electron microscope, and have sought to correlate fine structural changes with synaptic function, the characteristic properties of neuromuscular junctions that make them ideal experimental preparations for the physiologist—such as their accessible location and relative hardiness *in vitro*—have become important to the neuro-anatomist as well.

This chapter will describe how neuromuscular preparations that are isolated from frogs, and subjected to a variety of experimental manipulations *in vitro*, and then fixed by immersion in standard chemical fixatives for electron microscopy (or, more recently, frozen by application of a block of supercooled metal), can be stimulated to undergo structural alterations that help to elucidate the role of synaptic vesicles and other presynaptic organelles in the secretion of acetylcholine from the motor nerve terminal.

II Methodological considerations

The most difficult part of these morphological investigations has been to find neuromuscular junctions within the experimental muscles once

they have been fixed and embedded in plastic. The sparse distribution
of junctions—usually one to each muscle fibre—may be an advantage
to the electrophysiologist who wishes to study one synapse in isolation,
and can poke around in a muscle with a microelectrode until he has
found one, or more recently can use Nomarski light-microsopic optics
to directly visualize living ones; but for the morphologist who must cut
slowly and blindly through a muscle which has been made opaque by
impregnation with heavy metals that are needed to provide contrast in
the electron microscope, the paucity of neuromuscular junctions
within a muscle is the major technical obstacle.

Some investigators have circumvented this problem by studying the
electric organs of fishes, which have an abundance of synapses that are
essentially evolved from neuromuscular junctions. But these organs are
much more difficult to isolate or maintain in a good physiological state
in vitro, which is a prerequisite for the correlative studies we have in
mind which involve controlled stimulation and monitoring.

In the face of this difficulty, it is fortunate that neuromuscular

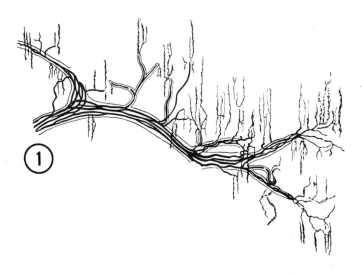

Fig. 1. Drawing from Kuhne (1888) of the terminal ramifications of a bundle of
motor axons in a frog twitch muscle, which illustrates how adjacent neuromuscular
junctions overlap.

junctions in the frog have a peculiar and characteristic shape. This was shown over a hundred years ago by Kuhne (1887) and Von Kolliker (1863) who used a gold-chloride stain, that selectively impregnates the nerves within muscles, to show that where a motor axon joins a muscle fibre in the fast twitch muscles of the frog (Fig. 2), it sprouts into a tiny bush with delicate branches that are oriented precisely parallel to the muscle fibre and, fortunately for the morphologist, extend for a considerable distance along the fibre, often up to 0·5 millimetres. We shall see later that the precise orientation and cylindrical shape of these nerve terminals has been a great help in interpreting the morphological changes found in them.

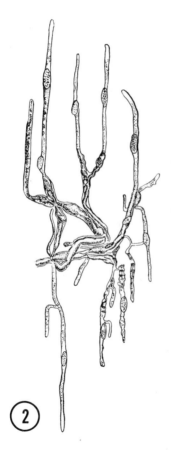

Fig. 2. Drawing from Kuhne (1887) of the sort of terminal bush that each individual motor axon forms at the frog neuromuscular junction.

The immediate advantage of their elongated form, however, is that the chances of cutting into one are greatly increased when frog muscles are sectioned transversely, or perpendicular to the long axis. Indeed, since junctions on adjacent muscle fibres tend to occur in the vicinity of each other, a single transverse section through a region of frog muscle that contains the terminal arborization of a bundle of motor axons, such as Fig. 1, will often strike a number of different junctions, and as Fig. 3 illustrates, will often provide views of several different random points within individual nerve terminal bushes, since each bush often consists of multiple branches that run parallel to each other. Thanks to this, transverse sections through frog muscles provide a quick and accurate assessment of how many junctions are affected by any experimental treatment and a general idea of what structural changes they have undergone.

On the other hand, the chances of finding such an elongated junction

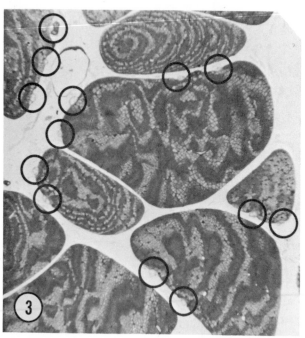

Fig. 3. Transverse 1 micron thick-section through the edge of a frog sartorius muscle in a region rich in neuromuscular junctions, such as Fig. 1, which provides a random sample of several different junctions and several different branches within the individual junctions.

in a lengthwise section through a frog muscle, cut along the axis of the muscle fibres, is correspondingly worse. When finally one is struck, however, it can provide a panoramic view of the interior of the nerve terminal which immediately gives a clear indication of the distribution and extent of the structural changes that have occurred within the individual synapse, such as can be seen by comparing Fig. 5 with Fig. 4. Taken together, transverse and longitudinal sections can thus provide an accurate assessment of structural changes in the sort of synapse which is, of course, also uniquely suitable for physiological analysis.

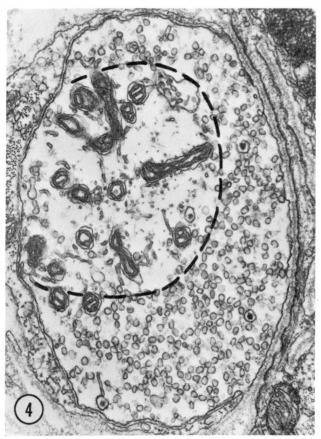

Fig. 4. Electron micrograph of a single terminal branch cut transversely, which illustrates its central filamentous core which is an extension of the pre-terminal axon and contains many mitochondria (surrounded by dashes), and the cytoplasmic expansion which characterizes its secretory region and is filled with synaptic vesicles. × 20 000.

III Organization of the nerve terminal: vesicles and presynaptic specializations

In random cross-section views of frog neuromuscular junctions such as Fig. 4, we can see that where the motor axon contacts the muscle, it possesses a cytoplasmic expansion that is filled primarily with hordes of synaptic vesicles. When electron microscopy first revealed that this secretory region of the nerve terminal was filled with vesicles (Palade, 1954), del Castillo and Katz (1955) proposed that these tiny organelles were packets of the transmitter, acetylcholine, which they had predicted might exist within the nerve terminal, when they found that acetylcholine was secreted from it in discrete multimolecular pulses or quanta (del Castillo and Katz, 1954). In one sense, the whole point of this chapter will be to illustrate how strong the morphological evidence now is in favour of this vesicle hypothesis.

Also in the random cross-section such as Fig. 4, we can discern a backbone that runs throughout the length of each terminal branch and consists of a direct continuation of all the organelles found in the axon before it reached the muscle, especially the long bundles of neurofilaments and the extensive reticulum of narrow membranous tubes which occur all along the axon, as well as a few microtubules and other variegated membrane forms. We shall see how these structures represent the link between the synaptic terminal and cell body that is manifest by axoplasmic flow (Rambourg and Koenig, 1975). Couteaux (1960) clearly explained how the fibrous parts of this backbone were probably what were stained in the old gold-chloride preparations that revealed the unique branching pattern of frog motor nerves so clearly. In addition, as in all presynaptic terminals, mitochondria are abundant in cross-section views of the frog terminals. As Fig. 4 illustrates, they are most often encountered back away from the presynaptic surface, at the margin of the vesicle-rich cytoplasmic expansion, and mixed up in the filamentous backbone of the terminal.

A closer look at the contents of the presynaptic cytoplasm reveals that every so often, in about 1 to 10 random cross-sections or once in every micron in longitudinal sections, the presynaptic membrane possesses a row of electron-dense tufts of vaguely granular or filamentous material that extend up into the cytoplasm for a few hundred Angstroms before fading out of view (Fig. 5). In the immediate vicinity of these tufts, synaptic vesicles appear to reach their highest concentration in the cytoplasm, and—most important—appear to establish their closest proximity to the presynaptic membrane. Such a distribution of vesicles around presynaptic dense tufts was first observed in CNS synapses (Palay, 1958; Gray, 1963), and has long been interpreted to mean that these specializations demarcate zones on the presynaptic membrane where synaptic vesicles probably discharge transmitter.

Early circumstantial evidence that the presynaptic dense tufts in frog nerve terminals could indeed be sites of transmitter discharge was their close physical register with the postsynaptic folds (Fig. 5). When Birks *et al.* (1960) first showed the presynaptic densities in the frog by staining them with osmium tetroxide and phosphotungstic acid, they found that each occurred directly above a deep transverse fold in the surface of the muscle, which they imagined would be an admirable arrangement for increasing the receptive area of the postsynaptic membrane immediately opposite from where transmitter might be discharged from the nerve terminal.

Their micrographs showed that much of the muscle membrane beneath the nerve terminal, including some of the membrane in the folds, was also underlaid with an electron-dense fuzz of about the same thickness, and just as indistinct, as the presynaptic density. Recently it has been possible to show that this specialized postsynaptic membrane can be selectively labelled with toxins that bind to the acetylcholine receptor (Fertuch and Salpeter, 1974; Daniels and Vogel, 1975; Lentz *et al.*, 1975) and displays in freeze-fracture (as seen in Fig. 8) a high concentration of punctate deformations or particles in the plane of fracture through the lipid interior of the membrane (Peper *et al.*, 1974; Heuser *et al.*, 1974), of the sort that have been correlated in other membranes with the location of surface receptors (Marchesi *et al.*, 1972) and transmembrane ion channels (McNutt and Weinstein, 1970). This is all good evidence that acetylcholine receptors are concentrated in the electron-dense regions of the postsynaptic membrane, and thus supports the original notion that the folds increase the receptive area of the postsynaptic membrane.

IV   Freeze-fracture evidence for vesicle discharge

A   APPEARANCE OF THE "ACTIVE ZONE"

What is so fortunate about the precise and characteristic transverse orientation of these pre- and postsynaptic specializations could not be fully appreciated until the frog neuromuscular junction was visualized

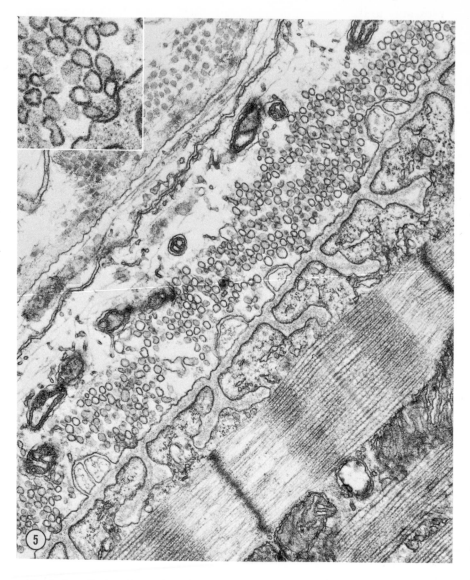

Fig. 5. Lengthwise thin section through a nerve terminal from a resting frog neuro-muscular junction that was fixed in osmium tetroxide, which shows the characteristic clusters of synaptic vesicles around electron-dense tufts that overlie the presynaptic membrane just opposite deep folds in the muscle plasma membrane, one of which is shown at higher magnification in the inset. × 35 000.

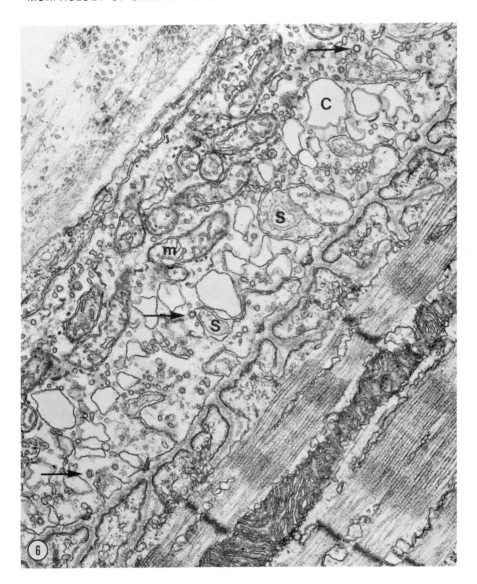

Fig. 6. Thin section of a nerve terminal that was stimulated at 10 Hz for 15 min and then fixed exactly as in Fig. 5, which shows that prolonged acetylcholine secretion results in a severe depletion of synaptic vesicles and expansion of the plasmalemma into deep folds that contain Schwann cell processes (S), and a proliferation of numerous coated vesicles (arrows) and cisternae (C) in place of the missing synaptic vesicles. Also this treatment results in a swelling of the mitochondria (m) which is reversible after stimulation. × 30 000.

in freeze-fracture. The pioneering application of this technique by Dreyer *et al.* (1973) afforded broad panoramic views of the pre-synaptic surface of the nerve terminal, such as Fig. 7, on which the presynaptic specializations could immediately be recognized as the transverse ridges on the fracture face which are elevated towards the opposite muscle fold and demarcated by two double rows of unusually large intramembranous particles, of the sort—as mentioned previously —thought to be functional protein molecules, possibly ion channels, embedded in the inner lipid core of the membrane (Branton, 1971). What also characterizes these presynaptic specializations in freeze-fracture is that just beside each of them occur many discrete plasma-lemmal deformations, which look like circular dimples or craters in the presynaptic membrane, depending on the freeze-fracture view (cf. Fig. 7 with Fig. 17). Such deformations have been found in other sorts of tissues, from CNS synapses (Pfenninger *et al.*, 1972) to capillary endothelial cells (Simonescu *et al.*, 1974), where it can be seen that they are sites at which underlying vesicles are fused to the plasmalemma. Thus it was immediately assumed that the ones found just beside the presynaptic specializations of frog nerve terminals represented examples of synaptic vesicles fused with the plasmalemma.

In fact, prior to this pioneering freeze-fracture study, Couteaux and Pecot (1970) showed that in longitudinal thin-sections of frog muscles fixed with aldehyde, they could find many examples of pockets in the presynaptic membrane which were also located just beside the elongated presynaptic densities. Couteaux recognized that if these were indeed synaptic vesicles fused with the plasmalemma, their restricted distribu-tion seemed to indicate that vesicle discharge could occur only beside the presynaptic specializations; and with this in mind he renamed these regions the "active zones" of the nerve terminal.

B EVIDENCE THAT VESICLES DISCHARGE TRANSMITTER

Critical evidence in favour of the idea that the plasmalemma deforma-tions seen in thin-sections and freeze-fractures do indeed represent discharging synaptic vesicles is that they cannot be found on resting or unstimulated nerve terminals such as Fig. 15 (Heuser *et al.*, 1974). To illustrate this, we had to carefully remove the neuromuscular prepara-tions from frogs without stimulating them unduly, soak them in Ringer's solution containing magnesium instead of calcium in order to

Fig. 7. Freeze-fracture view of a nerve terminal fixed with glutaraldehyde during 10 Hz stimulation, which illustrates the active zones that occur opposite each muscle fold (arrow). In this fracture, the outer half of the plasma membrane was stripped away to reveal the interior of the presynaptic membrane as seen from outside the nerve terminal. The small particles that are scattered over the fractured surface and occur in distinct rows at the active zones are thought to be functional proteins embedded in the membrane. The larger dimples are thought to be the deformations produced by underlying synaptic vesicles that were caught in the act of fusing with the plasma membrane to discharge acetylcholine during fixation. × 40 000.

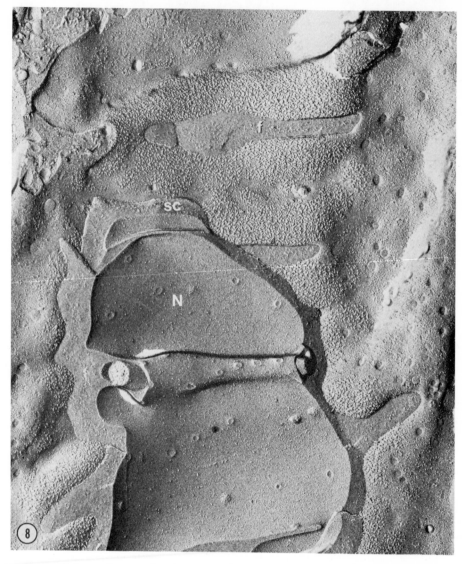

Fig. 8. A freeze-fracture that has scooped away all of the nerve terminal except a patch of the outer leaflet of its membrane (N) and has jumped across the synaptic cleft (sc) to reveal the internal structure of the postsynaptic membrane as it looks from outside the muscle. The large clusters of distinctive intermembranous particles which characterize the muscle membrane in these regions and extend for some distance into the postsynaptic folds (f) correspond to the location of acetylcholine receptors. × 45 000.

prevent any transmitter secretion from occurring at the last moment when the electron microscopic fixative depolarizes the nerve, and finally make sure that the fixative did not exert other damaging osmotic or ionic effects directly on the nerve terminals as they were being fixed. Prepared in this way, the active zones could be recognized by their basic and unchanging elevation and rows of particles, but no plasmalemmal deformation could ever be found beside them (Fig. 15).

In contrast, when isolated neuromuscular preparations were tetanized by 20 Hz nerve stimulation in normal, calcium Ringer as they were fixed in aldehydes, plasmalemmal deformations occurred abundantly beside the active zones, as in Fig. 10, but not when the secretory response to this nerve activity was blocked by soaking them in magnesium during the stimulation, as in Fig. 9 (Heuser et al., 1974). Since then, several other sorts of nondepolarizing stimuli of acetylcholine secretion have also been found to produce the same sort of plasmalemmal deformations in the same location; including such stimulation as ionic lanthanum, black widow spider venom and high osmotic pressure. Thus such deformations are strictly correlated with transmitter secretion and, we believe, are examples of synaptic vesicles that have been caught in the act of fusing with the plasmalemma to discharge transmitter, which is the process termed "exocytosis" in other secretory cells (deDuve, 1963).

The muscles that were investigated in the pioneering studies mentioned previously were not intentionally stimulated during fixation and so it might seem inconsistent that they displayed any plasmalemmal deformations at all. But there is now reason to suspect that in both studies some nerve terminals may have been inadvertently stimulated, either by an unusually high osmolarity of the electron microscopic fixative (Couteaux and Pecot, 1970), or by the radical dissection originally thought necessary to obtain freeze-fracture samples (Dreyer et al., 1973). Now that we have seen nerve terminals which have been quick-frozen by pressing them against a copper block cooled to 4°K, without ever exposing them to electron microscopic fixatives or otherwise stimulating them, we are sure that absolutely no signs of synaptic vesicle fusion occur in the absence of stimulation (Heuser et al., 1976).

C QUANTITATIVE CONSIDERATIONS

Of course, we expect that the slow spontaneous trickle of quanta from these terminals which produce m.e.p.p.'s in the absence of any stimula-

Fig. 9. An active zone, viewed as in Fig. 7, from a nerve terminal which was stimulated at 10 Hz in Ringer's solution containing 10 mM magnesium in place of all
calcium as it was fixed with glutaraldehyde, which reveals one of the few examples of
vesicle fusion we could find (arrow) in terminals that are stimulated but unable to
secrete acetylcholine because of the lack of calcium. × 95 000.

Fig. 10. An active zone from a nerve terminal stimulated at 20 Hz in normal Ringer's
solution as it was fixed with glutaraldehyde, which shows how abundant images of
vesicle fusion can be, when nerve stimulation is accompanied by the normal levels of
acetylcholine secretion. × 95 000.

Fig. 11. An active zone from a terminal stimulated in normal Ringer's solution at the
same rate as the terminal in Fig. 10, but fixed instead with a low concentration of
formaldehyde, which shows an unusually large number of fusing vesicles are captured
by such a slow-acting fixative. × 95 500.

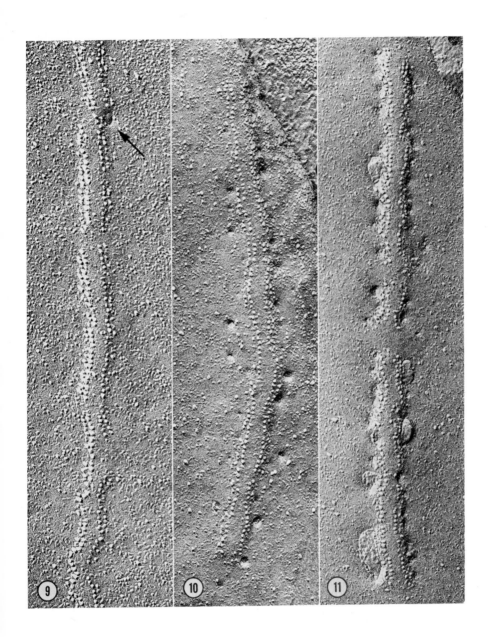

tion should also result from the same sort of vesicle discharge. But when we remember that nerve stimulation increases the instantaneous rate of quantal discharge by more than 100 000 fold (Katz, 1969), and that frog terminals can sustain a continuous rate of secretion at least 1000 times greater than the rate of spontaneous m.e.p.p. production (Heuser and Miledi, 1971), we realize how rarely we ought to see vesicles discharging in the absence of stimulation.

This depends on how long the membrane perturbations that result from each vesicle fusion remain visible, of course; but if we assume for the moment that each lasts 10 ms, which is an estimate we have reached from quick-freezing experiments discussed later, then the spontaneous resting release of 1 m.e.p.p. per s from an average neuromuscular junction would mean that a vesicle fusion should be visible somewhere over the terminal only one one-hundredth of the time. And since an average terminal can have some 500 active zones, we might expect to find a vesicle fusion only once in every 50 000 active zones; that rarely !

Unfortunately, any such attempt to correlate the number of vesicle fusion sites with the rate of quantal discharge will soon run into considerable difficulties, because it will lead us to doubt, as it led Birks *et al.* to doubt in their original study (1960), that we could ever catch synaptic vesicle discharge in the electron microscope. For example, if a terminal were secreting 1000 quanta per s, as it can do when activated with lanthanum or 10 Hz nerve stimulation, and the membrane changes resulting from each quantal discharge persisted about 10 ms, then we ought to find only 10 vesicles discharging at any moment in time throughout the whole terminal, and distributed among 500 active zones this would be only 1 vesicle fusion in every 50 active zones. How then can we explain the abundance of vesicle fusion sites we and others have found in stimulated active zones, such as Fig. 10? Should we conclude that they actually persist much longer than our estimate of 10 ms?

Some time ago we predicted that signs of the event might last 1 s, and were disturbed that we could not see it in terminals stimulated with lanthanum (Heuser and Miledi, 1971). That turned out to be because of problems with osmium tetroxide fixation (Heuser and Reese, 1975). When we fix instead with aldehydes during exposure to lanthanum, or during nerve stimulation, we often find as many as 10 vesicle fusion sites per active zone, which would suggest, working backward through the above simple calculations, that each membrane perturbation that results from a fused vesicle lasts about 10 s. This seems more and more unlikely as we learn about the fluid nature of biological membranes (Singer and Nicolson, 1973).

D THE EFFECT OF ALDEHYDE FIXATION

To make matters worse, we have shown that slowing down the rate of fixation during stimulation (by applying a low concentration of a poor crosslinking agent, formaldehyde, rather than the usual higher concentration of a better crosslinking agent, glutaraldehyde) yields an even greater abundance of images of vesicle fusion, often as many as 40 per active zone as in Fig. 11, which are as many as one zone can pack in at a time (Heuser et al., 1974). Rather than conclude that it is natural for each fusion reaction to last 40 s, we believe instead that during the last 40 s of fixation, just before the nerve terminals become completely fixed, nearly all the vesicles which discharged became trapped in the fused configuration and thereby remain visible to us. Perhaps this occurred because aldehyde fixatives stiffen the membrane or make the cytoplasm more viscous before they interrupt the molecular machinery that produces synaptic vesicle fusion with the presynaptic membrane.

We and several other groups have now reported that it is possible to record intracellularly from frog muscle fibres long enough during aldehyde fixation to confirm that secretion continues for a minute or more after the fixative is applied (Heuser and Reese, 1974; Hubbard and Laskowski, 1972; Clark et al., 1976). Such recordings have given no indication that aldehyde fixatives cause a final massive burst of transmitter release during stimulation, which could have been an alternative explanation for the unusual abundance of fusing vesicles that we are concerned about.

A further indication that the membrane perturbations produced by vesicle discharge are more transient than the abundance of vesicle fusions in fixed tissues would predict, is that we have been completely unable to find any instances of vesicle fusion, at all, in several dozen muscles that we have quick-frozen during stimulation at 20 Hz or even higher frequencies (Heuser et al., 1976). This method of freezing has been developed to arrest all function and all structural changes as fast and as synchronously as possible, and best estimates are that freezing is complete in less than 10 ms after the muscle is pressed against a copper block cooled to 4°K with liquid helium (Van Harreveld and Crowell, 1964; Van Harreveld *et al.*, 1974). But even when we carefully trigger a single nerve impulse, or apply a massive field-stimulation to the entire muscle, in the last few milliseconds before the muscle contacts the cold copper block, we can find no images of vesicle fusion. One conclusion might be that vesicle fusion does not occur *at all* during life, only during stimulation in fixatives. But for the moment we prefer to conclude simply that the presynaptic membrane changes that accompany each vesicle fusion normally persist less than 10 ms and are greatly prolonged and finally arrested midstream during aldehyde fixation.

V   Towards a model of vesicle discharge

A   THE ORIGINAL MODEL

Such problems with the fixative have somewhat discouraged efforts to scrutinize the detailed structure of the plasma membrane deformations for clues to exactly how synaptic vesicles undergo exocytosis. But what exactly are the molecular events that lead up to and cause vesicle fusion? We can begin to analyse this basic question by recalling the details of the vesicle hypothesis Katz originally proposed (1969). He imagined that synaptic vesicles are free to mill around inside the nerve terminal in Brownian motion, and as a result are constantly colliding with the presynaptic membrane, but cannot fuse with it and discharge transmitter unless they strike a "reactive site" on the membrane which can somehow start off the exocytotic process. He further supposed that the calcium which enters the nerve terminal during depolarization greatly increases the number of reactive sites in the presynaptic membrane and thus, for a brief time, until the calcium is somehow got rid of, allows many more vesicle collisions to be "effective" in discharging transmitter. Let us see how each aspect of this hypothesis fares in light of the morphological data that is currently available.

B   MODIFICATIONS REQUIRED BY THE STRICT LOCALIZATION OF VESICLE DISCHARGE

First, it must be modified to take account of the extreme concentration of vesicles and images of vesicle discharge around the active zones. In particular, Couteaux and Pecot (1974) were the first to appreciate that it is common to find whole rows of intact vesicles in the cytoplasm just exactly where images of vesicle fusion always occur. Clearly, this structural organization dictates that a vesicle must first come to lie within one of these rows before it can fuse with the plasma membrane and discharge acetylcholine.

The vesicles in these rows are usually so closely spaced, and so precisely aligned beside the dense tufts of filaments that extend up from the presynaptic membrane at the active zone, that it looks like they are not free to mill around throughout the terminal, but are held down by some attachment to the tufts. To see if such attachments really exist we will need better methods of fixing cytoplasmic filaments (cf. Gray, 1975). But even if nothing actually holds these vesicles in the correct position for discharge, it is clear that they are somehow concentrated around the active zone in close proximity to the plasmalemma; so vesicle collisions with the presynaptic membrane must be much more frequent in these areas.

C FURTHER IMPLICATIONS OF VESICLE LOCALIZATION: A READILY-RELEASABLE FRACTION

A further implication of the fact that a vesicle must lie beside the active zone before it can discharge is that the number of vesicles that can achieve this position at any one time is limited by the total length of the active zones. For example, a typical nerve terminal has around 500 active zones, each of which can accommodate about 40 vesicles when it is bounded by two complete rows (or "double cartridge belts" as Couteaux calls them), for a total of about 20 000 vesicles that can be in a favourable position for discharge at any one time, which is less than 10% of all the vesicles in the nerve terminal.

There is also biochemical and electrophysiological evidence that the amount of acetylcholine which is ready to be released at any moment in time is limited to a fraction of the total store of transmitter in the synapse. The evidence is that acetylcholine release promptly declines during continuous stimulation, long before the total stock of acetylcholine in the synapse declines, but soon reaches a steady plateau (Brown and Feldberg, 1936; Birks and MacIntosh, 1961). This has been interpreted to mean that acetylcholine is drawn from a small readily-releasable pool of transmitter which is promptly depleted during repetitive stimulation, and releases progressively less transmitter, until the rate at which acetylcholine is released from it equals the rate at which acetylcholine from the total store in the synapse is mobilized into it, at which time its size stabilizes and the rate of transmitter release plateaus.

The idea of such a readily releasable pool is not consistent with the original hypothesis that all vesicles were free to mill around in the synapse and collide with the presynaptic membrane, but it does fit with the existence of a limited population of vesicles that are in a favourable position for discharge.

If we consider the rows of vesicles beside the active zones to be the morphological counterpart of the readily releasable pool, then mobilization of transmitter into the pool could be thought of as movement of vesicles into spots along active zones vacated when other vesicles discharge, probably by random Brownian movement of vesicles free in the cytoplasm.† In this way, the presynaptic dense tufts could hold a stock of vesicles that would sustain transmitter release at a higher rate than could be sustained simply by random movements of vesicles up to the active zones on the presynaptic membrane. But this stock would be limited to some 20 000 vesicles, or about enough for 100 end-plate potentials, or a couple of seconds of jumping.

D A FURTHER RESTRICTION ON VESICLE DISCHARGE

However, even though these 20 000 vesicles may constitute a readily-releasable fraction, it would appear that something *further* limits how many of these vesicles can discharge with each nerve impulse. Usually, only 200 or so quanta discharge with each nerve impulse, and yet the variations in how many discharge from one impulse to the next are so small that they fit binomial statistics (Wernig, 1975). In his original hypothesis, Katz imagined that there were so many vesicles bombarding the presynaptic membrane that even during the peak of transmitter output after the nerve impulse, the chance that any individual collision would be effective in discharging transmitter was still very low, so the output would fit the Poisson approximation to binomial statistics (Katz and Miledi, 1967; Katz, 1969). And even if there are only 20 000 vesicles

---

† Possibly, the conditions that have been reported to increase the rate of transmitter mobilization, such as prolonged stimulation or potassium depolarization (Hubbard, 1963; Braun *et al.*, 1966), could do so by raising internal calcium, and thereby lowering the viscosity of the axoplasm (Hodgkin and Katz, 1949), which would be expected to increase such Brownian movement of vesicles (Einstein, 1905; Shea and Karnovsky, 1966).

quivering near the presynaptic membrane ready to discharge, repeated discharges of around 200 of them at a time ought still to fit Poisson's statistics.

But what we will have to conclude from the reports that transmitter release fits binomial statistics is that only around 500 quantal packets are immediately available for discharge at any moment in time (Wernig, 1975), or only a fraction of the 20 000 or so vesicles that look like they are all in equally favourable positions for discharge. The number 500 is so close to the total number of active zones in the terminal that it immediately leads us to wonder if it is not highly probable that one quantum will discharge from each active zone and highly improbable that more than one could discharge with each nerve impulse (A. R. Martin, personal communication). But why only one, when more than 40 vesicles are lined up beside each active zone?

Will we discover that each quantum actually represents the synchronous discharge of many vesicles from an individual active zone (Kriebel and Gross, 1974)? If this were so, it would be difficult to conceive of how the basic constancy of quantal amplitude could be maintained during prolonged transmitter release, which is one of the cornerstones of the quantal hypothesis (del Castillo and Katz, 1956). Also, it is unlikely for several other reasons, including—as we shall see later—the fact that during several sorts of stimulation, the number of vesicles which disappear and later reappear is not any greater than the number of quanta that are discharged (Heuser and Reese, 1973). Moreover, there are now several reports of treatments that will produce unusually large m.e.p.p., and in all cases these large quanta have been correlated with unusually large vesicles in the affected nerve terminals, rather than with synchronized discharge of smaller vesicles (Pecot and Couteaux, 1972, 1975; Heuser, 1974). So there is no compelling reason at the moment to doubt the notion that one quantum results from the discharge of one vesicle, and unless we learn differently when we catch all the vesicle fusions that result from one single nerve impulse by quick-freezing, we shall have to look more closely at the 40 vesicles along each active zone for subtle structural differences that might explain why only one could discharge with each nerve impulse.

The way in which transmitter release at the neuromuscular junction declines almost exponentially when three or four nerve impulses are given in rapid succession (so called "early tetanic rundown") indicates that the number of quanta immediately available for discharge rapidly declines from the number of 500 with successive nerve impulses (Elm-qvist and Quastel, 1965b). Takeuchi (1958) has shown that this number recovers exponentially after each nerve impulse with a time constant of 4–5 seconds. The low Q_{10} Takeuchi found for this time constant suggested that recovery depends on diffusion of large molecules (or vesicles) rather than a chemical reaction. This may illustrate not only that each active zone is limited to discharging one quantum at a time, but that after it discharges one, it is refractory to further discharge for a short period, until something which depends on diffusion is recon-stituted.

However, we see that recent data from several approaches suggests that discharge probably does not result from a statistically improbable event such as the collision of a free vesicle milling around in the synapse with a reactive site on the presynaptic membrane, but rather from a statistically probable event such as one particularly well-situated vesicle at an active zone interacting with an adjacent spot of presynaptic membrane.

VI   The mechanism of vesicle discharge

A   CALCIUM CONTROL OF VESICLE AND PLASMA MEMBRANE CONTACTS

If we now ask what is the exact nature of the interaction of vesicles with the plasma membrane, and how calcium alters these interactions to initiate exocytosis, we must first take account of the prevailing ideas about the nature of membrane/membrane interactions in general. It is widely held that most natural membranes have a negative charge on their surfaces, due to the polarized orientation of charged proteins and phospholipids in them. It is further supposed that such negative surface charges produce an electrostatic repulsion between two membranes when they approach each other, the strength of which should increase exponentially as they come closer and closer, and thus should constitute an energy barrier that would prevent them from coming into direct contact (Parsegian and Gingell, 1972). Synaptic vesicles have been shown to be negatively charged by electrophoresis (Vos et al., 1969; McLaughlin et al., 1973), so it is reasonable to think that when they collide with the presynaptic membrane they are prevented from coming too close to it by such an electrostatic repulsion. When synapses can be properly quick-frozen, it should become possible to show that favourably located vesicles near the active zones do not come any closer to the presynaptic membrane than a certain minimum distance, and thus provide evidence for the existence of such an electrostatic energy barrier.

Experiments with artificial membranes have illustrated that divalent cations such as calcium can overcome such an energy barrier and induce negatively charged membranes to contact each other and stick, either by screening their opposed negative charges or else by linking them together and thus acting as a bridge between them (Liberman and Nenashev, 1970; Neher, 1974; Lansman and Haynes, 1975). The calcium which enters the synapse with each nerve impulse may exert such an effect on vesicle and plasma membranes (Blioch et al., 1970). A spot on the inside of the presynaptic membrane whose negative charges become screened by calcium might thus be receptive to direct vesicle contact, or considered "reactive" in Katz' terminology.

During impulse transmission, such contacts may go on promptly to vesicle fusion; but fortunately they do not all progress to fusion so quickly that some cannot be captured by electron microscopic fixatives, because the ability to see such vesicle contacts in the electron microscope has provided some evidence that they are, indeed, induced by divalent cations. Palade (1975) has obtained several micrographs of secretory vesicles in the pancreas that are in direct contact with the plasma membrane, and reports that such contacts are easier to find in tissues soaked in elevated concentrations of calcium before fixation (personal communication). Boyne et al. (1974) have also observed intimate, contacts between synaptic vesicles and presynaptic membranes of nerves in Torpedo electric organs,† and found that such contacts were more abundant in organs that were fixed in the presence of 90 mM calcium. But what is particularly interesting about their results is that such close vesicle contacts were also more abundant in organs fixed in 90 mM magnesium, which supports the idea that there is indeed an

---

† Originally, Boyne et al. proposed that these contacts might be permeable, like gap junctions and might allow acetylcholine to be released from the vesicle without further membrane fusion. However, we have observed such contacts in freeze-fractured frog nerve terminals as well as Torpedo synapses, and find that in the region of contact such as is illustrated in Fig. 13, vesicle and plasma membranes are completely devoid of their usual complement of intramembranous articles (Heuser and Reese, 1975a). This illustrates how intimate the membrane contact is, but does not support the idea that it is permeable, because gap junctions possess a great abundance of intramembranous particles which probably represent ion channels through the opposed membranes (McNutt and Weinstein, 1970).

electrostatic repulsion between the vesicle and plasma membranes that can be overcome by various divalent cations.

We have found that similar vesicle and plasma membrane contacts are abundant, as Fig. 12 illustrates, in frog motor nerve terminals soaked in isotonic magnesium during life (Heuser *et al.*, 1971), and presume that this treatment produced an elevation in internal magnesium concentration that must have been sufficient to neutralize the negative surface charges on vesicle and plasma membranes. Furthermore, with

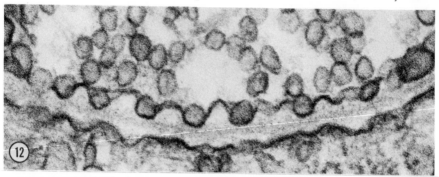

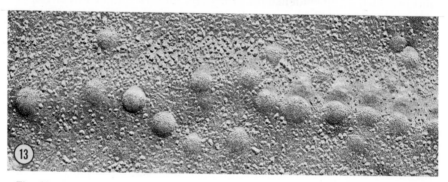

Fig. 12. Thin-section view of the close contacts between vesicles and plasma membranes which are induced by soaking frog muscles in isotonic magnesium chloride for 8 h at 4°C. These vesicle contacts are abnormally extensive and cannot go on to fusion and transmitter discharge, but are evidence that magnesium as well as calcium can neutralize an electrostatic repulsion between vesicle and plasma membranes and cause such intimate contacts to occur. × 105 000.

Fig. 13. Freeze-fracture appearance of the close vesicle contacts produced by isotonic magnesium as in Fig. 12, which illustrates that all of the intramembranous particles in the presynaptic membrane are excluded from the regions of contact. Such close opposition of the phospholipid backbones of the adjacent membranes is thought to be a prerequisite for vesicle fusion. × 120 000.

this treatment it is quite clear that the vesicle contacts persist for a long time, because they occur in equally great abundance after all sorts of fixation, fast or slow, and even after quick-freezing, which shows exactly how many exist at one moment in time and involves no artifactual induction of such contacts by fixation. Thus such contacts do not proceed immediately to vesicle fusion. In fact, frog junctions soaked in isotonic magnesium like this become totally unable to discharge quanta, and no longer even display spontaneous m.e.p.p.'s, until they are returned to normal Ringer solution, when they recover some degree of normal function (Heuser et al., 1971). It would appear that something else must happen after vesicle contact, something that magnesium cannot trigger but calcium can, before vesicles will fuse with the plasma membrane and discharge acetylcholine.

B WHAT HAPPENS DURING VESICLE FUSION?

It is difficult to conceive of what else this might be, when the general understanding of membrane fusion is currently so weak. There is no direct information about what molecular rearrangements occur during fusion. The occasional vesicle has been found in electron micrographs to form a contact with the plasma membrane which appears to possess fewer than the usual number of dark lines that would signify complete integrity of its membrane (Palade and Bruns, 1968), and from these Palade (1975) has proposed that fusion involves a progressive elimination of membrane layers from the area of contact. Our views of vesicle contacts in freeze-fracture (Fig. 13) would add that one component of the membrane, the intramembranous particles, is eliminated from the area of contact as soon as it forms, before elimination of the phospholipid backbones begins. Other than this, there is no information on how vesicle and plasma membranes fuse at the point of contact and produce an orifice that releases acetylcholine from the vesicle.

Nevertheless, most membrane biologists agree that fusion probably results from some sort of increased freedom of motion of the phospholipids in the area of contact between vesicle and plasma membranes, which leads to phospholipid exchanges or diffusional flip-flops between them, that somehow go on to complete fusion (Papahodjopoulos et al., 1974; 1975; Pagano and Huang, 1975; Van der Bosch and McConnell, 1975). Such phospholipid rearrangements undoubtedly slow down in the cold, which may be reflected in the dramatic way that transmitter release can be delayed by cooling (Katz and Miledi, 1965).

Some authors think that contact itself leads to close range interactions among the phospholipid backbones of the adjacent membranes that progress inevitably to fusion (Poste and Allison, 1973). But fusion clearly does not follow inevitably when vesicle contact is promoted by magnesium, as we have just seen. Nor does it follow inevitably when artificial membranes are prompted to contact by divalent cations (Liberman and Nenashev, 1970; Neher, 1975; Lansman and Haynes, 1975). Calcium must do more to promote vesicle fusion.

C CALCIUM'S ROLE IN VESICLE FUSION

For example, calcium might promote phospholipid exchanges by causing a small patch of phospholipids in one or both of the membranes to gel, which would generate a phase boundary between the gelled patch and surrounding fluid regions of the membrane (Papahadjo-poulos and Poste, 1975; Jacobson and Papahadjopoulos, 1975). At such phase boundaries, it has now been shown that phospholipids have increased freedom of motion (Van der Bosch and McConnell, 1975). The lack of intramembranous particles that we find at points where vesicles contact the plasmalemma (Fig. 13) could represent such a phase change, since intramembranous particles are typically pushed aside when patches of phospholipids gel within a membrane (Verklej et al., 1972; Kleemann and McConnell, 1974). But such bare patches of plasma membrane also occur at vesicle contacts induced by magnesium, which do not go on to fuse. Probably intramembranous particles are pushed aside whenever vesicles approach the plasma membrane very closely, so that the bare patches do not reflect whatever more basic phospholipid rearrangements are involved in calcium's action.

At this point we should recall that secretion from many sorts of cells is prevented by metabolic inhibitors (Jamieson and Palade, 1971) and may directly require ATP (Wooden and Weineke, 1964). But how the hydrolysis of ATP could be involved in membrane phospholipid rearrangements is not known. Early studies speculated that the products of hydrolysis might simply "destabilize" vesicle and plasma membranes by chelating calcium in their vicinity (Wooden and Weineke, 1964).

The recent discovery of actin and myosin-like proteins in nerve terminals (Berl et al., 1973) has raised the alternative possibility that ATP hydrolysis might be part of a calcium-triggered contractile event that leads to vesicle fusion. If so, it is *not* likely to be contraction of a filament linking the vesicle and plasma membranes, even if one were shown to exist; because we have just seen that contact between them is also promoted by magnesium and thus appears to result from a simple screening of negative charges. It is more likely that the calcium-dependent contractile event, if there is one, involves a conformational change of proteins within the plasma membrane that affects the organization of phospholipids.

D INVOLVEMENT OF INTRAMEMBRANOUS PARTICLES IN VESICLE FUSION

In looking for such functional molecules within the plasma membrane, we should consider the unique intramembranous particles that delineate the active zones in the presynaptic membrane of frog terminals. Their very proximity to the event would implicate them in vesicle fusion. Plasma membrane particles of a similar size and shape have been found to occur in distinctive clusters at sites where secretory vesicles undergo fusion in certain protozoa (Satir *et al.*, 1973). In fact, those particles actually surround the exocytotic orifice when it appears, and spread out as the orifice grows, so they are thought to be intimately involved in the molecular rearrangements leading to membrane fusion in this organism. But whether the movements of these membrane particles represents an active contractile or conformational change in them which leads to fusion, or simply a passive displacement as a result of the vesicle fusion, remains to be seen.

At the frog neuromuscular junction the physical relationship between particles and vesicle fusion looks somewhat different. At least in alde-hyde fixed terminals, images of vesicle discharge are found in the immediate vicinity of the intramembranous particles, never actually in their midst (Fig. 10); and as a result, discharge does not appear to involve any obvious displacement of the particles as it does in protozoa.†

---

† This has been a difficult point to determine, because we have found instances in which the rows of particles are disorganized after prolonged stimulation and we can completely disorganize them by prolonged soaks in high magnesium (Fig. 13). But in these cases, the particle movements seem to be part of a much more wide-spread disorganization of the internal structure of the synapse that results from such treatments. In rat neuromuscular junctions, however, images of vesicle fusion do appear in the *midst* of the short rows of intramembranous particles (Rash and Ellisman, 1974).

Of course, the sluggish way in which aldehyde fixatives arrest membrane changes, including particle movements (McIntyre *et al.*, 1974), probably would not be adequate to capture any particle changes that occur on the rapid time scale of synaptic transmitter discharge. So whether or not these particles move, or change in any other way during vesicle fusion, ought to remain an open question until this process can be visualized in quick-frozen synapses.

Even if they don't visibly change in any way, we ought to consider what other sorts of functional proteins these particles might be. It has been proposed that phospholipases in secretory vesicle membranes could increase the motional freedom of phospholipids in the plasma membrane and induce fusion by enzymatically digesting them (Lucy, 1969). But the evidence for the existence of such an enzyme in secretory membranes has now been called into severe question (Meldolesi *et al.*, 1971; Parkes and Fox, 1975) and none has ever been found in synaptic vesicle membranes (von Hungen *et al.*, 1968; Morgan *et al.*, 1973; Morgan *et al.*, 1974).

But whatever the particles might be, the important point is that the membrane changes which lead to vesicle fusion occur, during nerve stimulation at least, only in one location on the presynaptic membrane, and that is adjacent to these large particles. Even when these particles form ectopic rows that are not located over postsynaptic folds, we find that vesicle fusion sites still occur beside them (Fig. 5 in Heuser *et al.*, 1974). Thus, even if the particles are not directly involved in the molecular processes that lead to membrane fusion, they must somehow define where this process can occur. For example, they could represent receptors for vesicles, or anchoring points for filaments attached to the vesicles, that determine the initial location at which vesicles contact the plasma membrane. Also, by being so stably located at the active zones of the presynaptic membrane, they could thus indirectly produce the precise localization of vesicle fusion which we find there.

Alternatively, we have speculated that they could be the calcium channels in the presynaptic membrane themselves (Heuser *et al.*, 1974). If calcium channels were this localized, the distribution of calcium which entered with each nerve impulse ought to be so restricted that only vesicles near the active zones could be induced to discharge. After all, calcium has less than 0·5 ms to exert its effect (and actually closer to 0·2 ms when we realize that it takes some time for the calcium channels to open during a nerve impulse; Llinas and Heuser, 1976), and probably has such a low mobility in axoplasm that it can diffuse only a few hundred Angstroms in that brief time (Blaustein, 1974). Thus a highly localized injection of calcium ought to produce just as localized an effect on vesicle fusion.

If the location of calcium channels did determine the location of vesicle discharge in this way, however, we should expect that any treatment which increased the concentration of intracellular calcium diffusely ought to produce a diffuse vesicle discharge, rather than the localized discharge we see after nerve stimulation; but we seem to have the contrary result. All the alternative methods that we have used to induce transmitter release so far (including lanthanum, black widow spider venom, hypertonicity and isotonic magnesium), all of which may stimulate by raising internal calcium concentrations (Alnaes and Rahamimoff, 1975), produced a predominantly localized vesicle discharge. Unfortunately, it is impossible to be certain that they do not produce any diffuse vesicle discharge outside the active zones, because—as we shall see in the next section—they all inevitably induce endocytosis as well, which can occur all over the nerve terminal and can look just like exocytosis, and so would obscure any diffuse exocytosis that might have occurred. But, for the moment, we must conclude that vesicle discharge remains localized to the active zones regardless of how it is stimulated, and we have no morphological evidence to support the idea that a localization of calcium entry determines the localization of vesicle discharge. Of course it remains attractive to think that calcium might be injected into the nerve parsimoniously, only at the active zones where it is needed.

Support for this possibility awaits some morphological method to determine the localization of calcium channels, such as the availability of an agent which binds specifically to them and can be seen in the electron microscope. Possible candidates for such an agent include lanthanum and Botulinum toxin, both of which may arrest transmitter release by directly blocking the calcium channels (Harris and Miledi, 1971; Miledi, 1971). Lanthanum is electron dense and could be visualized directly if it were present in sufficiently large amounts, and Botulinum toxin could be coupled with horseradish peroxidase and located histochemically (Nakane and Kawaoi, 1974).

E SUMMARY: ROLE OF THE ACTIVE ZONE IN VESICLE FUSION

We are left with the impression that probably all of the structural specializations, that we see at the active zones, are part of a broad system of functional molecules that all act in concert to produce some control over where vesicles are supplied to the presynaptic membrane and, specifically, produce some convergence of this supply on to patches of the plasma membrane, where they also control and maintain a concentration of membrane molecules that are critical for the vesicle fusion reaction, thereby assuring that we will find vesicle discharge concentrated beside the active zones.

F   DEVELOPMENT OF THE ACTIVE ZONES

An indication of how this whole system of molecules becomes organized comes from the appearance of *developing* active zones on regenerating nerve terminals in denervated frog cutaneous pectoris muscles. M. Dennis, L. Evans and I have now examined several dozen regenerating terminals which we think had resided in contact with the muscle for less than one week before we took them for electron microscopy. We can see in freeze-fracture views, such as Fig. 14, what is also apparent in thin sections; namely that developing active zones are discontinuous structures, composed of individual short segments that each contain intramembranous particles and are surrounded by vesicles just like the adult active zones, and are aligned side by side in correct orientation over the muscle folds (which persist after the previous nerve degenerates). Adult active zones are rarely so discontinuous, as Fig. 9 illustrates quite dramatically, though they can occur in widely varying lengths, as they conform to vagaries in the postsynaptic pattern of folds.

The individual short segments that compose developing active zones show some variations that may reflect how the intramembranous particles in the presynaptic membrane become aligned above muscle folds, but in general look sufficiently well organized to support some vesicle discharge, albeit not as much as the long continuous active zones of the adult. So the synapse should gain in secretory capabilities as the active zones grow and fill in discontinuities, until they reach their adult capacity.

VII   The consequences of vesicle fusion

A   IS VESICLE FUSION REVERSIBLE?

In considering the outcome of vesicle fusion, we must first of all remember that no signs of vesicle fusion and no other alterations of the

Fig. 14. Freeze-fracture view of developing active zones in a regenerating nerve terminal, which illustrate that they consist of short segments which are more or less complete but are not as long and continuous as they are in the adult (Fig. 9). Most segments are already correctly aligned along the gentle ridges in the nerve surface that belie where postsynaptic folds were located in the muscle membrane which has been fractured away, but a few variants are present (arrows) which may represent earlier stages of development of the active zone. × 45 000.

presynaptic membrane are visible in muscles that have been quick-frozen within about 10 ms after fusion begins (Heuser *et al.*, 1976), so the fusion reaction must normally resolve quite quickly. In his original vesicle hypothesis, Katz (1969) imagined that the fusion between vesicle and plasma membranes would be a very transient event, lasting just long enough to permit acetylcholine discharge. This implied that the fusion reaction would be reversible, and that the vesicle simply closed off again after discharging a quantum. Indeed, the freeze-fracture appearance of vesicle fusion sites in CNS synapses (Pfenninger *et al.*, 1972) is somewhat consistent with such a model, because these fusion sites are all very small, and they comprise what look like open and closed forms that were in fact postulated to change back and forth reversibly when it was thought that they persisted longer than they do (Streit *et al.*, 1972).

We are cautioned from looking too closely at such images of vesicle fusion that aldehyde fixatives capture for the electron microscope, because they may be so distorted during the final moments of secretion and the early stages of fixation. We could only hope to see what actually happens to vesicles after discharge if we could catch the process with quick-freezing. However, what we can see in aldehyde-fixed neuro-muscular junctions such as Figs 11 and 17, is that vesicle fusion sites vary considerably in size and shape, and include some forms that look like very wide open or collapsed vesicles. These variations should not occur if vesicles open only a little and then immediately close off again.

Even more important, if vesicle fusion were a transient, reversible event, vesicles should never disappear during stimulation. For example, if all the vesicles in the first row beside the active zone just sat there and opened and closed occasionally, their numbers need never diminish as they discharge transmitter. But what is found is that even very brief periods of stimulation produce a transient depletion of synaptic vesicles in this local area,[†] and longer stimulation produces a more persistent depletion of vesicles throughout the nerve terminal (Heuser and Reese, 1973).

---

† Such local depletion of vesicles has been shown so far only in ganglionic synapses in the frog (Dickensen and Reese, 1974), but should be easy to catch at the active zones of motor nerve terminals by quick-freezing frog muscles, even though it remains hard to catch the vesicle fusion reactions themselves.

B EVIDENCE THAT VESICLES COLLAPSE AFTER FUSION

We think that the missing vesicles have collapsed completely into the plasma membrane after fusion, until their membrane is no longer readily distinguishable from the rest of the plasmalemma. The evidence for this is that during prolonged nerve stimulation the surface area of the presynaptic terminal also expands, and does so, as closely as we could measure in synapses stimulated at 10 Hz and 10°C, to about the same extent as the amount of vesicle membrane inside the synapse declines, which implies a bulk shift of vesicle membrane into the plasma membrane (Heuser and Reese, 1973). Such plasmalemmal expansion has been seen during stimulation of all sorts of other secretory cells and would be expected to occur if vesicle fusions usually proceed to complete collapse, but not if fusion were immediately reversible (Palade, 1959; Amsterdam et al., 1969; Pysh and Wiley, 1974).

Recently we have obtained more direct evidence from freeze-fracturing that it is indeed vesicle membrane which accumulates on the surface of nerve terminals during secretion (Heuser and Reese, 1975b). Thanks to the fortunate circumstance that synaptic vesicles display one or two unusually large intramembranous particles on the fracture face of the cytoplasmic leaflet of their membranes (Fig. 23), we have been able to treat this morphological feature as a marker of synaptic vesicle membrane and count how many such particles appear on the surface of the nerve terminals during stimulation. The counts become quite large after prolonged stimulation (cf. Fig. 16), and are consistent with the extent of vesicle collapse into the presynaptic membrane that we would have expected from the previous measurements of membrane surface areas in thin sections.†

---

† Each vesicle has two or three such particles and a membrane surface area of 0.004 μ^2, for a concentration of 500 large particles per square micron. At a time when stimulation would be expected to have increased the surface area of nerve terminals by 25%, we find that the concentration of large intramembranous particles in the presynaptic membrane reaches nearly 100 per square micron, compared to background counts of 25 such particles per square micron at rest. If at this time, 80% of the surface of the nerve terminal was covered by old plasma membrane with 25 particles per square micron, and 20% was covered by vesicle membrane with 500 particles per square micron, the average concentration of particles would be 0.8 × 25 plus 0.2 × 500 = 120 particles per square micron, which is quite close to that observed (Heuser and Reese, 1975a).

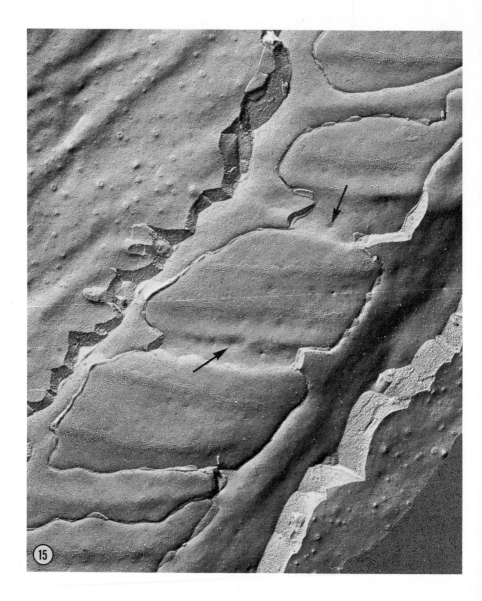

Fig. 15. Freeze-fracture view of a resting nerve terminal to illustrate that the smooth contours of the nerve are periodically interrupted by processes of the Schwann cells that repeatedly embraces it and separates adjacent active zones. Where Schwann fingers have been broken away during fracture, shallow dimples can be seen in the surface of the nerve terminal (arrows) which are examples of the type of endocytosis that becomes so abundant after stimulation (see Fig. 16). × 28 000.

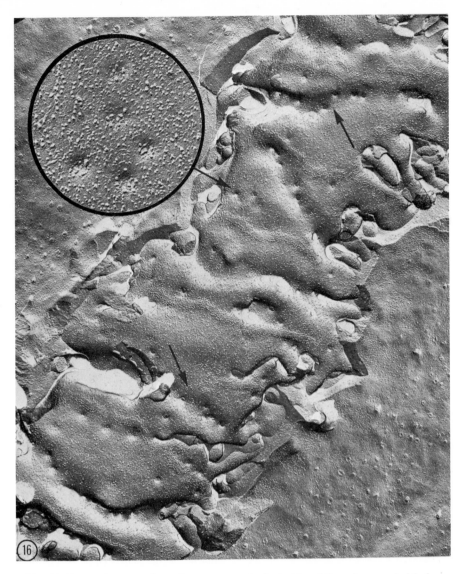

Fig. 16. Freeze-fracture view of a nerve terminal stimulated for a long period before fixation, to illustrate the consequences of exocytosis of synaptic vesicles, including the extreme degree of surface distortion that can result from adding large amounts of vesicle membrane to the plasma membrane, and the increased concentration of large intramembranous particles in the plasma membrane, which are thought to be a distinctive component of the vesicle membrane. Also present are numerous small dimples which look like the sites of vesicle fusion that occur beside the active zones (Fig. 7), but *here* occur all over the presynaptic surface, and particularly in the shallow grooves formed by Schwann cell processes that were ripped away during fracture (arrows). At higher magnification, these dimples can be seen to bear a number of unusually large intramembranous particles, which serve to identify them as the freeze-fracture image of coated vesicles caught in the process of budding off from the plasma membrane, and distinguish them from synaptic vesicle exocytosis.

Of course, since we know nothing about the biochemical nature of these intramembranous particles, we cannot be sure that the ones which appear on the surface are the same as those originally in vesicles, they just look the same. In this regard, it is disappointing that the first attempt to produce antibodies to vesicle proteins, and to demonstrate by immunocytochemical methods that these vesicle proteins appear on the surface of the nerve terminal during stimulation, have so far yielded negative results (E. Heilbronn, personal communication).

Fig. 17. A freeze-fracture view of synaptic vesicle exocytosis that is complementary to Fig. 11. Here the nerve terminal has been scooped out during fracture, along with the synaptic vesicles that were fused to the plasma membranes, so that the outer leaflet of the presynaptic membrane which remains (which we view as if we were inside the nerve terminal looking out) displays crater-like plasmalemmal elevations at points where fused synaptic vesicles were avulsed from it. The large intramembranous particles which delineate the active zones mostly adhere to the inner half of the membrane which was pulled away, but leave faint imprints on this half of the presynaptic membrane which serve to identify the location of three active zones in this field, and illustrate that synaptic vesicle exocytosis occurs only in the immediate vicinity of these zones. The wide variety in sizes and shapes of these vesicle fusion sites may be different stages of vesicle collapse after fusion. × 70 000.

Fig. 18. A view like Fig. 17 of a terminal stimulated for a long period before fixation (like the one in Fig. 16 was stimulated) to illustrate that endocytosis of coated vesicles can occur all over the surface of the nerve, in sharp contrast to exocytosis of synaptic vesicles, which would be concentrated beside the two active zones visible in this field. If it were not for this characteristic difference in location, it would be hard to distinguish vesicle discharge from vesicle retrieval in such views of the presynaptic membrane, since the large particles which cluster within nascent coated vesicles mostly adhere to the inner half of the membrane and leave only faint imprints on this view (arrow). × 63 000.

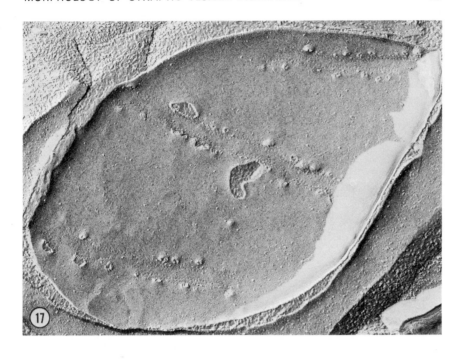

C THE RESULTING NEED FOR MEMBRANE RETRIEVAL

Of course, such collapse of vesicles after discharge could not occur indefinitely, in the absence of some sort of retrieval of vesicle membrane from the surface, or else the synapse would promply blow up like a balloon and completely run out of synaptic vesicles; as in fact it does when stimulated with black widow spider venom (Clark *et al.*, 1972), which may indeed block vesicle membrane retrieval, as we shall see. Moreover, there is other strong evidence that vesicles must be used over again, because it is now quite clear that frog neuromuscular junctions can secrete many times more quanta than the number of vesicles they typically contain at rest (Heuser and Miledi, 1971; Ceccarelli *et al.*, 1972, 1973), and can do so within a period of time too brief to imagine that many new vesicles could have been provided by axoplasmic flow, even if the axons had not been severed during the dissection of the neuromuscular preparation. On the other hand, they apparently cannot maintain secretion so long when black widow spider venom is used as a stimulus (Ceccarelli *et al.*, 1973). In response to this venom, frog motor nerve terminals secrete just about as many quanta as the number of vesicles they typically contain at rest, and then end up swollen and completely empty of vesicles (Clark *et al.*, 1972). Something in the venom must block the retrieval of vesicle membrane from the presynaptic surface and thus prevent vesicle reuse. Ordinarily, however, vesicle retrieval must be able to keep pace with vesicle discharge, and this must be why so many of the early efforts to deplete synaptic vesicles by nerve stimulation failed.

There is currently some disagreement in the literature about how synaptic vesicles reform from the plasma membrane. On the one hand, some authors think that vesicles might be able to reform in the same spot where they collapsed, if the molecular processes involved in exocytosis could be completely reversed (Ceccarelli *et al.*, 1973). Shown often as evidence for this is a micrograph of a synaptic vesicle fused to the plasma membrane at an active zone, which contains the extracellular tracer horseradish peroxidase (HRP) (Ceccarelli *et al.*, 1972). This micrograph is evidence only that during fusion, synaptic vesicles establish a wide open continuity with the extracellular space—which is, after all, the definition of exocytosis—but the particular vesicle in question was forever fixed in the fused position and we can never know whether it was about to pinch off from the surface again or had not yet collapsed into it.

Somehow, extracellular tracers such as HRP can gain access to the interior of synaptic vesicles that are free inside the nerve terminal. A number of studies have now shown that nerve stimulation in the presence of HRP, as well as other sorts of stimulation that cause transmitter discharge, dramatically increases the proportion of vesicles inside the synapse that contain this extracellular tracer (Holtzman *et al.*, 1971; Ceccarelli *et al.*, 1973; Heuser and Reese, 1973). Undoubtedly, the tracer is taken up when vesicle membrane is retrieved from the plasma-lemma. But where and how this occurs remains the question, and as in all such attempts to establish the route and timetable of a membrane redistribution, the answer has been sought by applying a brief pulse of the HRP tracer.

This we accomplished by fixing frog neuromuscular junctions within seconds or a minute after we began to stimulate in HRP, and then determining which organelles contain the first wave of HRP that is carried into the synapse (Heuser and Reese, 1973). We found that HRP definitely did not appear first in the vesicles that were lined up near the active zones, as would have been expected if they could reform directly from the plasma membrane after they discharged. Instead, as Fig. 19 shows, HRP appeared first in vesicles and larger vacuoles or flattened cisternae that were located at some distance away from the active zones. And among the labelled vesicles, most could be distinguished from synaptic vesicles because they were surrounded by a delicate web of cytoplasmic filaments that identified them as "coated vesicles".

D MEMBRANE RETRIEVAL BY COATED VESICLES

We should not have been surprised to find HRP in coated vesicles, since such vesicles have been shown to take up extracellular proteins in all sorts of cells (Bowers, 1964), including neurons (Rosenbouth and Wissig, 1964), and even the mammalian neuromuscular junction (Zacks and Saito, 1969). But finding that nerve stimulation could activate the formation of coated vesicles was new, even though we had suspected as much when we saw that coated vesicles were more abundant in lanthanum-stimulated terminals when we fixed them with osmium tetroxide and then stained them with uranyl acetate, which is a combination that dramatically accentuates the prominence of cytoplasmic filaments in the nerve terminal and is required to see coated vesicles clearly (Heuser and Miledi, 1971).

Coated vesicles have been found in all sorts of nerve terminals since Gray (1961) first saw them in mossy fibre terminals of the cerebellum (Gray and Willis, 1970). As it became clear that they formed from coated regions of the plasmalemma that dimpled and eventually pinched off from the surface, several investigators hypothesized that the membrane they take up from the plasmalemma could be used to make synaptic vesicles (Palay, 1963; Andres, 1964; Westrum, 1965). When our results showed that coated vesicle formation was induced by stimulation, it became possible to seriously consider that this form of vesicle retrieval or endocytosis could be coupled to synaptic vesicle exocytosis and could thus serve as a major route for the retrieval of vesicle membrane from the plasmalemma (Heuser and Reese, 1973). As with other aspects of the vesicle discharge process we are evaluating, this may also be the case in other sorts of secretory cells (Grynzpan-Winograd, 1971; Nagasawa et al., 1971; Orci, 1974).

From freeze-fracture we have learned a good deal more about this endocytosis of coated vesicles. We can recognize it in freeze-fracture views of the presynaptic plasmalemma by certain structural details that I will describe later. We find it can occur in considerable abundance after a prolonged period of transmitter secretion, such as in Fig. 16, in which case we see that it can occur almost anywhere on the presynaptic surface (though not usually at the active zones themselves, where there is probably no space among the tight rows of synaptic vesicles for a coated vesicle to form). When endocytosis is less prevalent, most examples of it occur beneath the Schwann cell processes that girdle the nerve terminals between nearly every active zone, and occur especially often at the edge of the Schwann cell's extension over the presynaptic plasmalemma, where perhaps there begins a special external environment maintained by the Schwann cells. Also when endocytosis is rare, as in the resting terminal in Fig. 15, it usually occurs beneath Schwann processes, but then happens often to involve bigger individual patches of membrane, which would form bigger coated vesicles than those seen during stimulation (Heuser and Reese, 1973; Heuser et al., 1974).

The relatively diffuse occurrence of endocytosis, however, in spots often far removed from the precise localization of exocytosis, means that if the same vesicle membrane constituents which join the plasmalemma at the active zone are to be retrieved, as they must be retrieved if the vesicle and plasma membranes are to maintain their chemical individuality (Von Hungen et al., 1968; Morgan et al., 1974; Gurd et al., 1974), they would have to diffuse laterally in the membrane for some distance. It is reasonable that they could do so (Singer and Nicolsen, 1973), but it would undoubtedly produce some scrambling of the constituents, which we may be seeing in freeze-fracture, because the large particles that

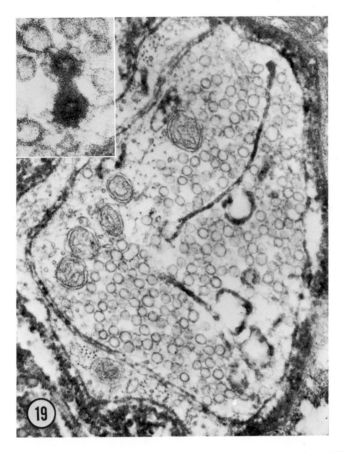

Fig. 19. Cross section of a frog terminal bathed in horseradish peroxidase (HRP), the location of which is identified by the black histochemical reaction product which surrounds the termal. This terminal was stimulated at 10 Hz for 60 s just before fixation, so the intercellular vacuoles and cisternae which contain HRP represent a form in which membrane is first retrieved from the plasma membrane after vesicle discharge. Note that the synaptic vesicles near the presynaptic surface do not contain the tracer, so they do not form directly from the presynaptic membrane. In the inset is a view of two coated vesicles fused to each other which also contain HRP, from a different terminal stimulated for 60 s, to illustrate that these compartments also form by endocytosis. × 50 000.

appear on the surface during stimulation and presumably come from vesicle membrane are invariably scattered randomly all over the presynaptic surface by the time we fix it (Fig. 16).

E   COATED VESICLES MAY BE SELECTIVE

Possibly the coat of coated vesicles bears multiple recognition sites for vesicle membrane constituents, or at least for those constituents with

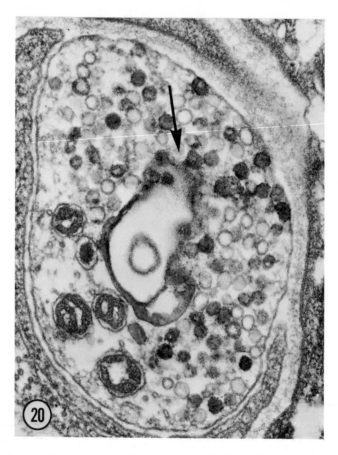

Fig. 20. Cross section of a terminal stimulated at 10 Hz for 15 min in the presence of HRP, and then rested and washed for 1 h, until the cisternae which took up HRP initially have nearly all disappeared and delivered the extracellular tracer into a new generation of synaptic vesicles. The one remaining cisternae in this field may have been forming new synaptic vesicles at the arrow. × 60 000.

cytoplasmic representation, which must become occupied before it can detach from the plasma membrane. This would serve to collect the vesicle membrane constituents together again before retrieval. Freeze-fractures such as Fig. 16 have provided some evidence for such selectivity in retrieval, by showing that the membrane of nascent coated vesicles possesses an unusually high concentration of intramembranous

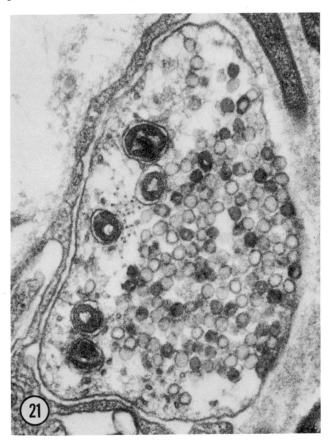

Fig. 21. Cross section of a terminal stimulated for 15 min in HRP and then rested, to illustrate that this period of stimulation turns over about half of all the synaptic vesicles, which is about how many quanta discharged during the stimulation. Though the reformed vesicles are here filled with dense reaction product, in life they probably contain only a very few HRP molecules and must be functionally normal, because in the absence of further stimulation, they remain in the terminal for at least 18 hours—the longest we have looked—but they promptly disappear if the terminal is stimulated again. × 65 000.

particles, some of which are particularly large ones on the cytoplasmic leaflet of its membrane, and thus could be the same sort of distinctive particles that appear to originate in synaptic vesicles and scatter over the surface during stimulation, finally in the process of being selectively retrieved.

There is a precedent for thinking that the coat might be able to confer some selectivity to membrane retrieval, because at the level of transfer of materials from the rough endoplasmic reticulum to the Golgi apparatus of pancreatic secretory cells, where distinct biochemical differences exist between the membranes of the two organelles, the transfer operates apparently by a shuttle of vesicles back and forth which somehow avoid mixing the distinct constituents of the two membranes. These "transition vesicles" have a distinct coat of cyto-plasmic filaments, much like coated vesicles in the synapse (Jamieson and Palade, 1968; Palade, 1975).

The coat might reflect a selectivity of retrieval in another regard, also. Coated vesicles are notorious for taking up soluble proteins and may do so with some degree of specificity, probably by selectively retrieving surface receptors that bind the protein specifically (Bowers, 1964). There are soluble proteins in cholinergic synaptic vesicles (Whittaker et al., 1972a), which ought to be discharged along with the transmitter during exocytosis of synaptic vesicles, but which apparently —according to the most recent work—do not become depleted from the vesicles during repetitive discharge the way the transmitter itself does (Whittaker et al., 1972b; Zimmerman and Whittaker, 1974), and thus would have to be almost quantitatively retrieved after discharge and used over again. It is hard to imagine how endocytosis could retrieve a soluble protein quantitatively, but it may help that coated vesicle formation occurs most abundantly around the periphery of the pre-synaptic membrane (Heuser and Reese, 1973), in regions where protein may be retarded from leaving the synaptic cleft.

Faced with the problem of conserving this soluble protein inside vesicles as they discharge acetylcholine and other small molecules, Whittaker et al. (1972b) proposed that exocytosis might be only partial, that the vesicle might fuse with the plasma membrane only enough to let out small molecules and not large ones, which harks back to the idea that vesicle fusion is a limited, transient event and that vesicle attachment sites are permanent structures which simply open and close. As we have seen, such a mechanism of discharge would never lead to the vesicle depletion, surface expansion, or proliferation of coated vesicles and uptake of extracellular tracers that are now known to occur in response to nerve stimulation in several sorts of synapses besides the frog neuromuscular junction. We prefer to think that such vesicle proteins are discharged during vesicle exocytosis, at least to the extent that they are soluble, but that they are recaptured by selective retrieval into coated vesicles. What is not captured may eventually leak out of the synaptic cleft and appear as protein in the muscle perfusate (Musick and Hubbard, 1972).

F MEMBRANE RETRIEVAL BY CISTERNAE

The cisternae which also begin to form promptly during stimulation in HRP (Fig. 19) invariably contain this extracellular marker, and yet often are clearly isolated structures and not just folds in the plasmalemma; so they too must be produced by some sort of endocytosis of the plasmalemma, and represent another form of membrane retrieval (Heuser and Reese, 1973).

Because we often found cisternae in contact with coated vesicles, and often apparently fused to many coated vesicles, we initially thought that cisternae might form entirely by coalescence of coated vesicles (Heuser and Reese, 1973). We did not consider how fast cisternae could appear in the HRP uptake studies, and how many coated vesicles would have had to pinch off from the plasmalemma and coalesce together in just a few seconds to produce them. Nor at the time had we viewed cisternae in freeze-fracture, which has recently revealed that their membranes contain a very odd and variable distribution of intramembranous particles. As Fig. 22 illustrates, over broad stretches cisternae often have no particles at all, which is highly unusual for most biological membranes except those of myelin (Branton, 1971), and in only a few places do they ever have a population of particles approaching that of coated vesicle membranes.

Thus it now seems more likely that cisternae do not form secondarily from coated vesicles. Probably they form directly from invaginations of the plasma membrane. We have observed a few examples of plasmalemmal invaginations in stimulated terminals that also lack any intramembranous particles and could have been in the process of separating from the plasma membrane to form cisternae when they were fixed.

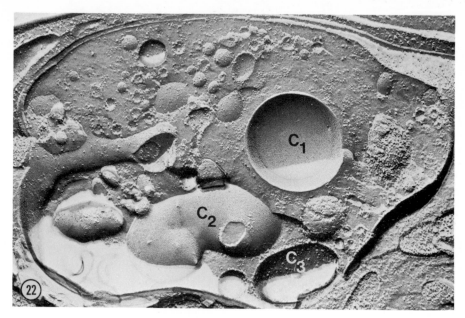

Fig. 22. A freeze-fracture that has broken across a nerve terminal stimulated at 10 Hz for 15 min at 10°C, like that in Fig. 6, to illustrate that large areas of the membranes of cisternae often contain no intramembranous particles (C_1, C_2), and thus are equivalent to myelin figures, while other areas possess more nearly normal particle populations (C_3). This odd particle distribution has led to the idea that cisternae do not form entirely by coalescence of coated vesicles, which are rich in particles, but by direct invagination of the presynaptic membrane during the period of rapid surface expansion and distortion consequent to vesicle exocytosis. × 70 000.

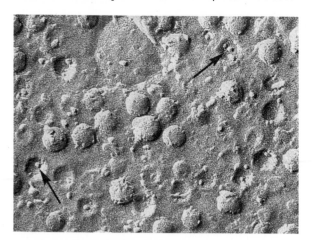

Fig. 23. High magnification view of freeze-fractured synaptic vesicles to illustrate that those which are scooped out leave one to three unusually large intramembranous particles behind on the remaining cytoplasmic leaflet of their membranes (arrows). When vesicles open on the surface of the nerve terminal, these particles should appear on the outer views of the presynaptic membrane such as in Fig. 16. × 140 000.

In contrast to endocytosis of coated vesicles, such a bulk uptake of very variable patches of membrane with very abnormal particle distributions would not be very selective for vesicle membrane retrieval. Often it might mistakenly entrap molecules that belong properly to the presynaptic membrane, which would explain why vesicles often interact with cisternae just like they do with the plasma membrane. For example, when high magnesium induces vesicles to form extensive contacts with the presynaptic membrane, they do so against cisternae as well, but not against other intracellular organelles (Fig. 13 in Heuser et al., 1971).

However, it remains to be seen how much nonspecific uptake of large blobs of membrane goes on during more physiological rates of transmitter secretion. Few synapses in the CNS ever show many large vacuoles or cisternae, even though they must have been in widely different states of activity when they were fixed.

G THE FATE OF CISTERNAE

What has led us to believe that cisternae are not a complete artifact of excessive stimulation, however, is that once they have accumulated in large numbers, their membrane does seem to be used over again to produce new vesicles. That is, when stimulation is stopped and frog terminals allowed to rest, cisternae soon become progressively less abundant while vesicles become more abundant, and the average size of the remaining cisternae declines (Fig. 24); so somehow cisternae can become partitioned again into smaller vesicles (Heuser and Reese, 1973). While this is occurring, cisternae may even begin to fill with acetylcholine and discharge occasionally, since we and others consistently find that unusually large and variable m.e.p.p.'s occur after intense stimulation, especially about an hour after this stimulation as shown in Fig. 24, when cisternae are almost back to vesicle size (Pecot-Dechavassine and Couteaux, 1972; Heuser, 1973). How cisternae with their unusual membranes could be partitioned into synaptic vesicles with a normal complement of intramembranous particles is a serious problem. Perhaps the coated vesicles we often find in contact with cisternae could help in this regard, if they could contribute necessary vesicle particles which they have specifically retrieved.

But some cisternal membrane may be hopelessly disorganized or for other reasons may be discarded from the terminal. Evidence that this occurs comes from a consideration of the studies that use HRP to trace pathways in the nervous system. These studies rely on the fact that when HRP is taken up by synapses at the site of injection, some of it is transported back up the axons to the cell bodies (Kristensson and Olsson, 1971). Moreover, this retrograde transport occurs within membrane-bounded organelles that are similar in size and shape to the cisternae and vacuoles that form during rapid stimulation (LaVail and LaVail, 1973). So it was not surprising to learn that the amount of HRP that is transported back up frog motor axons is practically nil when the nerve terminals are resting, but dramatically increases after

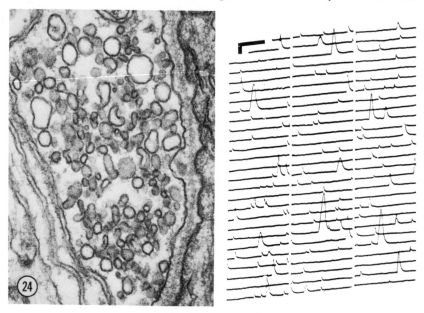

Fig. 24. Cross section of a frog terminal stimulated at 10 Hz for 15 min at 10°C (until it must have looked like Fig. 6) and then rested at 10° for 1 h before fixation, until vesicle numbers had returned toward normal and cisternae had been reduced to short lengths. The wide range of vesicle sizes which is characteristic of this period during recovery is reflected in the wide range of m.e.p.p. amplitudes that typically begin to appear about an hour after stimulation, shown on the right. (At the low amplification of 2 mv and 50 ms shown here, the little blips represent normal-sized m.e.p.p.'s and the large ones are "giants".) Presumably, by this time during recovery, cisternae have diminished in size enough to reach the active zones and discharge whatever acetylcholine they have already accumulated to produce the "giant" m.e.p.p.'s. × 140 000.

nerve stimulation (Litchy, 1973); presumably because some of the coated vesicles and cisternae that form during stimulation to retrieve vesicle membrane, and take up HRP in the process, are not used over again but are transported back up the axon. Apparently such discarded organelles end up as secondary lysosomes in the cell bodies and are eventually digested (LaVail and LaVail, 1973).

H HOW ENDOCYTOSIS OCCURS

Faced now with two possible types of endocytosis at the frog neuromuscular junction, we should consider what sorts of forces might be involved in producing each. The force for production of coated vesicles has long been thought to reside in the coat itself. Originally, the coat looked in thin sections like it was composed of a cluster of lollipop-shaped molecules attached to the membrane. This led to the idea that the force for deforming the plasma membrane into a sphere could be repulsion among the heads of these molecules (Roth and Porter, 1964). Later, when it became clear from negative-staining of isolated coated vesicles that the coat was, instead, a delicate basket of cytoplasmic filaments interconnected into polygons, it was imagined that re-arrangements among the connections of these filaments might provide the force for deforming a patch of membrane into a vesicle (Kanaseki and Kadota, 1969; Kadota and Kadota, 1973a).

Presumably energy must be required for the function of this coat, since it has been shown that the shuttle of coated vesicles that operates between the rough endoplasmic reticulum and the Golgi in the pancreas is stopped by metabolic inhibitors (Jamieson and Palade, 1968). Unfortunately, it has not been possible to show that coated vesicle production at the frog neuromuscular junction is also stopped by metabolic inhibitors, because these agents invariably disrupt transmitter release as well, either because energy is required for vesicle discharge in the first place (Beani et al., 1966), or because it is required to maintain a level of intracellular calcium compatible with the normal triggering of vesicle discharge (Glagoleva et al., 1970).

It would be very helpful to determine whether the coat requires energy to perform endocytosis at the synapse, because a positive answer would support the possibility that energy is required to deform a flat patch of membrane into a new vesicle. Also, if the plasma membrane did possess such a surface tension, then the opposite process—that of collapsing the deformation formed by a fused vesicle—should be energetically favourable. We would then have a simple thermodynamic explanation for why we find that vesicle formation is morphologically not just a mirror image of vesicle discharge; and why vesicles ordinarily collapse into the presynaptic membrane after they have discharged.

Exactly how the coat might use energy will not be answered without detailed biochemical analysis of its molecular composition. It could contain enzymes for transducting ATP hydrolysis into molecular motion, like myosin; or molecules for transmitting such molecular rearrangements to the membrane, like actin. Such molecules are present in synapses, as they are in all other cells, and the long-term maintenance of transmitter release does appear to depend on their function (Berl *et al.*, 1973; Thoa *et al.*, 1973b).

Even though it is possible to separate coated vesicles from other organelles by tissue fractionation (Kanaseki and Kadota, 1969; Kadota and Kadota, 1973), and to isolate their coats as a relatively pure protein (Pearse, 1975), in life their coats may actually be continuous with a much more widespread reticulum of filamentous and contractile proteins (Gray, 1972), which is now thought to be present throughout most of the cytoplasm of all cells. But what remains unique about their membranes is that some such filamentous material from the cytoplasm adheres to them tenaciously through the thick and thin of fixation or fractionation. This filamentous material would seem to be attached to the membrane strongly enough that it might be able to transmit a pull from the cytoplasm into a deformation of the membrane. Intramembranous portions of such attachments may form some of the many intramembranous particles we find in coated regions of the plasma membrane in freeze-fracture (Fig. 16). (For that matter, the large particles that distinguish synaptic vesicles might also turn out to be points of attachment for cytoplasmic filaments, which, if they all converged to form the presynaptic dense tufts, as Gray (1975) imagines, might draw the vesicles up to the active zones.)

The force that would produce larger deformations of the presynaptic membrane capable of forming cisternae is, however, another matter. Since cisternae form while the membrane is most rapidly expanding, we can only speculate that possibly during this period the phospholipids in it are threatened with compression and the resulting negative surface tension drives the membrane into large distortions or redundancies. The large areas devoid of intramembranous particles may arise because membrane phospholipids could flow more rapidly into such distortions that intramembranous particles, whose larger size would slow their rate of diffusion relative to phospholipids (Keleman *et al.*, 1974; Scandella *et al.*, 1972).

I A CENTRAL CLEARING-HOUSE OF MEMBRANE RECYCLING

Rather than attempt to fit all the various sorts of membrane compartments we find inside stimulated synapses into tight flow-sheets, perhaps we should simply think of them as all part of a multi-faceted clearing-house in the nerve terminal that is involved in the turnover of vesicle membrane. As Fig. 25 illustrates, membrane would be fed into this clearing-house by two sources: by flow down the axon of new membrane, in the form of smooth endoplasmic reticulum, and by retrieval of vesicle membrane from the plasmalemma, in the form of coated vesicles and cisternae. When large amounts of membrane were supplied rapidly by any of these sources, it would accumulate in variegated forms in the clearing-house, before slow points in its output processes, which would lead either to reformation of new synaptic vesicles or else, ultimately, to retrograde disposal of larger membrane forms (since only a few larger membrane compartments have been found to exist in resting nerve terminals, even when they are completely serial-sectioned (Andersson-Cedergren, 1959)). Anything that slowed vesicle reformation or otherwise reduced output from the clearing house would also exacerbate the accumulation of membrane in it. This may be why we found that stimulation in the cold caused the variegated membrane forms to become progressively more abundant, and compose a larger and larger fraction of all the internal membranes we measured (Heuser and Reese, 1973), until they eventually become the predominant form of membrane that remained inside severely exhausted nerve terminals stimulated in the cold with lanthanum (Heuser and Miledi, 1971).

A critical aspect of this model is that in the process of reforming synaptic vesicles, their individual membranes can at several stages become considerably disorganized and intermixed before they become reorganized into new vesicles. For this sort of turnover of vesicle membrane, we chose the term "recycling", in order to clearly distinguish it from the alternative scheme in which the integrity of the individual vesicle containers would at all times be conserved and the vesicles would simply be refilled with transmitter.

J THE IMPORTANCE OF LOCAL VESICLE RECYCLING FOR THE AUTONOMY OF THE SYNAPSE

Since a number of membrane constituents are delivered to the synapse by fast axoplasmic flow (Koenig et al., 1973; Droz, 1973; Forman et al., 1972), we must question the relative importance of this supply of new membrane, versus the local supply of recycled vesicle membrane, to the total economy of synaptic vesicles at the terminal. The answer seems rather obvious: even if isolated nerve terminals could not continue to secrete as long as they can in vitro, the fact that they contain a whole complex biochemical system capable of synthesizing new transmitter as fast as it is discharged illustrates that they are autonomous. Indeed, if terminals used each container of transmitter only once, and depended on a constant supply of new vesicles from axoplasmic flow, they could just as well stock the vesicles with transmitter in the cell body and eliminate all their biosynthetic machinery. Recycling of synaptic vesicles simply represents a local turnover of the transmitter containers concomitant with the local turnover of transmitter that occurs in the nerve terminal.

The studies of fast axoplasmic flow have not so far given any direct indication of how fast the total stock of vesicle membrane in the synapse is replaced by new membrane (Koenig et al., 1973). Best estimates of the life-time of vesicle membranes in the synapse range from one to several days (Von Hungen et al., 1968; Koenig et al., 1973); which would be long enough for most synaptic vesicles to recycle locally hundreds of times.

K A POSSIBLE DISCREPANCY BETWEEN TURNOVER OF TRANSMITTER AND RECYCLING OF VESICLES

However there is one apparent discrepancy between our concept of local vesicle membrane recycling and the prevailing views about the turnover of acetylcholine itself. Biochemical data on acetylcholine metabolism and discharge in stimulated synapses, most easily obtained from cat sympathetic ganglia, but available for neuromuscular preparations as well (Potter, 1970), have been interpreted to mean that a small portion of the total stock of transmitter in these tissues turns over faster than the rest, i.e. it is first to become labelled when hot substrates of acetylcholine are provided outside (Collier and MacIntosh, 1969) and then is discharged in preference to the larger store of unlabelled transmitter (Collier, 1969). If we were tempted by this biochemical data to imagine that within each individual synapse, some vesicles turn over faster than others, it would not fit at all with the relatively slow time course of the membrane retrievals and rearrangements that we have seen are involved in reforming vesicles in the neuromuscular junction. It would fit better with the idea that the vesicles near the active zone just sit there and open and close repeatedly; which, again, we believe does *not* occur for all the reasons described above.

But we often cooled the preparations in which we examined the time course of these phenomena, and a more recent HRP uptake study that applied extremely brief, local pulses of HRP directly onto individual nerve terminals suggests that these processes can operate much faster (McMahan and Yee, 1975). We could imagine that variations in the location where these processes occur could introduce differences in vesicle turnover time, and that those which occurred in the proximity of the active zones might yield new vesicles in a preferential position for discharge; if it were not likely that the viscosity of the axoplasm is low enough so that rapid mixing of organelles and vesicles is going on all the time. But rather than speculate further on how variations in vesicle turnover–time might come about, we should remember that the biochemical data could just as well result from some whole synapses turning over faster than others.

L HOW VESICLES ARE RECHARGED WITH TRANSMITTER

In any case, it is hard to imagine how certain vesicles could long contain a pool of acetylcholine different from others in their surround. Vesicles are certainly not formed, as some electron microscopists thought initially, by a condensation of a totally impermeable membrane around a preformed packet of transmitter molecules; intracellular organelles simply do not form that way. Nor is the reformation of vesicles in any way dependent upon the availability of transmitter, at least at the neuromuscular junction. All the membrane changes we have found to be involved in vesicle retrieval and reformation also occur during stimulation in hemicholinium-3 (Heuser and Reese, 1973) which prevents transmitter synthesis by blocking the supply of choline (Schueler, 1960). So even when nerve terminals are nearly exhausted of acetylcholine by this treatment, a normal number of vesicles can be found (Heuser and Reese, 1973; Hurlbut and Ceccarelli, 1974; contrary to an earlier report of Jones and Kwanbunbumpen, 1970a), and normal numbers of quanta are discharged with each nerve impulse (Elmqvist and Quastel, 1965a), though the number of molecules in each quantum that is released is extremely small, so most of the vesicles must be nearly empty of acetylcholine.

As Katz (1969) has pointed out, there ought to be a period during stimulation in hemicholinium, before the terminals are completely exhausted, when undischarged vesicles are mixed with ones that have already discharged. If the ones that had not yet discharged could retain their stock of acetylcholine in the absence of any new synthesis, they should continue to produce full-sized m.e.p.p.'s, while the vesicles that had reformed in the absence of a new supply of acetylcholine should produce small m.e.p.p.'s. But no such split in m.e.p.p. amplitude has been found during hemicholinium treatment (Elmqvist and Quastel, 1965a; Jones and Kwanbunbumpen, 1970b); instead, all the quanta decline in unison, so it seems more likely that acetylcholine is constantly re-equilibrated from vesicle to vesicle.

Such a split in m.e.p.p. amplitude has been found after poisoning in Botulinum toxin and has been interpreted to mean that the affected terminals contain a mixture of full and empty vesicles (Boroff et al., 1974). However, the small m.e.p.p. in this condition, like those now found in a variety of other conditions, are not as sensitive as the normal-sized m.e.p.p. to the usual treatments that stimulate vesicle discharge at the active zones (Harris and Miledi, 1971; Spitzer, 1972), and they could be the result of a variety of phenomena other than discharge of empty vesicles (Dennis and Miledi, 1974). There is still no compelling reason to suspect that neighbouring vesicles can maintain persistent differences in acetylcholine content.

Presumably this is because the vesicles are all in equilibrium with a common cytoplasmic pool, into which new transmitter must also first be delivered, since it is synthesized by a soluble cytoplasmic enzyme (Fonnum and Malthe-Sorenssen, 1973). Probably vesicles accumulate acetylcholine from this pool until they achieve a constant concentration ratio with it, either by a membrane pump or by an intravesicular ionic exchange protein, which may be among the critical vesicle constituents that need to be retrieved after discharge. Of course it remains possible that, as vesicles approach the presynaptic surface and become lodged above the active zone, they could become subject to a final re-equilibration with materials recently derived from outside the axon, and this could lead to a preferential release of newly synthesized acetylcholine.

M   COULD ACETYLCHOLINE BE DISCHARGED FROM THE CYTOPLASM?

Treatments that apparently lower the cytoplasmic pool of acetylcholine, such as injections of soluble acetylcholine esterase into the cytoplasm (Tauc et al., 1974), or stimulation of endocytotic uptake of soluble acetylcholine esterase (Politoff et al., 1975), have been reported to depress acetylcholine secretion. We need not conclude from these reports that acetylcholine is directly discharged from this cytoplasmic pool into the extracellular space. If vesicles are in equilibrium with it, then lowering the cytoplasmic pool by any means should commensurately lower the amount of acetylcholine in all of the vesicles, the way that hemicholinium does when it blocks new acetylcholine synthesis. Nor is it necessary to imagine that acetylcholine is discharged from a cytoplasmic pool, simply because this pool can comprise a large proportion of the total stock of acetylcholine in the synapse when vesicles are reduced in number by prolonged stimulation (Birks, 1974). Under these conditions, transmitter release still declines as the vesicles become depleted, it does not increase as the cytoplasmic pool enlarges (Birks and MacIntosh, 1961).

Finally, earlier evidence from fractionation of Torpedo electric organs that the readily releasable pool of acetylcholine appeared to be located free in the cytoplasm of these nerve terminals (Dunant et al., 1972) has now, in the light of improved methods of fractionation (Babel-Guerin and Dunant, 1972; Heuser and Lennon, 1973), been reinterpreted to indicate that this pool is actually located in vesicles that are particularly labile or easily discharged during fractionation (Dunant et al., 1974), presumably those vesicles near the surface of the nerve terminal.

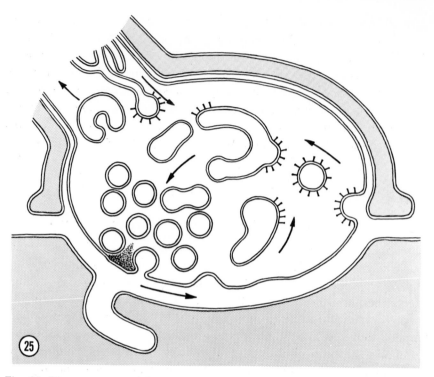

Fig. 25. Diagram summarizing our current view of vesicle recycling at the frog neuromuscular junction. A vesicle starts off milling around free in the cytoplasm. until it comes by chance into contact with sticky filaments that compose the presynaptic dense tufts. There it is held with a group of other vesicles that quiver close to the presynaptic membrane but cannot quite contact it because of an electrostatic energy barrier. When calcium enters with the nerve impulse, it neutralizes negatively charged spots on the vesicle and plasma membrane and allows the two to contact each other. In the presence of calcium, and perhaps with the help of intercalated protein molecules, their phospholipids mix at the point of contact until their membranes fuse and a perforation is formed that lets transmitter out of the vesicle. Then the vesicle collapses completely and its membrane diffuses into the presynaptic membrane. An equal amount of membrane is retrieved elsewhere, partly by a bulk uptake of phospholipids into intracellular cisternae when the plasma membrane is expanding fast, and partly by formation of coated vesicles which continue for some time after stimulation to retrieve certain specific vesicle membrane constituents and possibly also some soluble proteins discharged into the cleft. The retrieved membrane phospholipids and proteins mix inside with each other and with new vesicle constituents supplied by axoplasmic flow to produce a population of variegated membrane compartments. Somehow these compartments sort out the important vesicle constituents, begin to accumulate acetylcholine, and partition back to vesicle size again, thus forming a new generation of synaptic vesicles which can enter the cycle once again.

VIII Conclusion

In summary, all of the preceding anatomical considerations point to the vesicles as the prime actors in quantal transmitter discharge. Each in itself is not sufficient to prove that discharge is from vesicles, if proof could ever be had; but taken together, they provide a substantial and consistent body of evidence that the vesicle hypothesis is correct.

Our current view of how synaptic vesicles are discharged and their membrane recycled for reuse is summarized in Fig. 25.

IX Acknowledgements

I wish to thank Dr. Thomas Reese for his constant collaboration throughout this work and Ms. Louise Evans for her technical assistance. This work was supported in part by grants from the Muscular Dystrophy Association of America and NIH (NINDS) FNS 11979.

X References

Alnaes, E. and Rahamimoff, R. (1975). On the role of mitochondria in transmitter release from motor nerve terminals, *J. Physiol.* **248**, 285–306.

Amsterdam, A., Ohad, I. and Schramm, M. (1969). Dynamic changes in the ultrastructure of the acinar cell of the rat parotid gland during the secretory cycle. *J. Cell Biol.* **41**, 753–773.

Andersson-Cedergren, E. (1959). Ultrastructure of motor end-plates and sarcoplasmic components of mouse skeletal muscle fiber as revealed by three-dimensional reconstructions from serial sections. *J. Ultra. Res. Suppl.* **1**, 1–191.

Andres, K. H. (1964). Micropinozytose im zentralnervensystem. *Z. Zell Micr. Anat.* **64**, 63–73.

Babel-Guerin, E. and Dunant, Y. (1972). Entrie de calcium et liberation d'acetylcholine dans l'organe electrique de la Torpille. *C. R. Acad. Sci. (Paris)* D **275**, 2961–2964.

Beani, L., Bianchi, C. and Ledda, F. (1966). Effect of 2.4 dinitrophenol on neuromuscular transmission. *Brit. J. Pharmacol. Chemother.* **27**, 299–312.

Berl, S., Puszkin, S. and Nicklas, W. J. (1973). Actomyosin-like proteins in brain. *Science* **179**, 441–446.

Birks, R. I. (1973). The relationship of transmitter release and storage to fine structure in a sympathetic ganglion. *J. Neurocytol.* **3**, 133–160.

Birks, R., Huxley, H. E. and Katz, B. (1960). The fine structure of the neuromuscular junction of the frog. *J. Physiol.* **150**, 134–144.

Birks, R., and MacIntosh, F. C. (1961). Acetylcholine metabolism of a sympathetic ganglion. *Can. J. Biochem. Physiol.* **39**, 787–827.

Blaustein, M. (1974). The interrelationship between sodium and calcium fluxes across cell membranes. *Rev. Physiol. Biochem. Pharmacol.* **70**, 34–71.

Blioch, Z. L., Glagoleva, I. M., Liberman, E. A. and Nenasheo, V. A. (1968). A study of the mechanism of quantal transmitter release at a chemical synapse. *J. Physiol.* **199**, 11–35.

Boroff, D. A., del Castillo, J., Evoy, W. H. and Steinhardt, R. A. (1974). Observations on the action of type A botulinum toxin on frog nuromuscular junctions. *J. Physiol.* **240**, 227–253.

Bowers, B. (1964). Coated vesicles in the pericardial cells of the aphid. *Protoplasma* **59**, 351–367.

Boyne, A. F., Bohan, T. P. and Williams, T. H. (1974). Effects of calcium containing fixation solutions on cholinergic synaptic vesicles. *J. Cell Biol.* **63**, 780–795.

Branton, D. (1971). Freeze-etching studies of membrane structure. *Phil. Trans. Roy. Soc. Lond. B.* **261**, 133–138.

Braun, M., Schmidt, R. F. and Zimmerman, M. (1966). Facilitation of the frog neuromuscular junction during and after repetitive stimulation. *Pflugers Arch.* **287**, 41–55.

Brown, G. L. and Feldberg, W. (1936). The acetylcholine metabolism of a sympathetic ganglion. *J. Physiol.* **89**, 265–283

Ceccarelli, B., Hurlbut, W. P. and Mauro, A. (1972). Depletion of vesicles from frog neuromuscular junctions by prolonged tetanic stimulation. *J. Cell Biol.* **54**, 30–38.

Ceccarelli, B., Hurlbut, W. P. and Mauro, A. (1973). Turnover of transmitter and synaptic vesicles at the frog neuromuscular junction. *J. Cell Biol.* **57**, 499–524.

Clark, A. W., Hurlbut, W. P. and Mauro, A. (1972). Changes in the fine structure of the neuromuscular junction of the frog caused by black widow spider venom. *J. Cell Biol.* **52**, 1–14.

Clark, A. W., Hurlbut, W. P. and Mauro, A. (1976). Changes in the structure and function of neuromuscular junctions caused by variations in osmotic pressure. *J. Cell Biol.* (in press).

Collier, B. (1969). The preferential release of newly synthesized transmitter by a sympathetic ganglion. *J. Physiol.* **205**, 341–352.

Collier, B. and MacIntosh, F. C. (1969). The source of choline for acetylcholine synthesis in a sympathetic ganglion. *Can. J. Physiol. Pharmacol.* **47**, 127–136.

Couteaux, R. (1960). Motor end-plate structure. *In* "Structure and Function of Muscle" (G. H. Bourne, ed.) (1st edn.), p. 337. Academic Press, New York and London.

Couteaux, R. and Pecot-Dechavassine, M. (1970). Vesicules synaptiques et poches au niveau des zones actives de la jonction neuromusculaire. *C. R. Acad. Sci. (Paris)* D **271**, 2346–2349.

Couteaux, R. and Pecot-Dechavassine, M. (1974). Les zones specializes des membranes presynaptiques. *C.R. Acad. Sci. (Paris)* D **278**, 291–293.

Daniels, M. P. and Vogel, Z. (1975). Immunoperoxidase staining of α-Bungarotoxin binding sites in muscle end-plates shows distribution of acetylcholine receptors. *Nature Lond.* **254**, 339–341.

de Duve, C. (1963). The lysosome concept. *In* Ciba Symp. on Lysosomes (A. V. S. De Reuch and M. P. Cameren, eds.), pp. 126–216.

del Castillo, J. and Katz, B. (1954). Quantal components of the end-plate potential. *J. Physiol.* **124**, 560–573.

del Castillo, J. and Katz, B. (1955). Local activity at a depolarized nerve-muscle junction. *J. Physiol.* **128**, 396–411.

del Castillo, J. and Katz, B. (1956). Biophysical aspects of neuromuscular transmission. *Prog. Biophys. Biophys. Chem.* **6**, 121–170.

Dennis, M. J. and Miledi, R. (1974). Non-transmitting neuromuscular junctions during an early stage of end-plate reinnervation. *J. Physiol.* **239**, 553–570.

Dickinson, A. H. and Reese, T. S. (1973). Morphological changes after electrical stimulation of frog sympathetic ganglia. *Anat. Rec.* **175**, 306 (abst.).

Dreyer, F., Peper, K., Akert, K., Sandri, C. and Moor, H. (1973). Ultrastructure of the "active zone" in the frog neuromuscular junction. *Brain Res.* **62**, 373–380.

Droz, B., Koenig, H. L. and Di Gramberardino, L. (1973). Axonal migration of protein and glycoprotein to nerve endings. *Brain Res.* **60**, 93–127.

Dunant, Y., Gautron, J., Israel, M., Lesbats, B. and Manaranche, R. (1974). Changes in acetylcholine level and electrophysiological response during continuous stimulation of the electric organ of *Torpedo marmorata*. *J. Neurochem.* **23**, 635–643.

Einstein, A. (1905). Mean square displacement of particle in given direction by Brownian movement. *Ann. D. Physik.* **17**, 549.

Elmquist, D. and Quastel, D. M. J. (1965a). Presynaptic action of hemicholinium at the neuromuscular junction. *J. Physiol.* **177**, 463–482.

Elmquist, D. and Quastel, D. M. J. (1965b). A quantitative study of end-plate potentials in isolated human muscle. *J. Physiol.* **178**, 505–529.

Fertuch, H. C. and Salpeter, M. M. (1974). Localization of the acetylcholine receptor by I^{125} labeled α-Bungarotoxin binding at mouse motor end plates. *Proc. Nat. Acad. Sci.* **71**, 1376–1378.

Fonnum, F. and Malthe-Sorenssen, D. (1973). Membrane affinities and subcellular distribution of the different molecular forms of choline acetylase from rat. *J. Neurochem.* **20**, 1351–1359.

Forman, D. S., Grafstein, B. and McEwen, B. S. (1972). Rapid axonal transport of ^{3}H fucosyl glycoproteins in the goldfish optic system. *Brain Res.* **48**, 307–342.

Glagoleva, I. M., Liberman, E. A. and Khashayev, Z. Kh. (1970). Effect of uncoupling agents of oxidative phosphorylation on the release of acetylcholine from nerve endings. *Biofizika* **15**, 76–82.

Gray, E. G. (1961). The granule cells, mossy synapses, and Purkinje spine synapses of the cerebellum: light and electron microscope observations. *J. Anat.* **95**, 345–356.

Gray, E. G. (1963). Electron microscopy of presynaptic organelles of the spinal cord. *J. Anat.* **97**, 101–106.

Gray, E. G. (1972). Are the coats of coated vesicles artifacts? *J. Neurocyt.* **1**, 363–382.

Gray, E. G. (1975). Synaptic fine structure and nuclear, cytoplasmic and extracellular networks. The stereoframework concept. *J. Neurocyt.* **4**, 315–339.

Gray, E. G. and Willis, R. A. (1970). On synaptic vesicles, complex vesicles, and dense projections. *Brain Res.* **24**, 149–168.

Grynszpan-Winograd, O. (1971). Morphological aspects of exocytosis in the adrenal medulla. *Phil. Trans. Roy. Soc. Lond. B* **261**, 291–292.

Gurd, J. W., Jones, L. R. Mahler, H. R. and Moore, W. J. (1974). Isolation and partial characterization of rat brain synaptic plasma membranes. *J. Neurochem.* **22**, 281–290.

Harris, A. J. and Miledi, R. (1971). The effect of type D botulinum toxin on frog neuromuscular junction. *J. Physiol.* **217**, 497–515.

Heuser, J. E. (1974). A possible origin of the "giant" spontaneous potentials that occur after prolonged transmitter release at frog neuromuscular junctions. *J. Physiol.* **239**, 106–108.

Heuser, J., Katz, B. and Miledi, R. (1971). Structural and functional changes of frog neuromuscular junctions in high calcium solutions. *Proc. R. Soc. Lond. B* **178**, 407–415.

Heuser, J. E. and Lennon, A. (1973). Morphological evidence for exocytosis of acetylcholine during formation of synaptosomes from *Torpedo* electric organ. *J. Physiol.* **233**, 39–41.

Heuser, J. E. and Miledi, R. (1971). Effects of lanthanum ions on function and structure of frog neuromuscular junction. *Proc. Roy. Soc. Lond. B.* **179**, 247–260.

Heuser, J. E. and Reese, T. S. (1973). Evidence for recycling of synaptic vesicle membrane during transmitter release at the frog neuromuscular junction. *J. Cell Biol.* **57**, 315–344.

Heuser, J. E. and Reese, T. S. (1974). Morphology of synaptic vesicle discharge and reformation at the frog neuromuscular junction. *In* "Synaptic Transmission and Neuronal Interaction" (M. V. L. Bennett, ed.). Raven Press, New York.

Heuser, J. E. and Reese, T. S. (1975a). Redistribution of intramembranous particles from synaptic vesicles: direct evidence for vesicle recycling. *Anat. Rec.* **181**, 374.

Heuser, J. E. and Reese, T. S. (1975b). Structural changes at the "active zones" of synapses in stimulated Torpedo electric organs. *Biol. Bull.* **149**, 429.

Heuser, J. E., Reese, T. S. and Landis, D. M. D. (1974). Functional changes in frog neuromuscular junctions studied with freeze-fracture. *J. Neurocytol.* **3**, 109–131.

Heuser, J. E., Reese, T. S. and Landis, D. M. (1976). Preservation of synaptic structure by rapid freezing. *Cold Spr. Harb. Symp. Quant. Biol.* **40**, 17–24.

Hodgkin, A. L. and Katz, B. (1949). The effect of calcium on the axoplasm of giant nerve fibers. *J. Exp. Biol.* **26**, 292–294.

Holtzman, E., Freeman, A. R. and Kashner, L. A. (1971). Stimulation-dependent alteration in peroxidase uptake at lobster neuromuscular junction. *Science* **173**, 733–736.

Huang, L. and Pagano, R. E. (1975). Interaction of phospholipid vesicles with cultured mammalian cells I. Characteristics of uptake. *J. Cell Biol.* **67**, 38–48.

Hubbard, J. I. (1963). Repetitive stimulation at the mammalian neuromuscular junction and the mobilization of transmitter. *J. Physiol.* **169**, 641–662.

Hubbard, J. I. and Laskowski, M. B. (1972). Spontaneous transmitter release and ACh sensitivity during glutaraldehyde fixation of rat diaphragm. *Life Sci.* **11**, 781–785.

Hurlbut, W. P. and Ceccarelli, B. (1974). Transmitter release and recycling of synaptic vesicle membrane at the neuromuscular junction. *In* "Advances in Cytopharmacology" (B. Ceccarelli *et al.*, eds.), Vol. 2. Raven Press, New York.

Israel, M., Gautron, J. and Lesbats, B. (1970). Fractionnement de l'organe electrique de la torpille: localization subcellulare de l'acetylcholine. *J. Neurochem.* **17**, 1441–1450.

Jacobson, K. and Papahadjopoulos, D. (1975). Phase transitions and phase separations in phospholipid membranes induced by changes in temperature, pH, and concentration of bilavent cations. *Biochem.* **14**, 152–161.

Jamieson, J. D. and Palade, G. E. (1968). Intracellular transport of secretory proteins in the pancreatic exocrine cell. *J. Cell Biol.* **39**, 589–603.

Jamieson, J. D. and Palade, G. E. (1971). Condensing vacuole conversion and zymogen granule discharge in pancreatic exocrine cells: metabolic studies. *J. Cell Biol.* **48**, 503–522.

Jones, S. F. and Kwanbunbumpen, S. (1970a). The effects of nerve stimulation and hemicholinum on synaptic vesicles at the mammalian neuromuscular junction. *Physiol.* **207**, 31–50.

Jones, S. F. and Kwanbunbumpen, S. (1970b). Some effects of nerve stimulation and hemecholinium on quantal transmitter release at the mammalian neuromuscular junction. *J. Physiol.* **207**, 51–61.

Kadota, K. and Kadota, T. (1973a). Isolation of coated vesicles, plain synaptic vesicles, and flocculent material from a crude synaptosome fraction of guinea pig whole brain. *J. Cell Biol.* **58**, 135–151.

Kadota, K. and Kadota, T. (1973b). A nucleoside diphosphate phosphohydrolase present in a coated-vesicle fraction from synaptosomes of guinea-pig whole brain. *Brain Res.* **56**, 371–376.

Kanaseki, T. and Kadota, K. (1969). The "vesicle in a basket". *J. Cell Biol.* **42**, 202–220.

Katz, B. (1969). "The Release of Neural Transmitter Substances." Liverpool University Press, Liverpool.

Katz, B. and Miledi, R. (1965). The effect of temperature on the synaptic delay at the neuromuscular junction. *J. Physiol.* **181**, 656–670.

Katz, B. and Miledi, R. (1967). The release of acetylcholine from nerve endings by graded electric pulses. *Proc. Roy. Soc. (Lond.) B* **167**, 23–28.

Kleemann, W. and McConnell, H. M. (1974). Lateral phase separations in *Escherichia coli* membranes. *Biochim. Biophys. Acta* **345**, 220–230.

Kleemann, W., Grant, C. W. M. and McConnell, H. M. (1974). Lipid phase separations and protein distributions in membranes. *J. Supramol. Struc.* **2**, 609–616.

Kriebel, M. E. and Gross, C. E. (1974). Multimodal distribution of frog miniature endplate potentials in adult, denervated and tadpole leg muscle. *J. Gen. Physiol.* **64**, 85–103.

Koenig, H. L., Di Gramberardino, L. and Bennett, G. (1973). Renewal of proteins and glycoproteins of synaptic constituents by means of axonal flow. *Brain Res.* **62**, 413–417.

Kristensson, K. and Olsson, Y. (1971). Uptake and retrograde axonal transport of peroxidase in hypoglossal neurones. *Acta Neuropath. (Berl.)* **19**, 1–9.

Kuhne, W. (1887). Neue untersuchungen uber motorische nervenendigung. *A. Biol.* **23**, 1–148.

Kuhne, W. (1888). On the origin and the causation of vital movement. *Proc. Roy. Soc. Lond.* **44**, 427–448.

Lansman, J. and Haynes, D. H. (1975). Kinetics of a Ca^{++}-triggered membrane aggregation reaction of phospholipid membranes. *Biochim. Biophys. Acta* **394**, 335–347.

LaVail, J. H. and LaVail, M. M. (1973). The retrograde intraaxonal transport of horseradish peroxidase in the chick visual system: a light and electron microscopic study. *J. Comp. Neurol.* **157**, 303–357.

Lentz, T. L., Rosenthal, J. and Maxarkiewicz, J. E. (1976). Cytochemical localization of acetylcholine receptors by means of peroxidase-labelled α-Bungarotoxin. Neurosci. Abst., 5th Ann. Meeting, p. 627.

Liberman, Ye A. and Nenashev, V. A. (1970). Modelling of the interactions of cell membranes by artificial phospholipid membranes. *Biofizika* **15**, 1014–1021.

Litchy, W. J. (1973). Uptake and retrograde transport of HRP in frog sartorius nerve *in vitro*. *Brain Res.* **56**, 377–381.

Llinas, R. and Heuser, J. E. (1976). Depolarization-release coupling systems in neurons. *Neurosciences Research Program Bulletin*, M.I.T. Press (in press).

Lucy, J. A. (1969). Lysosomal membranes. *In* "Lysosomes in Biology and Pathology" (J. T. Dingle and H. Fell, eds), pp. 313–341, North Hill, Amsterdam.

Marchesi, V. T., Tillach, T. W., Jackson, R. L., Segrest, J. P. and Scott, R. E. (1972). Chemical characterization and surface orientation of the major glycoprotein of the human erythrocyte membrane. *Proc. Nat. Acad. Sci.* **69**, 1445–1449.

McIntyre, J. A., Gilula, N. B. and Karnovsky, M. J. (1974). Crypoprotectant-induced redistribution of intramembranous particles in mouse lymphocytes. *J. Cell Biol.* **60**, 192–203.

McLaughlin, J., Case, K. R. and Bosmann, H. B. (1973). Electrokinetic properties of isolated cerebral-cortex synaptic vesicles. *Biochem. J.* **136**, 919–926.

McMahan, U. J. and Yee, A. (1975). Rapid uptake and release of horseradish peroxidase (HRP) in motor nerve terminals of the snake. Neurosci, Abst., 5th Ann. Meeting, p. 624.

McNutt, N. S. and Weinstein, R. S. (1970). The ultrastructure of the nexus: a correlated thin-section and freeze-fracture study. *J. Cell Biol.* **47**, 666–688.

Meldolesi, J., Jamieson, J. D. and Palade, G. E. (1971). Composition of cellular membranes in the pancreas of the guinea pig. *J. Cell Biol.* **49**, 109–158.

Miledi, R. (1971). Lanthanum ions abolish the "calcium response" of nerve terminals. *Nature Lond.* **229**, 410–411.

Morgan, I. G., Zanetta, J. P., Breckenridge, W. C., Vincedon, G. and Gombos, G. (1973). The chemical structure of synaptic membranes. *Brain Res.* **62**, 405–411.

Morgan, I. G., Zanetta, J. P., Breckenridge, W. C., Vincedon, G. and Gombos, G. (1974). Adult rat synaptic vesicles: protein and glycoprotein composition. *Biochem. Soc. Trans.* **2**, 249–252.

Musick, J. and Hubbard, J. I. (1972). Release of protein from mouse motor nerve terminals. *Nature Lond.* **237**, 279–281.

Nagasawa, J., Douglas, W. W. and Schultz, R. A. (1971). Micropinocytotic origin of coated and smooth microvesicles ("synaptic vesicles") in neurosecretory terminals of posterior pituitary glands demonstrated by incorporation of horseradish peroxidase. *Nature Lond.* **232**, 341–342.

Nakane, P. K. and Kawaoi, A. (1974). Peroxidase-labelled antibody—A new method of conjugation. *J. Histochem. Cytochem.* **22**, 1084–1091.

Neher, E. (1974) Asymmetric membranes resulting from the fusion of two black lipid bilayers. *Biochem. Biophys. Acta* **373**, 327–336.

Orci, L. (1973). A portrait of the pancreatic B-cell. *Diabetologia,* **20**, 163–187.

Pagano, R. E. and Huang, L. (1975). Interaction of phospholipid vesicles with cultured, mammalian cells. II. Studies of mechanism. *J. Cell Biol.* **67**, 49–60.

Palade, G. E. (1954). Electron microscopic observations of interneuronal and neuromuscular synapses. *Anat. Rec.* **118**, 335 (abst.).

Palade, G. E. (1959). Functional changes in the structure of cell components. *In* "Subcellular Particles" (T. Hayashi, ed.), pp. 64–83. Ronald Press, New York.

Palade, G. E. (1975). Intracellular aspects of the process of protein synthesis. *Science* **189**, 347–358.

Palade, G. E. and Bruns, R. R. (1968). Structural modulations of plasmalemmal vesicles. *J. Cell Biol.* **37**, 633–649.

Palay, S. L. (1958). The morphology of synapses in the central nervous system. *Exp. Cell Res. Suppl.* **5**, 275–293.

Palay, S. L. (1963). Alveolate vesicles in Purkinje cells of the rat's cerebellum. *J. Cell Biol.* **19**, 89A (abstr.).

Papahadjopoulos, D. and Poste, G. (1975). Calcium-induced phase separation and fusion in phospholipid membranes. *Biophys. J.* **15**, 945–948.

Papahadjopoulos, D., Poste, G. and Mayhew, E. (1975). The interaction of phospholipid vesicles with mammalian cells *in vitro. Biochem. Soc. Trans.* **3**, 606–608.

Papahodjopoulos, D., Poste, G., Schaeffer, B. E. and Vail, W. J. (1974). Membrane fusion and molecular segration in phospholipid vesicles. *Biochem. Biophys. Acta* **352**, 10–28.

Parkes, J. G. and Fox, C. F. (1975). On the role of lysophosphatides in virus induced cell fusion and lysis. *Biochemistry* **14**, 3725–3729.

Parsegian, V. A. and Gingell, D. (1972). On the electrostatic interaction across a salt solution between two bodies bearing unequal charges. *Biophys. J.* **12**, 1192–1204.

Pearse, B. M. F. (1975). Coated vesicles from pig brain: purification and biochemical characterization. *J. Mol. Biol.* **97**, 93–98.

Pecot-Dechavassine, M. (1975). Mode of appearance of giant spontaneous potentials elicited by vinblastine on the isolated frog muscle preparation. *C. R. Acad. Sci. (Paris) D* **280**, 303–306.

Pecot-Dechavassine, M. and Couteaux, R. (1972). Potentials miniatures d'amplitude abnormale obtenus dans des conditions experimentales et changements concomitants des structures presynaptiques. *C.R. Acad. Sci. (Paris) D* **275**, 983–986.

Pecot-Dechavassine, M. and Couteaux, R. (1975). Modifications structurales des terminaisons motrices de muscles de grenouille soumis a l'action de la vinblastine. *C.R. Acad. Sci. (Paris) D* **280**, 1099–1101.

Peper, K., Dreyer, F., Sandri, C., Akert, Y. and Moor, H. (1974). Structure and ultrastructure of the frog motor endplate: a freeze-etching study. *Cell Tiss. Res.* **149**, 437–455.

Pfenninger, K., Akert, K., Moor, H. and Sandri, C. (1972). The fine structure of freeze-fractured presynaptic membranes. *J. Neurocytol.* **1**, 129–149.

Politoff, A., Blitz, A. L. and Rose, S. (1975). Incorporation of acetylcholinesterase into synaptic vesicles is associated with blockade of neuromuscular transmission. *Nature Lond.* **256**, 324–325.

Poste, G. and Allison, A. C. (1973). Membrane fusion. *Biochem. Biophys. Acta* **300**, 421–465.

Potter, L. T. (1970). Synthesis, storage and release of ^{14}C acetylcholine in isolated rat diaphragm muscles. *J. Physiol.* **206**, 145–166.

Pysh, J. J. and Wiley, R. G. (1974). Synaptic vesicle depletion and recovery in cat sympathetic ganglia electrically stimulated *in vivo*. *J. Cell Biol.* **60**, 365–374.

Rambourg, D. B. and Koenig, H. L. (1975). The smooth endoplasmic reticulum: structure and role in the renewal of axonal membrane and synaptic vesicles by fast axonal transport. *Brain Res.* **93**, 1–13.

Rash, J. E. and Ellisman, M. H. (1974). Macromolecular specializations of the neuromuscular junction and the nonjunctional sarcolemma. *J. Cell Biol.* **63**, 567–586.

Rosenbluth, J. and Wissig, S. L. (1964). The distribution of exogenous ferritin in toad spinal ganglia and the mechanism of its uptake by neurons. *J. Cell. Biol.* **23**, 307–325.

Roth, T. F. and Porter, K. R. (1964). Yolk protein uptake in the oocyte of the mosquito Aides aegypti L. *J. Cell Biol.* **20**, 313–332.

Satir, B., Schooley, C. and Satir, P. (1973). Membrane fusion in a model system: Mucocyst secretion in *tetrahymena, J. Cell Biol.* **56**, 153–176.

Scandella, C. J., Devaux, P. and McConnell, H. M. (1972). Rapid lateral diffusion of phospholipids in rabbit sarcoplasmic reticulum. *Proc. Nat. Acad. Sci.* **69**, 2056–2060.

Schueler, F. W. (1960). The mechanism of action of the hemicholiniums. *Int. Rev. Neurobiol.* **2**, 77–97.

Shea, S. M. and Karnovsky, M. T. (1966). Brownian motion: a theoretical explanation for the movement of vesicles across the endothelium. *Nature Lond.* **212**, 353–355.

Simonescu, M., Simonescu, N. and Palade, G. E. (1974). Morphometric data on the endothelium of blood capillaries. *J. Cell Biol.* **60**, 128–152.

Singer, S. T. and Nicolson, G. L. (1972). The fluid mosaic model of the structure of cell membranes. *Science (Wash.)* **175**, 720–731.

Spitzer, N. (1972). Miniature end-plate potentials at mammalian neuromuscular junctions poisoned by botulinum toxin. *Nature Lond.* **237**, 26–27.

Streit, P., Akert, K., Sandri, C., Livingston, R. P. and Moor, H. (1972). Dynamic ultrastructure of presynaptic membranes at nerve terminals in the spinal cord of rats. Anaesthetized and unanaesthetized preparations compared. *Brain Res.* **48**, 11–26.

Takeuchi, A. (1958). The long lasting depression in neuromuscular transmission of frog. *Jap. J. Physiol.* **8**, 102–113.

Tauc, L., Hoffmann, A., Tsuju, S., Hinzen, D. H. and Faille, L. (1974). Transmission abolished on a cholinergic synapse after injection of acetylcholinesterase into the presynaptic neuron. *Nature Lond.* **250**, 496–498.

Thoa, N. B., Wooten, G. F., Axelrod, J. and Kopin, I. J. (1972). Inhibition of release of dopamine-β-hydroxylase and norepinephrine from sympathetic nerves by colchicine, vinblastine, or cytochalasin-B. *Proc. Nat. Acad. Sci.* **69**, 520–522.

Wernig, A. (1975). Estimates of statistical release parameters from crayfish and frog neuromuscular junctions. *J. Physiol.* **244**, 207–221.

Westrum, L. E. (1965). On the origin of synaptic vesicles in cerebral cortex. *J. Physiol.* **179**, 4p–6p.

Whittaker, V. P., Essman, W. B. and Dowe, G. H. C. (1972a). The isolation of pure cholinergic synaptic vesicles from the electric organs of elasmobranch fish of the family Torpedinidae. *Biochem. J.* **128**, 833–846.

Whittaker, V. P., Dowdall, M. J. and Boyne, A. F. (1972b). The storage and release of acetylcholine by cholinergic nerve terminals,: recent results with non-mammalian preparations. *Biochem. Soc. Symp.* **36**, 49–68.

Woodin, A. M. and Weineke, A. A. (1964). The participation of Ca^{2+}, ATP, and ATPase in the extrusion of granule proteins from the polymorphonuclear leucocyte. *Biochem. J.* **90**, 498–509.

Van der Bosch, J. and McConnell, H. M. (1975). Fusion of dipalmitoylphosphatidylcholine vesicle membranes induced by concanavalin A. *Proc. Nat. Acad. Sci.* **72**, 4409–4413.

Van Harreveld, A. and Crowell, J. (1964). Electron microscopy after rapid freezing on a metal surface and substitution fixation. *Anat. Rec.* **149**, 381–386.

Van Harreveld, A., Trubatch J. and Steiner, J. (1974). Rapid freezing and electron microscopy for the arrest of physiological processes. *J. Microscopy* **100**, 189–198.

von Hungen, K., Mahler, H. R. and Moore, W. T. (1968). Turnover of protein and ribonucleic acid in synaptic subcellular fractions from rat brain. *J. Biol. Chem.* **243**, 1415–1423.

Verklej, A. J., Ververgaert, P. H. L., Van Deenen, L. L. M. and Elbers, P. F. (1972). Phase transitions of phospholipid bilayers and membranes of *Acholeplasma laidlawii B* visualized by freeze-fracturing electron-microscopy. *Biochem. Biophys. Acta* **288**, 326–332.

von Kolliker, A. (1863). Untersuchungen uber die letzten endigungen der nerven. *Z. Wiss. Zool.* **XII**, 149–164.

Vos, J., Kuriyama, K. and Roberts, E. (1968). Electrophoretic mobilities of brain subcellular particles and binding of γ-aminobutyric acid, acetylcholine, norepinephrine, and 5-hydroxytryptamine. *Brain Res.* **9**, 224–230.

Zacks, S. I. and Saito, A. (1969). Uptake of exogenous horseradish peroxidase by coated vesicles in mouse neuromuscular junctions. *J. Histochem. Cytochem.* **17**, 161–170.

Zimmerman, H. and Whittaker, V. P. (1974). Effect of electrical stimulation on the yield and composition of synaptic vesicles from the cholinergic synapses of the electric organ of Torpedo; a combined biochemical, electrophysiological and morphological study. *J. Neurochem.* **22**, 435–450.

4

The Role of Calcium in Transmitter Release at the Neuromuscular Junction

R. Rahamimoff

Dept. of Physiology, Hebrew University Medical School, Jerusalem, Israel

I Introduction

The dependence of neuromuscular transmission on extracellular calcium ($[Ca]_o$) has been recognized from the end of the last century (Locke, 1894) and has been confirmed and expanded many times since then (cf. Mines, 1911; Feng, 1936; del Castillo and Stark, 1952). If one examines the various processes from nerve activation to muscle mechanical response, it becomes clear that each of them is affected to a certain degree by partial or complete calcium deprivation. Thus, calcium influences the initiation of the nerve action potential and its propagation along the axon (Frankenhaeuser, 1957), the coupling between the depolarization of the nerve terminal by the action potential and the secretion of transmitter (Katz and Miledi, 1965b), the interaction between acetylcholine and the postsynaptic membrane (Takeuchi, 1963; Nastuk and Liu, 1966), the depolarization contraction coupling in muscle and finally the contraction of the skeletal muscle (cf. Bianchi, 1968). It seems however that the stage which is most sensitive to small changes in $[Ca]_o$ is the secretion of neurotransmitter from motor nerve terminals. This chapter discusses in detail the role of calcium in this last process.

II Extracellular calcium and transmitter release

A THE SITE OF ACTION OF CALCIUM IN EVOKED SYNAPTIC RESPONSE

Acetylcholine is liberated from motor nerve endings as multi-molecular packets or quanta (Fatt and Katz 1951, 1952; del Castillo and Katz, 1954b). At rest the rate of release of these quanta is low, approximately 1/s, and they appear at the postsynaptic membrane as the spontaneously occurring miniature end-plate potentials (m.e.p.p.'s). Arrival of an action potential greatly increases the rate of transmitter liberation; within a millisecond several hundred quanta are liberated (Katz, 1962), giving rise to the end-plate potential (e.p.p.). When this synaptic potential exceeds the electrical threshold of the muscle membrane, an action potential is created in the postsynaptic cell, thus completing the information transfer across the neuromuscular synapse.

The increase in the rate of transmitter release by the presynaptic action potential, depends on the presence of calcium ions in the extracellular medium; in their absence, stimulation of the motor nerve produces no e.p.p. This lack of response to nerve stimulation in the absence of calcium raises the question of whether it originates from a lack of invasion of the electrical signal into the nerve terminals or from lack of activation of the secretion process. The problem was solved several years ago by Katz and Miledi (1965b), by the use of focal extracellular recording from the motor nerve terminal. They observed that even in the absence of extracellular calcium, the action potential is still able to invade the nerve terminals, but fails to induce release of transmitter. Furthermore, the postsynaptic sensitivity is not greatly affected by calcium (cf. Takeuchi, 1963; Nastuk and Liu, 1966). Hence, these experiments clearly indicate that the activation of evoked secretion is the main calcium requiring process in the formation of the evoked end-plate response.

The presynaptic nerve terminal resembles in this respect many other secretory cells, where it has shown that there is a calcium requirement for release (cf. Matthews, 1970; Rubin, 1970; Case, 1973; Douglas, 1974; Thorn, 1974).

B   THE RELATION BETWEEN $[Ca]_o$ AND EVOKED TRANSMITTER RELEASE

Let us now examine the release process from the motor nerve terminal, when moderate changes are imposed on the extracellular calcium concentration. As a measure for transmitter release we shall use either the number of quanta released by a nerve impulse (quantal content—m), or the amplitude of the e.p.p. (the use of the latter parameter is justified only when there are no substantial changes in the postsynaptic sensitivity). The relation between $[Ca]_o$ and transmitter release is highly non-linear (Jenkinson, 1957; Dodge and Rahamimoff, 1967), with a sigmoidal start. This is illustrated in Fig. 1.1, where small

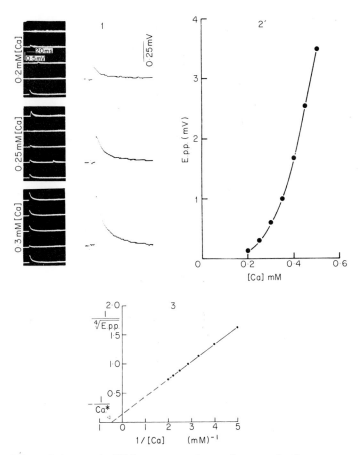

Fig. 1. Effect of changes in [Ca]₀ on transmitter release at the frog neuromuscular junction. 1. Sample records of e.p.p.s. (left) and automatically averaged response (right). Averaging step—160 μs. Three different calcium concentrations as shown in the figure. The same calibrations apply to all concentrations. 2. From the same end plate as Fig. 1.1. Relation between e.p.p. and [Ca]₀ on linear coordinates. The results include the three concentrations of [Ca]₀ shown in Fig. 1.1. 1m M Mg is present throughout the experiment. Note the more-than-linear relationship between e.p.p. and [Ca]₀. 3. Lineweaver-Burk plot of the results in Fig. 1.2. Ordinate: reciprocal of the fourth root of e.p.p. amplitude. Abscissa: reciprocal of [Ca]₀. The intersection with the abscissa gives − 1/Ca*; and $K_1 = Ca*/(1 + [Mg]/K_2)$. (From Dodge and Rahamimoff, 1967.)

changes in $[Ca]_o$ produce very large effects on liberation of acetylcholine. When this more-than-linear relation is plotted on doublelogarithmic coordinates a straight line with a slope of nearly four was obtained at the low range of $[Ca]_o$, which indicates that there is a power relation between $[Ca]_o$ and m. The effect of magnesium ions, which are competitive antagonists so calcium in transmitter release (del Castillo and Engbaek, 1954; del Castillo and Katz, 1954a; Jenkinson, 1957) is to cause a parallel displacement of the observed power relation.

The observation that at higher (and more physiological) calcium concentrations the slope of the log-log relation decreases (Fig. 2.1), led to a further analysis of the dependence of transmitter release on calcium. It was assumed that calcium ions can combine with sites (X) on the presynaptic membrane in a reversible fashion, forming a CaX complex:

$$Ca + X \overset{K_1}{\rightleftharpoons} CaX, \tag{1}$$

where K_1 is the dissociation constant. An additional parallel reaction is assumed to occur with magnesium ions:

$$Mg + X \overset{K_2}{\rightleftharpoons} MgX, \tag{2}$$

where K_2 is the dissociation constant for MgX.

If transmitter release would have been simply proportional to CaX then a rectangular hyperbola would have been the relation between $[Ca]_o$ and m. But since a sigmoidal relation was observed experimentally, is has been further assumed that several adjacent CaX complexes have to be formed in order to achieve release of transmitter. The simplest relation to describe this cooperative dependence is given by equation (3) and Fig. 2.4.

$$m = k \left(\frac{[Ca]/K_1}{1 + [Ca]/K_1 + [Mg]/K_2} \right)^n, \tag{5}$$

where n is the number of cooperating calcium ions, K_1 and K_2 are the dissociation constants of calcium and magnesium complexes respectively and k is a proportionality constant.

The slope of the power relation between $[Ca]_o$ and m suggested that the minimal number of calcium ions to cooperate in the release process is 4 $(n = 4)$. The other numerical values of equation (3) were obtained from the modified Lineweaver and Burk transformation (Fig. 1.3; Fig. 2.2). This cooperative dependence of transmitter release on $[Ca]_o$

was first analysed at the frog neuromuscular junction (Jenkinson, 1957; Dodge and Rahamimoff, 1967) and subsequently observed at the mammalian motor end-plate (Hubbard *et al.*, 1968b) and at other synapses (Katz and Miledi, 1970; Lester, 1970; Linder, 1973; Stjärne, 1973).

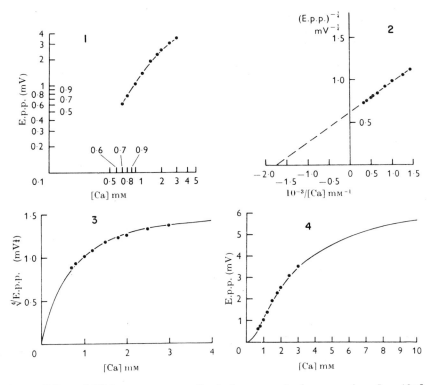

Fig. 2. Effect of $[Ca]_o$ on e.p.p. amplitude in a curarized preparation, 5×10^{-5} g/ml (+)-tubocurarine present throughout the experiment. 1. Double logarithmic coordinates. The first three points are averages of 128 responses (rate of stimulation 0·2/s). The other points are averages of 64 responses (rate of stimulation 0·1/s). The $[Ca]_o$ is higher than in Fig. 1 and the initial slope is only about 1·5 and decreases further with the increase in $[Ca]_o$. 2. Double reciprocal plot of the fourth root of the e.p.p. amplitude against $[Ca]_o$. From this plot K_1 was calculated. 3. Fourth root of the e.p.p. amplitude against $[Ca]_o$. The curve is a rectangular hyperbola following the equation $\sqrt[4]{e.p.p.} = 2\cdot7 [Ca]/(1 + [Ca]/0\cdot6)$. 4. E.p.p. amplitude against $[Ca]_o$ on linear coordinates. The points are the experimental results and the sigmoidal curve follows the equation: $e.p.p. = 2\cdot7\{[Ca]/(1 + [Ca]/0\cdot6)\}^4$. (From Dodge and Rahamimoff, 1967.)

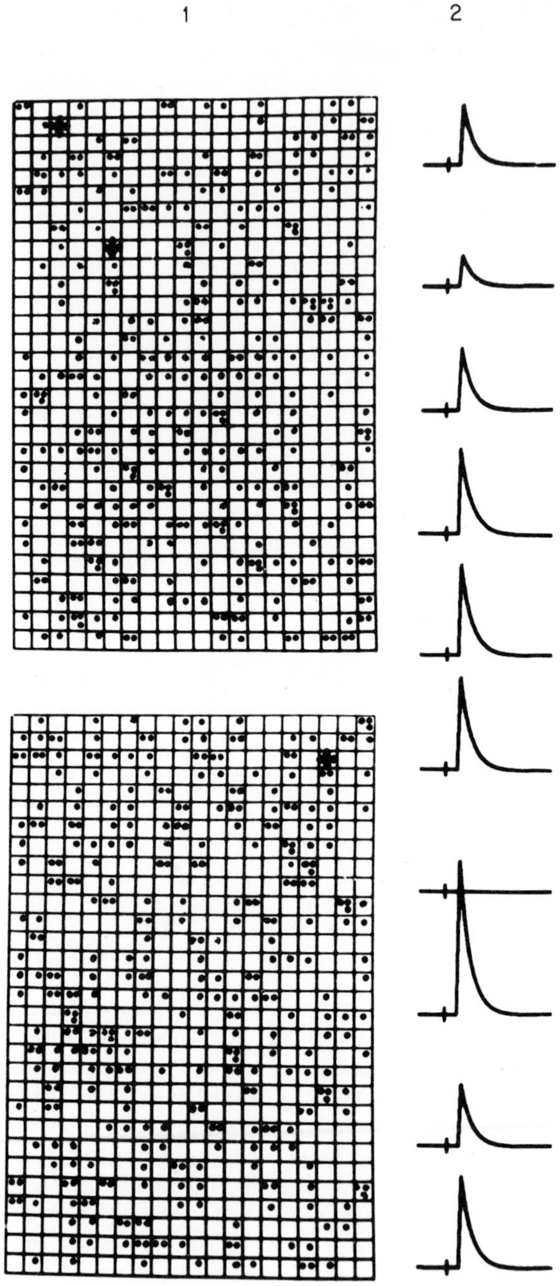
1
2

The cooperative action of calcium ions in transmitter release, can be visualized in a schematic fashion as illustrated in Fig. 3 (H. Meiri and Rahamimoff, 1972; Rahamimoff and Yaari, 1973; Rahamimoff, 1973). For this scheme several assumptions are necessary. Assume that there are many release locations on the presynaptic nerve terminal, each of them represented by a square, which are able to bind calcium ions (represented by dots); assume further that the binding of calcium ions is a random process and that release of transmitter can occur only when a certain minimal number of calcium ions will be found simultaneously in the same square (for the frog neuromuscular junction there must be at least four calciums in the square). Since the distribution of the calcium ions is a random process, then on various trials (simulated nerve impulses) a different number of squares will reach the required occupation (shown as crosses). Such simulation of the calcium activation of the release process can predict the observed relation between $[Ca]_o$ and transmitter release. Furthermore, the random distribution of calcium ions could be one of the causes of the fluctuating nature of synaptic response (del Castillo and Katz, 1954b; Martin, 1955; 1966). It should be pointed out that according to this schematic representation, only a small fraction of the attached calcium ions actually participate in release, but they are available when a subsequent stimulus is delivered (see "facilitation and calcium ions").

There is an interesting exception to the sigmoidal relation between $[Ca]_o$ and transmitter release—the neuromuscular junction of certain crustaceans, where a linear relation was observed between these two parameters (Bracho and Orkand, 1970; Ortiz and Bracho, 1972; Sarne, Parnas and Rahamimoff, 1972).

At two different preparations it was observed that this linear relation occurs at junctions with presynaptic inhibition (Dudel and Kuffler 1961). This raises the possibility that the linear relation is the combined

Fig. 3. Schematic representation of the random activation of release by calcium ions. The release process was simulated on a digital computer, by random distribution of 350 calcium ions on presynaptic location (1). It was assumed for simplicity that each location containing four calcium ions (+), leads to a release of a quantum. In the upper situation, two locations reached the required number; therefore the first simulated tracing in 2 has an "amplitude" of two quanta. The second simulation in 2 is also shown in detail in 1 and has an amplitude of one quantum. Note the fluctuations in amplitude.

result of the action of calcium on the excitatory and inhibitory nerve endings (Sarne, 1974; Parnas *et al.*, 1975).

There is an additional deviation from the high power relation, which can be observed at the frog neuromuscular junction when perfused with a medium containing very low $[Ca]_o$ (Crawford, 1974). This deviation will be discussed later on.

C EXTRACELLULAR CALCIUM AND SPONTANEOUS RELEASE OF TRANSMITTER

There is a large difference between spontaneous and evoked release of transmitter in regard to their dependence on extracellular calcium. First, when calcium ions are omitted from the medium, there is no evoked release of transmitter following the nerve action potential: this is not the case for the spontaneous release; even in the almost complete absence of calcium from the medium (and in the presence of the chelating agent—EGTA), there is still spontaneous release of acetylcholine (Miledi and Thies, 1971).

The second difference is regarding to the power of the relation between $[Ca]_o$ and transmitter release. As we have seen previously, there is a high power dependence on $[Ca]_o$ of the evoked release, in almost all synapses examined. For spontaneous release there is a species difference. At the amphibian motor end-plate there is only a very small dependence of the frequency of the m.e.p.p. on $[Ca]_o$ (Mambrini and Benoit, 1964). For the mammalian neuromuscular junction there is stronger dependence on $[Ca]_o$ but still smaller than that for the evoked release (Hubbard *et al.*, 1968a, b).

D THE TIMING OF EXTRACELLULAR CALCIUM REQUIREMENT IN EVOKED TRANSMITTER RELEASE

We have already seen that there is an absolute requirement for extracellular calcium in normal evoked transmitter release. Let us examine now when this calcium is required. This was studied by Katz and Miledi (1967c), who used a tetrodotoxin blocked preparation in which transmitter could be released by focal depolarization of the nerve terminal, provided that calcium ions were present in the extracellular medium. The supply of calcium ions could be restricted in time, by the use of a pipette filled with $CaCl_2$. By proper adjustment of the current passing through the calcium pipette it was possible to deliver the calcium ions at various intervals before, during or after the depolarizing pulse. By this technique it was demonstrated that calcium ions have to be present before and during the depolarizing pulse to induce release of transmitter. If calcium ions were supplied after nerve depolarization they were unable to trigger transmitter release. Hence, it can be concluded that calcium ions have to be present in the extracellular medium during the depolarization and prior to the actual transmitter release.

E   ACTIVATION OF EVOKED TRANSMITTER RELEASE BY STRONTIUM
AND BARIUM

We have seen up to now that calcium ions are able to activate transmitter release and that magnesium ions compete with calcium and inhibit the liberation of acetylcholine by nerve impulses. Other divalent ions are also able to act on this process. Two ions can replace calcium in evoked transmitter release—strontium and barium (Miledi, 1966; Dodge et al., 1969; Meiri and Rahamimoff, 1971; Alnaes et al., 1974; Meiri, 1975). With Sr there is a more-than-linear relation between the concentration of the divalent ion in the extracellular fluid and the number of quanta liberated by the nerve impulse suggesting that calcium and strontium activate the release process by a similar mechanism. If Ca and Sr ions are activating quantal release by two separate noninteracting processes, then one would predict, that the combined effect of Sr and Ca is at most the algebraic sum of the individual effects. The combined postsynaptic effect might be of course less than the algebraic sum of the individual effects, due to the action of divalent ions on the postsynaptic sensitivity and due to non-linear summation (del Castillo and Engback, 1954; del Castillo and Katz, 1954b; Martin, 1955). But, if the two ions were operating on the same mechanism, then in a certain range of concentrations, the presence of one of the ions would facilitate the release by the other, and vice versa. Figure 4 shows that the combined action of Ca and Sr is not equal to, or less than, the sum of their separate actions, but is larger. These results indicate that Ca and Sr activate transmitter release by the same mechanism and that there is an interaction in their effect. At higher concentrations of divalent ions, there is an inhibition, as expected for two agonists acting on the same saturable process.

The experiment in Fig. 4 also illustrates that Sr is much less effective than calcium in evoked transmitter release and approximately a seven times higher concentration of Sr is necessary in order to achieve the same effect as calcium. The effectiveness of barium is even smaller than that of strontium and it is only 4·6% of the effectiveness of calcium (Meiri, 1975).

Fig. 4. Non-linear summation of the effects of calcium and strontium ions on transmitter release at the frog neuromuscular junction. 1. Average response of 0·117 mV to 270 nerve stimuli. Mean quantal content 0·61. Bathing medium contains 0·3 mM Ca. 2. Average response of 0·119 mV to 135 nerve stimuli. Mean quantal content of 0·72. Bathing medium contains 2·0 mM Sr. 3. Average response of 0·355 mV to 135 nerve stimuli. Mean quantal content of 2·08. Bathing medium contains 0·3 mM Ca and 2·0 mM Sr.

1 mM Mg was present throughout the experiment. Voltage calibration applies to all three records. Averaging step of 125 µs. Stimulation rate 0·5/s in 1 and 0·25/s in 2 and 3. Note that when the two activating divalent ions were present simultaneously, transmitter release is 56% greater than the sum of the individual releases. (From Meiri and Rahamimoff, 1971.)

F IONS AS INHIBITORS OF EVOKED TRANSMITTER RELEASE

The effects of several other ions on neuromuscular transmission have been studied in detail in recent years. None of the ions examined were able to replace calcium as an activator of the release process and most of the ions were found to be inhibitory. I would like to illustrate this point first with two monovalent ions which occur naturally in the environment of the nerve terminal, namely sodium and hydrogen ions, and afterwards to mention in brief the action of some other ions.

Since sodium ions are abundantly present in the extracellular medium, their inhibitory action can be observed upon their partial withdrawal (Kelly, 1965; Rahamimoff and Colomo, 1967; Birks and Cohen, 1968). A reduction of $[Na]_o$ to 60% of the normal value, augments the end-plate potential amplitude. This increase in e.p.p. amplitude is accompanied by a reduction in the size of the m.e.p.p. and represents an increase in the number of quanta liberated by the nerve impulse. (In spite of the fact that the nerve impulse itself has a diminished amplitude due to the dependence of the action potential on sodium.) The withdrawal of sodium produces a linear shift in log $[Ca]_o$—log release relation (Fig. 5), suggesting that there is a competition between calcium and sodium in the release process (Colomo and Rahamimoff, 1968). This indicates that under normal physiological conditions, there is a partial suppression of the release process.

Qualitatively similar interaction has been found recently between calcium and hydrogen ions. Decrease of the pH of the external medium causes a subdued evoked release (Landau and Nachshen, 1975).

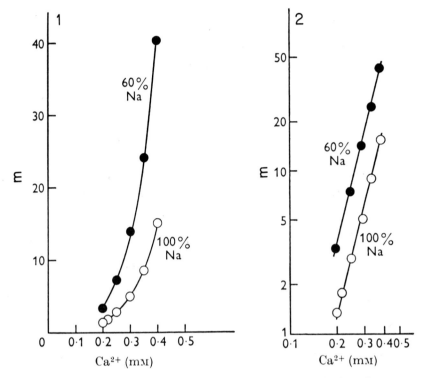

Fig. 5. Relation between the concentration of calcium ions and quantal content at 116 mM (○) and 70 mM (●) [Na]$_o$. Each point represents the average response to 64–256 stimuli. 1–linear plot; 2–double logarithmic plot. 1 mM Mg was present throughout the experiment. Note the increase in quantal content with partial withdrawal of sodium ions. (From Colomo and Rahamimoff, 1968.)

Within the 2A group of the periodic table, we have seen already that calcium, strontium and barium are activators of transmitter release and magnesium is an inhibitor. The first element in this group, beryllium, is also a strong inhibitor of evoked quantal output (Bliokh and Liberman, 1970). Several physical properties of the 2A group gradually change with the increase in atomic number (Be, Mg, Ca, Sr, Ba). Thus there is a progressive increase in the crystal ionic radius, a progressive decrease in the hydrated ionic radius (and a progressive decrease in the force of attraction of water molecules, shown as a progressive increase in the rate constants for H_2O substitution in the inner coordination sphere of the metal ions (Eigen, 1963)). However, there is sudden transition between an inhibitor (Mg) and a strong activator (Ca). We do not know at present, which physical parameter of an ion is responsible for its releasing properties. Cadmium for example is very similar to calcium in respect to crystal ionic radius (0·97 Å for Cd and 0·99 Å for Ca) and water substitution rate constant ($3.2 \cdot 10^8$ s^{-1} for Ca and $2.3 \cdot 10^8$ s^{-1} for Cd), but while calcium is a strong activator, cadmium is a strong inhibitor (Meiri, 1975).

Other cations that inhibit evoked transmitter release are nickel (Benoit and Mambrini, 1970), manganese (Meiri and Rahamimoff, 1972; Balvave and Gage, 1973), cobalt (Weakly, 1973; Crawford, 1974), lead (Manalis and Cooper, 1973), yttrium (Bowen, 1972), lanthanum (Heuser and Miledi, 1971) and praseodymium (Alnaes and Rahamimoff, 1974).

G FREQUENCY OF STIMULATION AND TRANSMITTER RELEASE RECEPTIVITY

If the motor nerve is stimulated at low frequency, for example once every 30 s, the mean quantal content remains constant over a long period of time. If however, the frequency of stimulation is increased, the quantal content becomes dependent on the time interval between successive stimuli. This becomes clear if a series of paired nerve action potentials are evoked and we compare the end-plate potential amplitude and the quantal content of the first response (m_1) to that of the second (m_2) at different time intervals (t) between them. If t is short (up to about 100 ms), m_2 will be larger than m_1, and the phenomenon is known as facilitation ($F = m_2/m_1$) (cf. Eccles et al., 1941; del Castillo and Katz, 1954c). In general, the value of F is larger, the smaller the t is; the decay of F with t is not monotonic however and shows a secondary rise at longer t's (Mallart and Martin, 1967).

When t is of the order of several hundred ms, m_2 will be slightly smaller than m_1; this neuromuscular depression will vanish gradually until m_2 will be again equal to m_1.

The two phenomena mentioned above can be demonstrated by the use of low frequency paired stimulation. There is an additional frequency dependent modulation of transmitter release which is manifested after tetanic activation of the motor nerve terminals. This is the post tetanic potentiation, where after a high frequency stimulation, there is a rather prolonged increase (up to minutes) of the quantal content of a test response.

In the past few years some of the properties of these stimulation frequency dependent modulatory processes have been attributed to calcium ions.

1   Neuromuscular facilitation and calcium ions

Neuromuscular facilitation depends on the presence of calcium ions in the extracellular medium. This was shown very clearly in the experiments of Katz and Miledi (1968). A neuromuscular preparation was immersed in a calcium free medium and the only source of calcium was a microelectrode. By adjusting the direction of the current passing through the micropipette, it was possible to deliver calcium at predetermined instances of time. They found that facilitation was observed only when calcium was delivered before the first and the second nerve action potential. If calcium was given only before the second nerve action potential, no substantial facilitation was detected. Hence it appears that the propagation of the impulse into the nerve terminals is not a sufficient condition for facilitation; it is necessary in addition to have calcium during the first impulse in order to have increased release of transmitter.

The value of F is not constant at any given time interval t, but depends also on the quantal content of the first response. If one examines F at relatively short time intervals, it becomes clear that the higher the value of m_1 the lower the value of F. This has been shown by two methods of changing m_1; by changing the calcium (Rahamimoff, 1968) or by changing the magnesium composition of the medium (Mallart and Martin, 1968).

All these experiments led to the formulation of the "residual calcium" hypothesis of facilitation (Katz and Miledi, 1968; Rahamimoff, 1968). It was assumed that the first nerve action potential produces certain attachment of calcium ions to critical sites on the presynaptic membrane (or calcium influx through the presynaptic membrane as it would be discussed later on). If the detachment of these calciums is not instantaneous after the first response, then at certain time intervals afterwards, some of these calcium ions will be attached to the critical sites. Due to the cooperative requirements of several calcium ions for the release of a quantum of transmitter, the effect of these residual ions on their own will be rather small and will appear as an increased frequency of "spontaneous" m.e.p.p. (delayed release—cf. Dodge *et al.*, 1969; Rahamimoff and Yaari, 1973). However, when a second nerve impulse

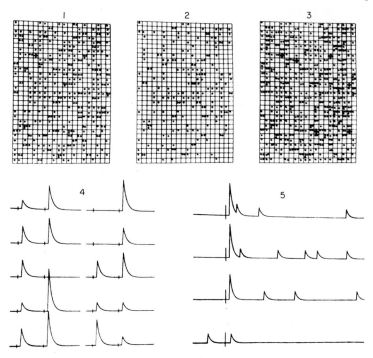

Fig. 6. Schematic representation and simulation of residual effects of Ca ions. 1. Schematic diagram of the presynaptic terminal after the first nerve action potential. Ca ions represented by black dots, are randomly distributed among the 600 release locations. A random number generator was used to select the location of each Ca ion attached. Only locations with four (or more) Ca ions (shown as crosses) can lead to the release of a quantum of transmitter. The simulation was done on a PDP 15/40 computer and photographed from the VP 15 oscilloscope screen. 2. Same as 1, but a certain time period afterwards; some of the Ca ions were dissociated or inactivated. 3. Same as 2, but after the arrival of a second nerve action potential. In addition to the residual Ca, new Ca ions are attached and contribute to the release of transmitter. 4. Simulation of number of trials as shown in 1, 2 and 3. It was assumed that for each location having at least four Ca ions attached, a quantum of transmitter is liberated. Note the fluctuation in amplitude of the first and second responses. The simulated responses were photographed from the VP 15 oscilloscope screen. 5. Simulation of end-plate potentials and delayed random release of transmitter. The e.p.p.s were simulated as shown in 1. For the delayed release, it was further assumed, that in addition to the residual Ca ions, new Ca ions attach at a very low rate. (From Rahamimoff and Yaari, 1973.)

arrives at the nerve terminal and produces an additional attachment of calcium to the critical sites (or an additional calcium influx), there will be calcium available for release from two sources—the left over and the newly available, leading to much higher rates of transmitter release. For graphical illustration of the residual hypothesis see Fig. 6.

Not only the initial phase of facilitation, but also its second component (Mallart and Martin, 1967) is dependent on calcium ions (Younkin, 1974).

2 Neuromuscular depression and calcium ions

When a pair of stimuli are given to the motor nerve, the amplitude of the second (test) response depends on the time interval that elapsed from the first response. At short time intervals (t—less than about 150 ms) the predominant phenomenon is that of facilitation and the second e.p.p. is larger than the first. At longer t's quite frequently the amplitude of the second e.p.p. is smaller than the first. This depression of the e.p.p. was described by Eccles *et al.* (1941) and by Lundberg and Quilish (1953). The origin of this depression is presumably pre-synaptic, since Otsuka *et al.* (1962) have shown that the sensitivity to applied acetylcholine is unchanged during the depression period.

Neuromuscular depression is an inherent property of the release process. It can be demonstrated when release is restricted to a short length of the nerve terminal and when transmitter release is initiated by electrotonic depolarization of the terminals, in a preparation in which the nerve action potentials are abolished by tetrodotoxin (Betz, 1970).

The magnitude of depression depends on the amount of transmitter released by the conditioning stimulation and thus on the concentration of calcium ions in the extracellular medium (Takeuchi, 1958; Thies, 1965). At higher $[Ca]_o$ more transmitter is liberated and depression is much more profound, while at lower $[Ca]_o$ depression is not prominent and the main phenomenon which is observed with paired stimulation is that of facilitation.

3 Post-tetanic potentiation and calcium ions

High frequency tetanic stimulation of the motor nerve produces an increase in the amplitude of the end-plate potential which can last up to several minutes after the end of the stimulation period (Feng, 1941; Liley and North, 1953; Gage and Hubbard, 1966, Braun et al., 1966; Rosenthal, 1969). The increase in e.p.p. amplitude is due to augmented liberation of transmitter from the motor nerve terminals (Hutter, 1952; del Castillo and Katz, 1954c; Liley, 1956). Since this post-tetanic potentiation in transmitter release can be obtained in a tetrodotoxin blocked preparation using electrotonic depolarization of the nerve terminals, it has been suggested that sodium ions are not the main cause of this phenomenon (Weinrich, 1971). Furthermore, post tetanic potentiation is observed in isotonic calcium solution, indicating that calcium ions are associated in its development. It seems therefore that the two phenomena in which there is an increase in evoked release following conditioning stimulation, namely facilitation and post tetanic potentiation, are closely related to calcium ions; however, evidence exists, that they do not share the same mechanism (Landau et al., 1973; Magleby, 1973; Magleby and Zengel, 1975).

III Calcium conductance of the presynaptic membrane and transmitter release

A CALCIUM ENTRY

Up to now we have discussed the effects of extracellular calcium ions on transmitter release. Several lines of evidence suggest, that the entry of calcium ions through the presynaptic membrane is directly correlated with evoked release of transmitter. In this section we shall examine a number of relevant experiments which relate the depolarization-induced influx of calcium to transmitter liberation. It should be noted that some of these experiments were performed on the motor nerve terminals, while others were done on the large presynaptic fibres of the squid where intracellular recording and intracellular injection of substances is possible.

The first suggestion that an inward movement of a positively charged compound across the presynaptic nerve membrane is associated with transmitter release, came from the experiments of Katz and Miledi. They found that a certain small synaptic delay, of approximately 0·5 ms exists between the depolarization of the nerve ending by the action potential and the release of transmitter (Katz and Miledi, 1965a). By the use of tetrodotoxin (TTX) (Katz and Miledi, 1967b), they were able to abolish the normal nerve action potential and to substitute it with artificial depolarizing pulses. This gives the freedom to change the amplitude and the duration of the depolarization, and to examine the effect of altering these parameters, on transmitter release. The basic observation in these conditions was that when the depolarizing pulses were rather small, there was an increase in the amount of transmitter liberated with increase in pulse amplitude and duration. In addition, it was observed that with prolongation of the depolarizing pulse there was a *latency shift*, manifested as an increase in synaptic delay (Katz, and Miledi, 1967c). The interpretation of this finding was that depolarizing pulses are able to oppose the inward movement of a positively charged substance, which is presumably calcium.

An even more striking example of the prevention of transmitter release by strong depolarizing pulses was obtained at the squid synapse. There, much larger and prolonged depolarizations were obtained after blocking the depolarization-induced increase in potassium conductance by intracellular electrophoretic application of tetraethyl ammonium (TEA). By combined use of TTX and TEA it was possible to polarize the presynaptic fibre by up to 200 mV for 18 ms and to show that transmitter release *during* the pulse (ON response) was suppressed; transmitter was liberated, however, *after* the end of the pulse (OFF response). These experiments were very important in the development of the idea that an influx of a positively charged particle is responsible for transmitter release and that this influx depends on the electrochemical gradient across the presynaptic membrane.

The tentative identification of this positively charged particle came later, when it was shown that depolarization can produce a regenerative inward current even when the voltage dependent sodium permeability was blocked by TTX (and the depolarization induced potassium permeability blocked by TEA). This regenerative inward current is sensitive to the calcium concentrations in the external medium and presumably represents a calcium spike (Katz and Miledi, 1969). Release of transmitter followed the occurrence of these spikes. Important developments took place after the introduction of calcium sensitive proteins in neurophysiological research. Shimomura *et al.* (1963) isolated from the jellyfish *Aequora forskalea* a protein named aequorin, which emits light when it reacts with calcium. The injection of aequorin inside nerve cells and the measurement of the changes in the rate of light emission following nerve activity, led to the estimation of calcium permeability during and after action potentials. It was found that there are two phases of calcium entry induced by depolarization in the squid giant axon (Baker *et al.*, 1971; Baker, 1972). The early phase of calcium entry is probably through the sodium channels. It parallels exactly the rise in Na permeability and it is abolished reversibly by tetrodotoxin. The late phase of calcium entry is TTX-insensitive but is blocked by the divalent ions manganese, magnesium and cobalt (Baker, 1972; Baker *et al.*, 1973). Pharmacological evidence suggests that this late phase of calcium influx is mainly responsible for transmitter release, since nerve terminal depolarization induces liberation of transmitter even in the presence of tetrodotoxin (Katz and Miledi, 1967b). Furthermore, transmitter release is blocked by magnesium, manganese and cobalt (Jenkinson, 1957; Meiri and Rahamimoff, 1972; Balnave and Gage, 1973; Weakly, 1973; Crawford, 1974). Similar intracellular injection of aequorin in Aplysia neurones showed that under physiological conditions, every action potential was accompanied by an influx of calcium (Stinnakre and Tauc, 1973). It is of interest to note, that repetitive action potentials produced an accumulation of calcium inside the cell, manifested as a progressive increase in light emission.

Therefore, it seems that depolarization of the presynaptic nerve causes an increase in the calcium conductance. Under normal conditions this increased conductance leads to calcium influx, since there is large electrochemical gradient. Very large depolarizations however would bring the membrane towards the calcium equilibrium potential and cancel the electrochemical gradient. In such conditions there is an increase in the calcium conductance but there is no net influx of calcium. This was shown in Aplysia neurons (Stinnakre and Tauc, 1973) and at the squid giant synapse (Llinas and Nicholson, 1975). The increase in calcium conductance outlasts the depolarization and upon return towards resting values of membrane potential, an influx of calcium occurs. This delayed calcium influx is presumably responsible for the OFF response.

B CALCIUM GRADIENT ACROSS THE PRESYNAPTIC MEMBRANE AND TRANSMITTER RELEASE

The concentration of free calcium ions inside the nerve terminal has not yet been determined accurately, but is probably around $3 \cdot 10^{-7} M$

(cf. Baker, 1972). Since the extracellular [Ca] is about $2 \cdot 10^{-3}$M, there is a very substantial concentration gradient across the nerve membrane. The electrochemical gradient across the membrane is even larger, since at rest the inside of the nerve membrane is negatively charged in respect to the outside. Therefore any increase of the calcium membrane conductance should produce an influx of calcium which is closely correlated to transmitter release. The increase in calcium conductance is produced presumably by the depolarization of the nerve terminal by the action potential, which is able to invade the nerve terminals also at very low extracellular calcium concentrations (Katz and Miledi, 1965b; Miledi and Thies, 1971). Hence, it was of interest to examine the effect of nerve impulses at very low extracellular calcium concentrations, when the increased calcium conductance will be presumably accompanied by a reversed electrochemical gradient of calcium across the presynaptic membrane. Very low $[Ca]_o$ cannot be achieved simply by omitting calcium from the extracellular medium since the muscle is a huge store of calcium which will leak out and elevate $[Ca]_o$ (Shanes and Bianchi, 1959); it is necessary to use calcium buffers such as EGTA (cf. Portzehl et al., 1964).

The experiment illustrated in Fig. 7 shows the effect of nerve stimulation on transmitter release in a medium containing no added calcium, 1 mM EGTA and 1 mM $MgCl_2$. The nerve was stimulated once every two seconds and the average frequency of appearance of m.e.p.p.s. after the nerve stimulation was estimated by the "moving bin" method (Rahamimoff and Yaari, 1973). The dashed line shows the mean frequency in the resting preparation. Immediately after the nerve stimulation there was reduction in the rate of release of the m.e.p.p. The average decrease was about 20% (Rotshenker et al., 1976). The return towards basic frequency was not a monotonic function but showed periodic fluctuations (Rahamimoff et al., 1975). The most plausible explanation of this inhibitory action of nerve impulses on transmitter release at very low calcium solutions is that, due to the reversed electro-chemical gradient for calcium, there is an efflux of calcium which reduces the $[Ca]_i$ near the sites where it is required for release. However an alternative explanation cannot be excluded at present, namely that the action potential increases the membrane conductance to EGTA, which enters the terminal and sequesters the calcium ions necessary for release. The latter explanation is not very probable in view of the low permeability of some biological membranes

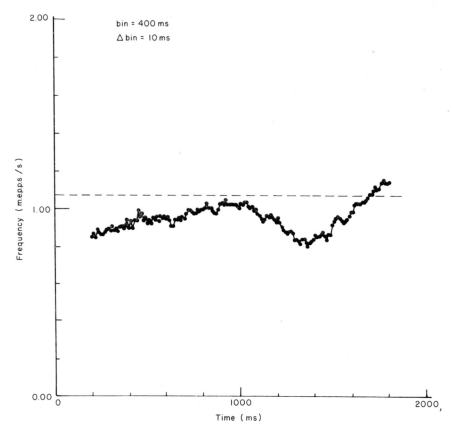

Fig. 7. Inhibitory effect of nerve stimulation on transmitter release at very low
$[Ca]_o$. 1 mM EGTA Ringer solution with no added calcium. 1 mM Mg. Stimulation
frequency 0·5/s. Moving bin method (Rahamimoff and Yaari, 1973) was used to
obtain a time histogram" of m.e.e.p. frequency. The testing frequency (dashed line)
was 1·076/s. Note that during the first second after nerve stimulation the frequency
is subnormal.

to EGTA (cf. Weber *et al.*, 1966). In any case this experiment suggests
that intracellular calcium is associated with transmitter release.

We have seen in this section that the increase in calcium conductance,
brought about by the nerve terminal depolarization and the resultant
calcium influx are presumably the trigger for evoked release of neurotransmitter. We shall now turn to examine more in detail the role of
intracellular calcium.

IV Intracellular calcium and transmitter release

A ACTIVATION OF TRANSMITTER RELEASE BY INTRACELLULAR CALCIUM

In the previous section, we have seen that evoked release of transmitter is associated with the influx of calcium ions through the presynaptic membrane. The next problem we would like to consider is whether the transmitter release process can be activated only by calcium ions coming from the outside, or whether calcium ions appearing from the inside of the nerve terminal can produce similar activation. Two different approaches were used to supply calcium ions directly to the inside of the nerve terminal, without crossing the presynaptic membrane. The first approach was to deliver calcium ions to the inside of the squid nerve terminal from a calcium filled micropipette; the second approach was to release calcium from intracellular stores.

The squid presynaptic nerve fibre is large enough to allow an introduction of an intracellular micropipette filled with calcium ion. By ejecting calcium ions from such a pipette, Miledi (1973) was able to demonstrate an increase in the background level of transmitter release. These experiments strongly suggested that calcium ions coming from the inside of the presynaptic membrane are able to reach the sites responsible for transmitter release.

B MITOCHONDRIA, INTRACELLULAR CALCIUM AND TRANSMITTER RELEASE

The free calcium ion concentration inside the nerve terminal can be increased by releasing calcium from intracellular stores. One of the subcellular organelles which is known to bind calcium reversibly is the mitochondrion. This section will dwell upon the evidence that mitochondria inside the nerve terminals play a role in the regulation of the free calcium ion concentration. This in turn will converge upon the problem of activation of transmitter release by calcium that does not cross the surface membrane, but comes from the inside of the nerve terminal.

In a large number of tissues it has been shown that the mitochondria are able to take up calcium against large concentration gradients (cf. Lehninger, 1970). They do that in a reversible fashion by a number of processes. Some of these processes, such as binding to high-affinity and

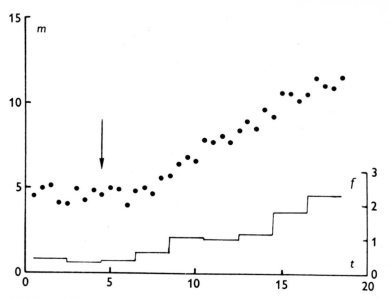

Fig. 8. Effect of dicoumarol, on spontaneous and evoked transmitter release. Each point denotes the average quantal content (m) of thirty consecutive nerve stimulations in high magnesium (10 mM) low calcium (0·7 mM) solution. Lower curve shows corresponding changes in the frequency of miniature end-plate potentials (right-hand ordinate). Arrow marks indicate addition of 80 μM dicoumarol. (From Alnaes and Rahamimoff, 1975.)

low-affinity sites do not require metabolic energy, whereas the transport against the electrochemical gradient is closely related to energy production (cf. Carafoli and Rossi, 1971). Therefore, it is expected that inhibitors of energy production (or its utilization), would impair calcium uptake by mitochondria, would release sequestered calcium and thus increase the concentration of free calcium ions inside the nerve terminals. If calcium ions inside the nerve terminal are able to activate the release process, then it is expected that the rate of spontaneous liberation of transmitter quanta would be increased. Figure 8 illustrates that this is the case when dicoumarol, an uncoupler of oxidative phosphorylation, is added to the perfusion medium; within several minutes there is a marked increase in m.e.p.p. frequency. Before accepting that the increase in m.e.p.p. frequency is due to release of calcium from intracellular stores, an alternative explanation should also be considered. The action of an uncoupling agent is not restricted to inhibition of calcium uptake by mitochondria, but it will also affect other energy requiring processes; one of them is the maintenance of the resting potential. Therefore it is conceivable that the increase in spontaneous release after dicoumarol is due to depolarization of the terminal membrane, which could increase calcium conductance and cause an influx of calcium. Although resting membrane potential cannot be measured at the frog motor nerve endings, indirect evidence suggests that the main action of dicoumarol is probably not due to depolarization. Depolarization-induced release depends to a large extent on the presence of calcium ions in the perfusion medium, however, even in the absence of extracellular calcium, there is a large increase in m.e.p.p. frequency following the addition of dicoumarol (Alnaes and Rahamimoff, 1975). This suggests that the source of calcium for dicoumarol-induced increase in spontaneous release is intracellular, and is presumably the mitochondrion. Therefore, spontaneous release can be activated by calcium ions arriving from the inside of the synaptic membrane.

Figure 8 shows an additional effect of mitochondrial inhibition. Not only is spontaneous release augmented by dicoumarol poisoning but also the release following the nerve impulse. A possible interpretation of this observation is that dicoumarol partially uncouples the mitochondria and causes some calcium to escape from them; the increased $[Ca]_i$ at rest sums up with the calcium entering the terminal due to the voltage-dependent conductance increase produced by the action potential, thus leading to higher $[Ca]_i$ and larger e.p.p.'s. If the evoked release of transmitter is a function of the sum of the resting $[Ca]_i$ and the amount entering due to the action potential then this may account for the deviation from fourth power kinetics at low $[Ca]_o$ (cf. Crawford, 1974). Presumably a more-than-linear relation exists between $[Ca]_i$ and transmitter release; otherwise it would be difficult to understand why evoked transmitter release increases by much more than the corresponding spontaneous transmitter liberation during the period of the end-plate potential. An alternative explanation would be that the entry of calcium is affected by $[Ca]_i$, much more than expected from the electrochemical gradient.

During the past ten years, the effect of a number of substances known to interfere with mitochondrial function has been tested on the neuromuscular junction (Hubbard and Løyning, 1966; Glagoleva et al., 1970; Katz and Edwards, 1973; Thomson and Turkanis, 1973; Rees, 1974). Figure 9 illustrates schematically some of the energy-linked

processes in mitochondria, and the probable sites of action of the mitochondrial inhibitors. In all cases where the effects of these inhibitors
were examined, they produced marked increase in m.e.p.p. frequency
(cf. Alnaes and Rahamimoff, 1975; Rahamimoff *et al.*, 1975). Of special
interest is the action of the inorganic dye Ruthenium red, which inhibits rather specifically calcium uptake by mitochondria at very low
concentrations (Moore, 1971; Vasington *et al.*, 1972). The current hypothesis is that Ruthenium red acts by combination with a glycoprotein

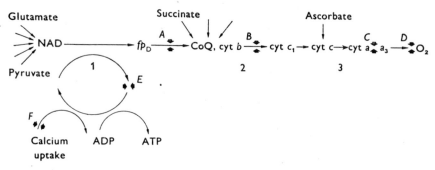

Fig. 9. Metabolic pathways of mitochondria. Electron transport proceeds along
upper horizontal arrows. One (marked (1)) of three ATP generating side branches
is drawn schematically. Calcium uptake is thought to occur facultatively to ADP–ATP
conversion. Probable sites of action of mitochondrial inhibitors: A. barbiturates,
rotenone; B. antimycin; C. cyanide; D. hypoxia; E. uncoupling agents (e.g. dicoumarol, DNP); F. specific calcium uptake antagonists (e.g. Ruthenium red). After
Lehninger, 1970; Carafoli and Rossi, 1971.)

molecule in the mitochondrion which participates in the transport of
calcium. Similar to other mitochondrial inhibitors, it increases the
frequency of the m.e.p.p.s; but instead of augmenting the quantal
content, it causes a reduction (Rahamimoff and Alnaes, 1973). This
reduction can be reversed by elevating $[Ca]_o$. Qualitatively similar
results were obtained with the lanthanide Praseodymium (Alnaes and
Rahamimoff, 1974). Lanthanides are known to block the transport of
calcium through the nerve membrane and to inhibit calcium uptake
by mitochondria (Mela, 1968). This raises the possibility that the calcium transport process across the presynaptic nerve membrane shares
some common properties with the calcium transport system in the
mitochondria.

The calcium binding capability of mitochondria from various origins, including the brain, has been well documented in the past. However, most experiments were performed at concentrations a few orders of magnitude larger than those which supposedly occur inside the nerve. Recently, uptake studies were performed at [Ca] levels of less than 10^{-6}M, on rat brain mitochondria which were separated from other subcellular fractions on ficoll gradients, incubated in calcium containing medium, and their rate of calcium uptake determined by the filtration method. The mitochondria were capable of lowering [Ca] to levels of less than 10^{-7}M (Rahamimoff et al., 1975a, b). The relation between the extramitochondrial [Ca] and the rate of calcium uptake is a non-linear process with a sigmoidal start (Rahamimoff and Rahamimoff, 1975; Scarpa, 1975).

The abundance of the mitochondria in the nerve terminal, the mitochondria occupy on an average 6.59% of the nerve terminal (Alnaes and Rahamimoff, 1975), and their ability to operate in the calcium concentration range supposed to exist inside the nerve terminal are strong enough reasons to assume that they play an important role in the regulation of the intracellular free calcium concentration and thus in the regulation of transmitter release.

In addition to the mitochondria, several other subcellular elements seem to participate in calcium binding. These include the nerve membrane (Hillman and Llinas, 1974; Oschman et al., 1974), synaptic vesicles (Politoff et al., 1974) and perhaps even a soluble protein. These fractions probably do not contribute more than a few percent to the calcium binding capacity of the terminal, but since even small changes in $[Ca]_i$ may have a large effect on transmitter release, they can play a significant role.

In the long run, the prime regulator of the intracellular calcium concentration is the transport across the nerve membrane. This process would be presumably responsible for the steady state concentration of calcium. The efflux has been studied extensively in recent years in the perfused squid giant axon (cf. Blaustein, 1974; Mullins and Brinley, 1975) and in isolated synaptosomes (Blaustein and Oborn, 1975) and it has been found that factors that decrease the calcium efflux are also known to increase transmitter release.

The *balance* between the calcium supplying and the calcium removing processes would determine the intracellular calcium concentration (Fig. 10). If the neurotransmitter release process "reads" $[Ca]_i$ and

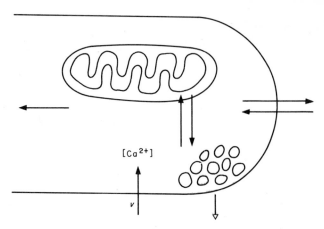

Fig. 10. Some of the factors which influence the intracellular free calcium level, which in turn may determine transmitter release. Shown by arrows are the resting and the voltage dependent calcium influx, the efflux, the uptake and the release by mitochondria (see text). (From Alnaes and Rahamimoff, 1975.)

acts accordingly, then any factor that changes the *balance* should affect neurosecretion and thus the efficiency of information transfer across the synapse. In this context, mitochondria can contribute to the de-crease in transmitter release after the e.p.p. and can participate, to-gether with other processes, in residual phenomena. Furthermore, calcium transport in and out of mitochondria may serve as a communica-tion channel between biochemical and electrophysiological processes at the synapse.

V   References

Alnaes, E. and Rahamimoff, R. (1974). Dual action of Praeseodymium (Pr$^3+$) on transmitter release at the frog neuromuscular synapse. *Nature Lond.* **247**, 478–479.

Alnaes, E. and Rahamimoff, R. (1975). On the role of mitochondria in transmitter release from motor nerve terminals. *J. Physiol.* **248**, 285–306.

Alnaes, E., Meiri, U., Rahamimoff, Hannah and Rahamimoff, R. (1974). Possible role of mitochondria in transmitter release. *J. Physiol.* **241**, 30p.

Baker, P. F. (1972). Transport and metabolism of calcium ions in nerve. *Prog. Biophys. Molec. Biol.* **24**, 177–223.

Baker, P. F., Hodgkin, A. L. and Ridgway, E. B. (1971). Depolarization and calcium entry in giant squid axons. *J. Physiol.* **218**, 709–755.

Baker, P. F., Meves, H. and Ridgway, E. B. (1973). Effects of manganese and other agents on calcium uptake that follows depolarization of squid axons. *J. Physiol.* **231**, 511–526.

Balnave, R. J. and Gage, P. W. (1973). The inhibitory effect of manganese on transmitter release at the neuromuscular junction of the toad. *Brit. J. Pharmacol.* **47**, 339–352.

Benoit, P. R. and Mambrini, J. (1970). Modification of transmitter release by ions which prolong the presynaptic action potential. *J. Physiol.* **210**, 681–695.

Betz, W. J. (1970). Depression of transmitter release at the neuromuscular junction of the frog. *J. Physiol.* **206**, 629–644.

Bianchi, C. P. (1968). "Cell Calcium," p. 131. Butterworths, London.

Birks, R. I. and Cohen, M. W. (1968). The influence of internal sodium on the behaviour of motor nerve endings. *Proc. R. Soc. B.* **170**, 401–421.

Blaustein, M. P. (1974). The interrelationship between sodium and calcium fluxes across cell membranes. *Rev. Physiol. Biochem. Pharmacol.* **70**, 33–82.

Blaustein, M. P. and Oborn, C. J. (1975). The influence of sodium on calcium fluxes in pinched-off nerve terminals in vitro. *J. Physiol.* **247**, 657–686.

Bliokh, Z. L. and Liberman, Y. A. (1970). Effect of beryllium ions on the end-plate potential and the frequency of miniature potentials at the neuromuscular junction of the frog. *Biofizika* **15**, 447–452.

Bowen, J. M. (1972). Effects of rare earths and yttrium on striated muscle and neuromuscular junction. *Can. J. Physiol. Pharmacol.* **50**, 603–611.

Bracho, H. and Orkand, R. K. (1970). Effect of calcium on excitatory neuromuscular transmission in the crayfish. *J. Physiol.* **206**, 61–71.

Braun, M., Schmidt, R. F. and Zimmerman, M. (1966). Facilitation at the frog neuromuscular junction during and after repetitive stimulation. *Pflugers Arch. Ges. Physiol.* **287**, 41–55.

Carafoli, E. and Rossi, C. S. (1971). Calcium transport in mitochondria. *Adv. Cytopharmacol.* **1**, 209–227.

Case, R. M. (1973). Calcium and Gastrointestinal Secretion. *Digestion* **8**, 269–288.

Christensen, B. N. and Martin, A. R. (1970). Estimates of probability of transmitter release at the mammalian neuromuscular junction. *J. Physiol.* **210**, 933–945.

Colomo, F. and Rahamimoff, R. (1968). Interaction between sodium and calcium ions in the process of transmitter release at the neuromuscular junction. *J. Physiol.* **198**, 203–218.

Crawford, A. C. (1974). The dependence of evoked transmitter release on external calcium ions at very low mean quantal contents. *J. Physiol.* **240**, 255–278.

Del Castillo, J. and Engbaek, L. (1954). The nature of the neuromuscular block produced by magnesium. *J. Physiol.* **124**, 370–384.

Del Castillo, J. and Katz, B. (1954a). The effect of magnesium on the activity of motor nerve endings. *J. Physiol.* **124**, 553–559.

Del Castillo, J. and Katz, B. (1954b). Quantal components of the end-plate potential. *J. Physiol.* **124**, 560–573.

Del Castillo, J. and Katz, B. (1954c). Statistical factors involved in neuromuscular facilitation and depression. *J. Physiol.* **124**, 574–585.

Del Castillo, J. and Stark, L. (1952). The effect of calcium on the motor end-plate potentials. *J. Physiol.* **116**, 507–515.

Dodge, F. A. and Rahamimoff, R. (1967). Co-operative action of calcium ions in transmitter release at the neuromuscular junction. *J. Physiol.* **193**, 419–432.

Dodge, F. A. Miledi, R. and Rahamimoff, R. (1969). Strontium and quantal release of transmitter at the neuromuscular junction. *J. Physiol.* **200**, 267–283.

Douglas, W. W. (1974). Exocytosis and the exocytosis-vesiculation sequence: with special reference to neurohypophysis, chromatin and mast cells, calcium and calcium ionphores. In "Alfred Benzon Symposium VII," Secretory mechanisms of exocrine glands (N. A. Thorn and O. H. Petersen, eds.), pp. 116–129. Munksgaard, Copenhagen.

Dudel, J. and Kuffler, S. W. (1961). The quantal nature of transmission and spontaneous miniature potentials at the crayfish neuromuscular junction. *J. Physiol.* **155**, 514–529.

Dunant, Y., Jirounek, P., Israel, M., Lesbats, B. and Manaranche, R. (1974). Sustained oscillations of acetylcholine during nerve stimulation. *Nature Lond.* **252**, 485–486.

Eccles, J. C., Katz, B. and Kuffler, S. W. (1941). Nature of the end-plate potential in curarized muscle. *J. Neurophysiol.* **4**, 362–387.

Eigen, M. (1963). Fast elementary steps in chemical reaction mechanisms. *In* "Coordination Chemistry, I.U.P.A.C.," p. 131. Butterworths, London.

Fatt, P. and Katz, B. (1951). An analysis of the end-plate potential recorded with an intracellular electrode. *J. Physiol.* **115**, 320–370.

Fatt, P. and Katz, B. (1952). Spontaneous subthreshold activity at motor nerve endings. *J. Physiol.* **117**, 109–128.

Feng, T. P. (1936). Studies on the neuromuscular junction. The universal antagonism between calcium and curarizing agencies. *Chin. J. Physiol.* **10**, 513–528.

Feng, T. P. (1941). Studies on the neuromuscular junction. XXXI. The changes of the end-plate potential during and after prolonged stimulation. *Chin. J. Physiol.* **16**, 341–372.

Frankenhaeuser, B. (1957). The effect of calcium on the myelinated nerve fibre. *J. Physiol.* **137**, 245–269.

Frankenhaueser, B. and Hodgkin, A. L. (1957). The action of calcium on the electrical properties of squid axons. *J. Physiol.* **137**, 217–243.

Gage, P. W. and Hubbard, J. I. (1966). An investigation of the post-tetanic potentiation of end-plate potentials at a mammalian neuromuscular junction. *J. Physiol.* **184**, 353–375.

Glagoleva, I. M., Liberman, Y. A. and Khashayev, Z. (1970). Effect of uncoupling agents of oxidation phosphorylation on the release of acetycholine from nerve endings. *Biofizika* **15**, 76–83.

Heuser, J. and Miledi, R. (1971). Effect of lanthanum ions on function and structure of frog neuromuscular junction. *Proc. R. Soc. B* **179**, 247–260.

Hillman, D. E. and Llinas, R. (1974). Calcium containing electron dense structures in the axons of the squid giant synapse. *J. Cell Biol.* **61**, 146–155.

Hubbard, J. I. and Løyning, Y. (1966). The effects of hypoxia on neuromuscular transmission in a mammalian preparation. *J. Physiol.* **185**, 205–223.

Hubbard, J. I., Jones, S. F. and Landau, E. M. (1968a). On the mechanism by which calcium and magnesium affect the spontaneous release of transmitter from mammalian motor nerve terminals. *J. Physiol.* **194**, 381–407.

Hubbard, J. I., Jones, S. F. and Landau, E. M. (1968b). On the mechanism by which calcium and magnesium affect the release of transmitter by nerve impulses. *J. Physiol.* **196**, 75–86.

Hutter, O. F. (1952). Post-tetanic restoration of neuromuscular transmission blocked by D-tubocurarine. *J. Physiol.* **118**, 216–227.

Jenkinson, D. H. (1957). The nature of the antagonism between calcium and magnesium ions at the neuromuscular junction. *J. Physiol.* **138**, 434–444.

Johnson, E. W. and Wernig, A. (1971). The binomial nature of transmitter release at the crayfish neuromuscular junction. *J. Physiol.* **218**, 757–767.

Katz, B. (1962). The Croonian lecture: the transmission of impulses from nerve to muscle and the subcellular unit of synaptic action. *Proc. R. Soc. B* **155**, 455–477.

Katz, B. (1969). The release of neural transmitter substances. The Sherrington lecture No. 10, Liverpool University Press, Liverpool.

Katz, N. L. and Edwards, C. (1973). Effects of metabolic inhibitors on spontaneous and neurally evoked tansmitter release from frog motor nerve terminals. Proc. 26th Meeting Soc. Gen. Physiol. *J. Gen. Physiol.* **61**, 259.

Katz, B. and Miledi, R. (1965a). The measurement of synaptic delay and the time course of acetylcholine release at the neuromuscular junction. *Proc. R. Soc. B.* **161**, 483–495.

Katz, B. and Miledi, R. (1965b). The effect of calcium on acetycholine release from motor nerve terminals. *Proc. R. Soc. B.* **161**, 495–503.

Katz, B. and Miledi, R. (1967a). The timing of calcium action during neuromuscular transmission. *J. Physiol.* **189**, 535–544.

Katz, B. and Miledi, R. (1967b). Tetrodotoxin and neuromuscular transmission. *Proc. R. Soc. B.* **167**, 8–22.

Katz, B. and Miledi, R. (1967c). The release of acetylcholine from nerve by graded electric pulses. *Proc. R. Soc. B.* **167**, 23–38.

Katz, B. and Miledi, R. (1967d). A study of synaptic transmission in the absence of nerve impulses. *J. Physiol.* **192**, 407–436.

Katz, B. and Miledi, R. (1968). The role of calcium in neuromuscular facilitation. *J. Physiol.* **195**, 481–492.

Katz, B. and Miledi, R. (1969). Tetrodotoxin resistant electric activity in presynaptic terminals. *J. Physiol.* **203**, 459–487.

Katz. B. and Miledi, R. (1969a). The effect of divalent cations on transmission in the squid giant synapse. *Publ. Staz. Napoli.* **37**, 303–310.

Katz, B. and Miledi, R. (1970). Further study of the role of calcium in synaptic transmission. *J. Physiol.* **207**, 789–801.

Kelly, J. S. (1965). Antagonism between sodium and calcium at the neuromuscular junction. *Nature Lond.* **205**, 296–297.

Kusano, K., Livengood, D. R. and Werman, R. (1967). Correlation of transmitter release with membrane properties of the presynaptic fiber of the squid giant synapse. *J. Gen. Physiol.* **50**, 2579–2601.

Landau, E. M. and Nachshen, D. (1975). The action of hydrogen ions in neuromuscular transmission. *J. Physiol.* (in press).

Landau, E. M., Smolinsky, A. and Lass, Y. (1973). Post-tetanic potentiation and facilitation do not share a common calcium-dependent mechanism. *Nature New Biol.* **244**, 155–157.

Lehninger, A. L. (1970). Mitochondria and calcium ion transport. *Biochem. J.* **119**, 129–138.

Lester, H. A. (1970). Transmitter release by presynaptic impulses in the squid stellate ganglion. *Nature Lond.* **227**, 493–496.

Liley, A. W. (1956). The quantal components of the mammalian end-plate potential. *J. Physiol.* **133**, 571–587.

Liley, A. W. and North, K. A. K. (1953). An electrical investigation of effects of repetitive stimulation on mammalian neuromuscular junction. *J. Neurophysiol.* **16**, 509–527.

Linder, T. M. (1973). Calcium and facilitation at two classes of Crustacean neuro-muscular synapses. *J. Gen. Physiol.* **61**, 56–73.

Llinas, R. and Nicholson, C. (1975). Calcium role in depolarization-secretion coupling, an aequorin study in squid giant synapse. *Proc. Nat. Acad. Sci.* **72**, 187–190.

Locke, F. S. (1894). Notiz uber den einfluss physiologischer kochsalz losung auf die erregbarkeit von muskel und nerv. *Zbl. Physiol.* **8**, 166.

Lundberg, A. and Quilisch, H. (1953). Presynaptic potentiation and depression of neuromuscular transmission in frog and rat. *Acta Physiol. Scand.* **30**, Suppl., 111, 121–129.

Magleby, K. L. (1973). The effect of tetanic and post-tetanic potentiation on facilitation of transmitter release at the frog neuromuscular junction. *J. Physiol.* **234**, 353–371.

Magleby, K. L. and Zengel, Janet, E. (1975). A dual effect of repetitive stimulation on post-tetanic potentiation of transmitter release at the frog neuromuscular junction. *J. Physiol.* **245**, 163–182.

Mallart, A. and Martin, A. R. (1967). An analysis of facilitation of transmitter release at the neuromuscular junction of the frog. *J. Physiol.* **193**, 679–694.

Mallart, A. and Martin, A. R. (1968). The relation between quantal content and facilitation at the neuromuscular junction of the frog. *J. Physiol.* **193**, 593–604.

Mambrini, J. and Benoit, P. R. (1964). Action du calcium sur la jonction neuro-musculaire chez la grenouille. *C. R. Soc. Biol. (Paris)* **158**, 1454–1458.

Manalis, R. S. and Cooper, P. C. (1973). Presynaptic and postsynaptic effects of lead at the frog neuromuscular junction. *Nature Lond.* **243**, 354–356.

Martin, A. R. (1955). A further study of the statistical composition of the end-plate potential. *J. Physiol.* **130**, 114–122.

Martin, A. R. (1966). Quantal nature of synaptic transmission. *Physiol. Rev.* **46**, 51–66.

Matthews, E. K. (1970). Calcium and hormone release. *In* "Calcium and Cellular Function" A. W. Cuthbert, (ed.), pp. 163–182. Macmillan, London.

Meiri, H. and Rahamimoff, R. (1972). Fluctuations in end-plate potential amplitude and calcium ions. *Israel J. Med. Sci.* **8**, 4.

Meiri, U. (1975). Divalent ions and synaptic transmission at the neuromuscular junction. Ph.D. Thesis. Hebrew University, Jerusalem.

Meiri, U. and Rahamimoff, R. (1971). Activation of transmitter release by strontium and calcium ions at the neuromuscular junction. *J. Physiol.* **215**, 709–726.

Meiri, U. and Rahamimoff, R. (1972). Neuromuscular transmission: Inhibition by manganese ions. *Science* **176**, 308–309.

Mela, L. (1969). Inhibition and activation of calcium transport in mitochondria: Effect of lanthanides and local anesthetic drugs. *Biochemistry* **9**, 2481–2486.

Miledi, R. (1966). Strontium as a Substitute for calcium in the process of transmitter release at the neuromuscular junction. *Nature Lond.* **212**, 1233–1234.

Miledi, R. (1973). Transmitter release induced by injection of calcium ions into nerve terminals. *Proc. R. Soc. B.* **183**, 421–425.

Miledi, R. and Thies, R. (1971). Tetanic and post-tetanic rise in frequency of miniature end-plate potentials in low-calcium solutions. *J. Physiol.* **212**, 245–257.

Mines, G. R. (1911). On the replacement of calcium in certain neuromuscular mechanisms by allied substances. *J. Physiol.* **42**, 251–266.

Moore, C. L. (1971). Specific inhibition of mitochondrial Ca^{2+} transport by Ruthenium red. *Biochem. Biophys. Res. Commun.* **42**, 298–305.

Mullins, L. Y. and Brinley, F. J. (1975). The sensitivity of calcium efflux from squid axons to changes in membrane potential. *J. Gen. Physiol.* **63**, 135–152.

Nastuk, W. L. and Liu, J. H. (1966). Muscle postjunctional membrane: Changes in chemosensitivity produced by calcium. *Science, N.Y.* **154**, 266–267.

Ortiz, C. L. and Bracho, H. (1972). Effect of reduced calcium on excitatory transmitter release at the crayfish neuromuscular junction. *Comp. Biochem. Physiol.* **41A**, 805–812.

Otsuka, M., Endo, M. and Nonomura, Y. (1962). Presynaptic nature of neuromuscular depression. *Jap. J. Physiol.* **12**, 573–584.

Parnas, I., Rahamimoff, R. and Sarne, Y. (1975). Tonic release of transmitter at the neuromuscular junction of the crab. *J. Physiol.* **250**, 275–286.

Politoff, A. L., S. Rose and Pappas, G. D. (1974). The calcium binding sites of synaptic vesicles of the frog sartorius neuromuscular junction. *J. Cell Biol.* **61**, 818–823.

Portzehl, H., Caldwell, P. C. and Ruegg, J. C. (1964). The dependence of contraction and relaxation of muscle fibres from the crab Maia squinado on the internal concentration of free calcium ions. *Biochim. Biophys. Acta* **79**, 581–591.

Rahamimoff, H. and Rahamimoff, R. (1975). Calcium uptake by brain mitochondria. *Israel J. Med. Sci.* **11**, 68.

Rahamimoff, R. (1967). The use of the Biomac 500 computer for estimating facilitation at single end-plates. *J. Physiol.* **191**, 12–148.

Rahamimoff, R. (1968). A dual effect of calcium ions on neuromuscular facilitation. *J. Physiol.* **195**, 471–480.

Rahamimoff, R. (1973). Modulation of transmitter release at the neuromuscular synapse. *In* "The Neurosciences Third Study Program," F. O. Schmitt and F. G. Worden, eds.), pp. 943–952. MIT Press, Cambridge, Mass. U.S.A.

Rahamimoff, R. and Alnaes, E. (1973). Inhibitory action of Ruthenium Red on neuromuscular transmission. *Proc. Nat. Acad. Sci. U.S.A.* **70**, 3613–3616.

Rahamimoff, R. and Colomo, F. (1967). The inhibitory action of sodium ions on transmitter release at the motor end-plate. *Nature Lond.* **215**, 1174–1176.

Rahamimoff, R. and Meiri, H. (1975). Effect of calcium ions on the statistical parameters of transmitter release. *Israel J. Med. Sci.* **11**, 1, 68.

Rahamimoff, R. and Yaari, Y. (1973). Delayed release of transmitter at the frog neuromuscular junction. *J. Physiol.* **228**, 241–257.

Rahamimoff, R., Erulkar, S. D., Alnaes, E., Meiri, H., Rotshenker, S. and Rahamimoff, H. (1975). Modulation of transmitter release by calcium ions and nerve impulses. *Cold Spr. Harb. Symp. Quant. Biol.* **40** (in press).

Rahamimoff, R., Rahamimoff, H., Binah, O. and Meiri, U. (1975). Control of neurotransmitter release by calcium ions and the role of mitochondria. *In* "Calcium Transport in Contraction and Secretion" (E. Carafoli, F. Clementi and A. Margretti, eds.). Elsevier, North Holland, Amsterdam (in press).

Rees, D. (1974). The effect of metabolic inhibitors on the cockroach nerve-muscle synapse. *J. Exp. Biol.* **61**, 331–343.

Rosenthal, J. (1969). Post-tetanic potentiation at the neuromuscular junction of the frog. *J. Physiol.* **203**, 121–133.

Rotshenker, S., Erulkar, S. D. and Rahamimoff, R. (1976). Reduction in the frequency of miniature end-plate potentials by nerve stimulation in low calcium solutions. *Brain Research* **101**, 362–365.

Rubin, R. P. (1970). The role of calcium in the release of neurotransmitter substances and hormones. *Pharmacol. Rev.* **22**, 389–428.

Sarne, Y. (1974). Synaptic inhibition in a neuromuscular system. Ph.D. Thesis, The Hebrew University, Jerusalem.

Sarne, Y., Parnas, I. and Rahamimoff, R. (1972). Tonic inhibition in crustacean nerve-muscle synapses. *Israel J. Med. Sci.* **8**, 1771.

Scarpa, A. (1975). Kinetic and thermodynamic aspects of Ca^{+2} transport in mitochondria. *In* Symposium on calcium transport in contraction and secretion, p. 59. Carlo Erba Foundation, Milan.

Shanes, A. M. and Bianchi, C. P. (1959). The distribution and kinetics of release of radiocalcium in tendon and skeletal muscle. *J. Gen. Physiol.* **42**, 1123–1137.

Shimomura, O., Johnson, F. H. and Saiga, Y. (1963). Microdetermination of calcium by aequorin luminescence. *Science* **140**, 1339–1340.

Stinnakre, J. and Tauc, L. (1973). Calcium influx in active aplysia neurons detected by injected aequorin. *Nature New Biol.* **242**, 113–115.

Stjärne, L. (1973). Michaelis–Menten kinetics of sympathetic neurotransmitter as a function of external calcium, effect of graded alpha-adrenoceptor blockade. *Naunyn-Schmiedebergs Arch. Pharmacol.* **278**, 323–327.

Takeuchi, A. (1958). The long-lasting depression in neuromuscular transmission of frog. *Jap. J. Physiol.* **8**, 103–113.

Takeuchi, A. and Takeuchi, N. (1960). On the permeability of end-plate membrane during the action of transmitter. *J. Physiol.* **154**, 52.

Takeuchi, N. (1963). Effects of calcium on the conductance change of the end plate membrane during the action of transmitter. *J. Physiol.* **167**, 141–155.

Thies, R. (1965). Neuromuscular depression and the apparent depletion of transmitter in mammalian muscle. *J. Neurophysiol.* **28**, 427–442.

Thomson, T. D. and Turkanis, S. A. (1973). Barbiturate-induced transmitter release at a frog neuromuscular junction. *Brit. J. Pharmacol.* **48**, 48–58.

Thorn, N. A. (1974). Role of calcium in secretory processes. *In* "Alfred Benzon Symposium." VII. Secretory mechanisms of exocrine glands (N. A. Thorn and O. H. Petersen, eds.), pp. 305–326. Munksgaard, Copenhagen.

Vasington, F. D., Gazotti, P., Tiozzo, R. and Carafoli, E. (1972). The effect of Ruthenium Red on Ca^{2+} transport and respiration in rat liver mitochondria. *Biochim. Biophys. Acta* **256**, 43–54.

Weakly, J. N. (1973). The action of cobalt ions on neuromuscular transmission in the frog. *J. Physiol.* **234**, 597–612.

Weber, A., Herz, R. and Reiss, I. (1966). Study of the kinetics of calcium transport by isolated fragmented sarcoplasmic reticulum. *Biochem. Z.* **345**, 329–369.

Weinreich, D. (1971). Ionic mechanism of post tetanic potentiation at the neuromuscular junction of the frog. *J. Physiol.* **212**, 431–446.

Wernig, A. (1972a). Changes in statistical parameters during facilitation at the crayfish neuromuscular junction. *J. Physiol.* **226**, 751–759.

Wernig, A. (1972b). The effects of calcium and magnesium on statistical release parameters at the crayfish neuromuscular junction. *J. Physiol.* **226**, 761–768.

Younkin, S. G. (1974). An analysis of the role of calcium in facilitation at the frog neuromuscular junction. *J. Physiol.* **237**, 1–14.

5

Desensitization at the Neuromuscular Junction

L. G. Magazanik and F. Vyskočil

Sechenov Institute of Evolutionary Physiology and Biochemistry, Academy of Sciences of the U.S.S.R., Leningrad, Soviet Union and *Institute of Physiology, Czechoslovak Academy of Sciences, Prague, Czechoslovakia.*

I Introduction

The reaction of an excitable tissue to applied stimulation depends on the duration of the stimulus. Generally speaking, the response decreases with time and this phenomenon is called adaptation, accommodation, inactivation, tachyphylaxis, etc. according to the type and characteristics of the tissue in any particular case. However, various specific cellular and molecular mechanisms are apparently involved in this process.

The term *desensitization* (Thesleff, 1955) is commonly used for the special case of lowered sensitivity to the transmitter and its analogues in chemosensitive regions of an excitable cell, which develops as the result of long-lasting contact between the cell and its agonist. This phenomenon has been observed during the action of acetylcholine (ACh) on the postsynaptic muscle membrane of vertebrate skeletal muscles, mollusc neurones, Renshaw cells in the mammalian spinal cord, electric organs of the fish, smooth muscles of the ileum, leech muscles and elsewhere. The progressive decrease of sensitivity of particular cells to receptor agonists has also been demonstrated during protracted application of glutamate, histamine, catecholamines, GABA, angiotensin, oxytocine and other biologically active compounds.

In different tissues, (and for different agonists as well), the mechanisms of the decreased sensitivity may be different in nature; the term "desensitization" is therefore used for the description of molecular processes which have a common general pattern.

The mechanism of desensitization (DS) is not clear even in such a well studied synapse as the skeletal muscle end-plate. At present considerable progress has been made in revealing the molecular organization of the muscle postjunctional membrane (PJM) and in understanding its activation by ACh and other cholinomimetics. Desensitization represents the opposite pole of this process, i.e. inactivation of the PJM and naturally the question arises, how both processes of activation and inactivation are internally coupled in this complex functional system.

II General features of desensitization

The effect of ACh on the end-plate is connected with increased ionic permeability, which progressively diminishes during the prolonged presence of the drug. When ACh application is stopped, the sensitivity recovers to the original level.

When studying the kinetics of DS, it is at first necessary to measure the initial sensitivity of the tissue to the agonist. After this the lowering of sensitivity during long-lasting application of this agonist is assessed and this is finally followed by the recovery after the washing out of the drug. The actual rates of DS onset and recovery of original sensitivity depend to a great extent on the experimental method employed, namely, on the method of ACh application to the chemosensitive area and the technique for recording the response.

When the muscle is bathed in a solution containing ACh (or other agonist), one can, upon removal of the drug, measure the rate of muscle fibre repolarization in the end-plate zone, i.e. the decrease of originally developed depolarization. The rate constants of DS onset and recovery are of the order of minutes, when the agonist is applied by this method. However, in this case the changes in internal ionic composition caused by long-lasting depolarization may complicate the estimation of DS (Jenkinson and Terrar, 1973) and it is therefore better to use a voltage-clamp technique by which only ionic currents are registered in the activated end-plate zone.

The DS develops more rapidly when agonists are applied more locally, e.g. when the end-plate zone is superfused with the agonist from a macropipette (Manthey, 1966), or when ionophoretic application through a micropipette is used (Katz and Thesleff, 1957). In the latter

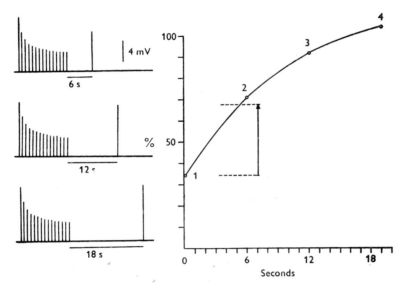

Fig. 1. Desensitization and recovery of sensitivity on the post-junctional membrane after application of a train of constant ionophoretic impulses from a single acetylcholine micropipette. On the left, oscillographic records of three trains of ACh potentials evoked at the same site at 60 s intervals. After each train the period of rest (6, 12 and 18 s) was followed by a single test pulse, indicating the recovery of sensitivity. On the right, the curve of recovery: abscissa, the time in seconds between last pulse in the train and test pulse; ordinate, amplitude of the last pulse in the train (1), and of test pulses at 6 s (2), 12 s (3) and 18 s (4) after end of train (100% = amplitude of the first ACh potential in the series). Arrow indicates the time at which the recovery reaches 50% of the difference between desensitization level (1) and maximal sensitivity (100%, point 4) (half-time of recovery, *t*). Temperature 22°C. (From Magazanik and Vyskočil, 1975; with permission of the *J. Physiol., Lond.*)

case the time course of the DS is often in the order of seconds. Because the phenomenon of DS with such local application may be induced repeatedly on the same cell, this method appears to be the most suitable.

A very important question is connected with the procedure of obtaining reliable data about the rate of recovery after washing out of the agonist. It is desirable to measure both the onset and the recovery rates in the same experiment. This requirement is met, to a great extent, by using the method of repetitive ionophoretic pulse application of ACh at a frequency of 0·2–2 Hz. Such a train of ACh pulses leads to a progressive decrease of subsequent responses, i.e. to DS. When the intervals between pulses are progressively prolonged to 5, 10, 15 s, etc. recovery of sensitivity can be observed until it reaches the original level (Fig. 1).

However, the success of this method depends on the susceptibility of the given tissue to DS. When DS develops very slowly, the use of some potentiating drug is desirable.

The rate at which the response diminishes under such conditions apparently reflects several simultaneously acting processes. Every dose of ACh evokes a response, during which DS develops and persists. If the next pulse comes after a sufficiently short period of time, the PJM is in a partially desensitized state and its amplitude drop is a measure of the degree of prevailing DS. But the second pulse deepens DS until an equilibrium during the train of pulses is eventually reached reflecting the situation, when the rates of DS onset and recovery are approximately equal. In such an experiment, one measures the rate at which an equilibrium is established. This process is apparently slower than the actual rate of DS onset during each ACh-pulse. The estimation of the rate of recovery is less complicated. Although each test-pulse used after the cessation of a train aggravates DS, it does not significantly affect the time course of recovery when the test pulse is used only once during the measurement (Fig. 1).

The dependence of absolute rates of DS onset and recovery on the mode of ACh application, diffuse or local, still remains obscure. The difference is so great (up to 100 times), that it leads one to speculate about the possibility of two different mechanisms of DS: a fast one, which can be revealed only by ionophoretic microapplication of agonists and a slow one, observed during more diffuse application (Rang and Ritter, 1970a). But many factors affecting the kinetics of DS are effective during both types of application (see Magazanik and Vyskočil, 1973; Terrar, 1974). The hypothesis about two separate types of DS is therefore less attractive. It seems that during longer-lasting contact with the agonist, additional factors play a role and make the situation more complex. To clear up this controversy appears urgent, since it could help to elucidate the blocking effect of "desensitizing" muscle relaxants such as decamethonium or succinylcholine, which are used in clinical practice and are administered parenterally, i.e. diffusely.

III   Some features of desensitization kinetics

Kinetics of DS in the end-plate area have many important properties which help us to understand its mechanism (Katz and Thesleff, 1957).

1. The rate of DS onset is directly proportional to the amount (concentration of ACh in prolonged contact with the PJM.

2. The rate of recovery after washing out of the agonist is constant and independent either of the concentration of agonist or the degree of DS.

It is therefore possible to arrange the initial conditions so that the rate of DS onset is slower than the rate of recovery.

A process with such kinetic properties undoubtedly has two successive reactions with independent rate constants: inactivation of the system and its restitution. Two separate processes underly these reactions and are coupled by a common intermediate product. This idea about the relative independence of DS onset and recovery can be supported by comparing the effect of different factors on the kinetics of DS. As shown in Table I, there exist factors, which selectively increase the rate of DS onset, but have no effect on its recovery. In experiments on different muscle preparations the rate of onset differs during the same level of long-lasting depolarization (Axelsson and Thesleff, 1958; Rang and Ritter, 1970a), but the rate of recovery remains relatively constant. At

TABLE 1

Influence of different drugs on kinetics of activation and
desensitization of PJM (Magazanik, 1976)

Drug	Activation		Desensitization	
	Decay of e.p.c.	Life-span of channels	Onset	Recovery
Procaine	biphasic		accelerate	
Lidocaine	biphasic		accelerate	
d-tubocurarine	slightly shortened	no-effect	absent	absent
atropine	shortened	shortened	absent	absent
serotonine	slightly shortened		absent	absent
pentobarbitone	shortened		accelerate	
SKF–525A	absent		accelerate	absent
chlorpromazine	absent		accelerate	absent
N-butanol	prolonged	prolonged	accelerate	

e.p.c. = end-plate current

$20°C$ the half-time of recovery is 5–7 s in different types of muscles (Vyskočil, 1976).

The special properties of DS kinetics can be illustrated by the cyclic scheme, proposed by Katz and Thesleff (1957).

$$A + R \underset{K_{-1}}{\overset{K_1}{\rightleftharpoons}} AR$$
$$K_R \Big\updownarrow K_{-R} \qquad K_{-D} \Big\updownarrow K_D \qquad (1)$$
$$A + R' \underset{K_{-2}}{\overset{K_2}{\rightleftharpoons}} AR'$$

According to this scheme (1), the receptor can exist in the active (R) and inactive (R') state, (2), both forms can combine with the agonist (A), the change from one state to the other is much slower than the formation and dissociation of the agonist-receptor complex, i.e. K_1, K_{-1}, K_2, K_{-2} $\gg K_D$, K_{-D}, K_R, K_{-R}. The equilibrium state may then be described as

$$\frac{K_2 . K_{-1}}{K_{-2} . K_1} = \frac{K_D K_R}{K_{-D} K_{-R}}.$$

For the scheme to be valid and to correspond to actual rates measured in experiments, one must assume the following.

1. $K_D > K_{-D}$; if this were not so, DS would not occur.
2. $K_R > K_{-R}$; in the opposite case most receptors would be in the desensitized state even in the absence of the agonist.

From these presumptions one can draw the following important conclusions:

1. $\dfrac{K_2}{K_{-2}} > \dfrac{K_1}{K_{-1}}$,

i.e. affinity of the desensitized receptor to the agonist should be higher than that of the normal receptor, and

2. $\dfrac{K_2 . K_{-1}}{K_{-2} . K_1} > \dfrac{K_D}{K_R}$,

i.e. the ratio of affinities should be higher than the ratio of rate constants for DS and recovery.

This scheme can be simplified when we assume that $K_R \gg K_{-R}$ and $K_D \gg K_{-D}$, i.e. that there are practically no desensitized receptors in the absence of the agonist and that a change of the complex AR' to AR is extremely improbable.

Then

$$A + R \underset{K_{-1}}{\overset{K_1}{\rightleftharpoons}} AR$$

$$K_R \uparrow \qquad \qquad \downarrow K_D \qquad \qquad (2)$$

$$A + R' \underset{K_{-2}}{\overset{K_2}{\rightleftharpoons}} AR'$$

Furthermore, if this scheme is to be in accordance with experimental data, it needs some assumptions concerning the ratio between the rate constants. In experiments with small concentrations of the agonist, the rate of onset could be lower than the rate of recovery. It would then be necessary to accept that $K_2/K_{-2} > K_1/K_{-1}$, i.e. the affinity to the desensitized receptor is higher than that to the normal one.

There exist fundamental reasons for giving preference to scheme (1) rather than to scheme (2). The non-equilibrium scheme (2) requires an energy supply, whereas the changes of receptors described in scheme (1) should occur without an extra energy source (Katz and Thesleff, 1957). It is now known that DS can take place in systems where there is no energy source (see below). Nevertheless, one has to take into account the above mentioned limiting assumptions about the possibility that K_D and K_R prevail over K_{-D} and K_{-R}, respectively.

IV   The connection between desensitization and activation of the postjunctional membrane

It is quite clear that DS develops in the presence of the agonist and relatively rapidly disappears after its removal. This indicates that the formation of the agonist-receptor complex is of primary importance for both processes, activation as well as DS. The question naturally arises, how both processes are connected: is activation a necessary prerequisite for the development of DS and vice versa and can DS affect, to a certain extent, the process of activation?

It has been suggested, on the basis of general considerations, that activation of the PJM undergoes two stages (del Castillo and Katz, 1957):

$$ACh + R \underset{K_{-1}}{\overset{K_1}{\rightleftharpoons}} AChR \underset{\alpha}{\overset{\beta}{\rightleftharpoons}} AChR^* \qquad (3)$$

where R is the resting state of the receptor and R^* is the active state when ion channels are open.

A detailed investigation of the time course of postjunctional currents (Magleby and Stevens, 1972a, b) and elementary fluctuations of the postjunctional current ("acetylcholine noise") (Katz and Miledi, 1972; Anderson and Stevens, 1973; see Chapter 2) has made it possible to postulate that a change of the AR complex into the active form is slower than the formation of the complex itself: $K_1, K_{-1} \gg \beta, \alpha$. This means that the ionic gate controlling mechanism is the limiting factor during activation-desensitization relationship.

In connection with this, one can suggest two possibilities for the activation-densensitization relationship.

1. DS always follows activation.

$$A + R \underset{K_{-1}}{\overset{K_1}{\rightleftharpoons}} AR \underset{\alpha}{\overset{\beta}{\rightleftharpoons}} AR^* \tag{4}$$

$$K_R \big\updownarrow K_{-R} \qquad\qquad K_{-D} \big\updownarrow K_D$$

$$A + R' \underset{K_{-2}}{\overset{K_2}{\rightleftharpoons}} AR'$$

2. For DS to develop the formation of the $A + R$ complex is sufficient The complex can afterwards transform itself into the active form with a higher probability or into the desensitized state (with a lower probability).

$$A + R \underset{K_{-1}}{\overset{K_1}{\rightleftharpoons}} AR \underset{\alpha}{\overset{\beta}{\rightleftharpoons}} AR^* \tag{5}$$

$$K_R \big\updownarrow K_{-R} \qquad K_{-D} \big\updownarrow K_D$$

$$A + R \underset{K_{-2}}{\overset{K_2}{\rightleftharpoons}} AR'$$

When the magnitudes of the rate constants of the slowest processes in this scheme are compared, both possibilities may occur. With ACh, α is less than 2 ms^{-1}. We do not know exactly the absolute values of β, but even if it were smaller than α, the difference between them cannot be too great. In the opposite case, activation of the complex should be less probable, assuming that the preceding reaction is very fast. The equilibrium constant of AR' formation which depends, to a great extent, on K_D is in the order of seconds, or even minutes, in the case of diffusely applied agonist. It then follows that in both schemes (4) and (5) the transformation AR (AR^*) into AR' is the limiting step for DS. It is therefore understandable why DS has practically no effect on the parameters of the elementary events (Katz and Miledi, 1972; Anderson and Stevens, 1973).

Many factors are known to affect the kinetics of activation and many other influence DS (see Table 1). Changes in the time course of the end-plate current can apparently be ascribed to changes of constant α and the effect on DS should be connected with other stages in the cyclic scheme. It is also known that hyperpolarization slows the kinetics of activation (Kordaš, 1969, 1972; Magleby and Stevens, 1972a, b; Anderson and Stevens, 1973), but increases the onset of DS (Magazanik and Vyskočil, 1970, 1973, 1975) without altering the rate of recovery of sensitivity (Magazanik and Vyskočil, 1975). Lowering of temperature unambiguously affects both processes, slowing activation (Kordaš, 1972; Magleby and Stevens, 1972) and DS onset as well as recovery (Magazanik and Vyskočil, 1973, 1975).

It can therefore be concluded that up to now no apparent correlation exists between the action of different factors on the kinetics of activation and on DS. However, this does not mean that such correlation is not masked by differences in the time course of both processes.

V The ability of different cholinomimetics to induce desensitization

There exists considerable differences in the ability of various cholinomimetics to activate the PJM. This ability is commonly estimated from the dose-response curves, which provide information about the affinity for the receptor and also about the intrinsic activity of the drug. The membrane current fluctuations in the presence of cholinomimetics provide us with a new tool for obtaining parametra connected with their efficacy. There are two such parameters—the mean conductivity of a single activated channel and the time constant of channel relaxation (Katz and Miledi, 1972; 1973; Anderson and Stevens, 1973; Colquhoun et al., 1975). For an understanding of the mechanism of DS and in particular for the elucidation of a possible connection between activation and DS, it should be important to compare the ability of different cholinomimetics to activate and to desensitize the PJM. Unfortunately, only fragmentary and scant information is so far available.

Thesleff (1955) compared the desensitizing effect of diffusely applied ACh, decamethonium, succinylcholine and nicotine and found no substantial differences between equipotent concentrations in inducing DS. The local application of other agonists such as carbachol and succinyldicholine (Katz and Thesleff, 1957) or butyrylcholine, tetramethylammonium, decamethonium and suberyldicholine (Magazanik, 1968) led to the same conclusion. Rang and Ritter (1970a), using the contractile response of the tissue, found that in chick muscles 10 out of the 14 compounds studied possessed a comparable ability to desensitize as carbachol and four compounds desensitized the muscle more effectively. In the frog muscle they observed, similarly to Gissen and Nastuk (1966), that the most potent desensitizing agent is phenyltrimethylammonium and heptyltrimethylammonium. Both these compounds were equally potent on producing DS as carbachol when tested on chick muscle. The rate of recovery from DS was found to be almost identical for all the compounds studied. However, all agonists, which were able to desensitize more effectively than carbachol exhibited a lower activating potency (8–20 times) in the frog muscle than carbachol itself. No such correlation was found in the chick muscle. Moreover, among "strong" and "normally" desensitizing agents substances were found which are known to be only partial agonists.

Quite a different scale of DS activity was found in isolated ganglion neurones of the snail *Limnea stagnalis* (Bregestovski *et al.*, 1975). The rate of membrane conductance decrease in the presence of constant equally activating concentrations of cholinomimetics were used as a measure of DS. Many of the studied agonists had a lower DS potency than ACh. The following sequence of potencies was found: dimethylamid of phosphodicholine ester > thiocholine > ACh > tetramethylammonium > dicholinester of glutaric acid > dicholinester of adipic acid > carbachol > suberyldicholine > sebacinyldicholine. No close correlation was found between the activating and the DS-inducing potency of these compounds, but some of the "strong" DS agents were found to be weak agonists and vice versa. Unfortunately, the rate of recovery from DS was not measured in these experiments.

Important information could be obtained from a comparison of the DS-inducing potency of the agonists and the kinetic parameters of the ionic channels opened by them. But "noise" has been recorded only in the presence of a small number of agonists. In several experiments with DS, we tested three of these: acetylthiocholine, which induces the short-lasting opening of the channels ($\tau = 0.12$ ms), ACh ($\tau = 1$ ms) and suberyldicholine, which activates the ionic channels with a longer life-span ($\tau = 1.65$ ms) (Katz and Miledi, 1973).

It was found that suberyldicholine and ACh do not differ significantly in their ability to desensitize the PJM (Fig. 2); but in the case of acetylthiocholine, when applied ionophoretically in equipotent dose, DS developed very poorly. This observation should be further analysed, namely by searching for other agonists which can open the ionic chan-

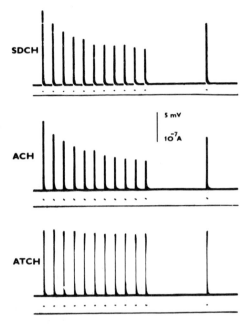

Fig. 2. Desensitization and recovery of sensitivity after application of a train (1 Hz) of constant ionophoretic pulses from single micropipettes filled with 2M suberyldicholine (SDCH), acetylcholine (ACH) or acetylthiocholine (ATCH) respectively. After a pause of 6 s, the sensitivity recovered approximately to 50% of the difference between the first and last pulse of the preceding train in the case of SDCH and ACH. The recovery of sensitivity to ATCH cannot be estimated because of the very weak desensitization induced by this drug. All records are from the same sartorius muscle fibre at 22°C. Chlorpromazine 5 × 10^{-7} M was present in the bath. (Record slightly retouched.)

nels for only a short period of time and to test their ability to desensitize the PJM.

Up to now it is difficult to put forward a plausible explanation for why the rate of DS onset differs with different agonists, whereas the rate of recovery is rather similar in all cases tested by the ionophoretic method.

VI   Factors influencing the kinetics of desensitization

It is undoubtedly important for an understanding of the mechanism of DS to study different factors which modulate the kinetics of this process. Up to now a number of such influences are known. The kinetics of

DS depend on the ionic composition of the medium. A change in $[Ca^{2+}]_o$ concentration has the most dramatic effect; an increase of $[Ca^{2+}]_o$ leads to an earlier onset of DS and vice versa (Manthey, 1966, 1970; Magazanik, 1968, 1969; Magazanik and Shekhirev, 1970; Magazanik and Vyskočil, 1970). It is noteworthy that DS of mollusc neuronal post-synaptic membrane is completely insensitive to changes in $[Ca^{2+}]_o$ (Bregestovski et al., 1975). The observed slowing of DS onset after an increase of $[K^+]_0$ (Manthey, 1966) can apparently be explained by the depolarizing effect of K^+, because it is known that depolarization of the muscle fibre membrane by excess K^+ causes a slowing of DS. This can be restored by artificial repolarization of the membrane (Magazanik and Vyskočil, 1970). Lowering of $[Na^+]_o$, similarly as the exchange of Na^+ for Li^+ decreases the DS onset and excess of Na^+ increases it (Manthey, 1966; Magazanik, 1968; Parsons et al., 1973). Some poly-valent cations (Ba^{2+}, Sr^{2+}, Al^{3+}, La^{3+}) are more active than Ca^{2+} in potentiating DS (Magazanik and Vyskolčil, 1970; Lambert and Par-sons, 1970). The rate of DS onset in the end-plate zone of the frog muscle fibre depends on the membrane potential level, hyperpolariza-tion increases and depolarization decreases it (Magazanik and Vyskočil, 1970, 1973, 1975). This phenomenon was neither observed in mollusc neurones (Bregestovski et al., 1975) where DS is independent of the membrane potential level. Changes in temperature affect the onset of DS as well as its recovery (Magazanik and Vyskočil, 1973; Magazanik and Vyskočil, 1975; see also Bregestovski et al., 1975). The lowering of temperature slowed both processes, but recovery was less affected (Fig. 3).

There exist a large number of chemical agents which affect the kinetics of DS. The majority of them only increase the rate of onset having no effect on recovery. SKF–525A (2-diethylaminoethyl di-phenylpropylacetate) and chlorpromazine (Magazanik, 1970, 1971a) belong to the most potent agents. A similar effect was found in the presence of local anaesthetics such as procaine, lidocaine and bar-biturates (amylobarbitone and pentobarbitone), antihistamines (di-phenylhydramin and prometazine), higher alcohols and propranolol (Magazanik, 1968, 1969, 1970, 1971a, b; Magazanik and Vyskočil, 1973, 1975).

It is characteristic for all these compounds to diminish the iono-phoretically evoked ACh-responses in concentrations so low that they do not affect the amplitude of natural miniature end-plate potentials

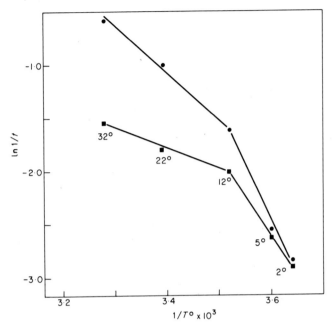

Fig. 3. Arrhenius plot for the rate of onset (●) and recovery (■) of desensitization (t_0 and t_r) during temperature changes between 32 and 2°C. Abscissa, reciprocal of the absolute temperature ($T°$). Ordinate, ln (1/t_0) or ln (1/t_r). (From Magazanik and Vyskočil, 1975; with permission of the *J. Physiol., Lond.*)

(m.e.p.p.s). Higher concentrations of some of these agents have a pre-synaptic effect, manifested by a reduced quantum content and changes in the frequency of m.e.p.p. Eventually when they are applied in even higher concentrations (50–200 times), the time course and the amplitude of m.e.p.p. is altered. Their ability to diminish the ACh-response in small doses may be ascribed to their DS potentiating effect (Magazanik, 1970, 1971a, b; Magazanik and Vyskočil, 1973, 1975). This idea is supported by the following facts.

1. The extent of ACh-potential decrease depends on the frequency of ACh application. After a short period of rest the amplitude returns to the original level.
2. The larger the amplitude and the longer the duration of the ACh-potentials, the more easily they are diminished during the train by DS.
3. The first ACh-potential in a train (after a sufficient period of rest)

has the same amplitude as before a DS potentiating drug has been added to the bath.

The following compounds seem to depress ACh-potentials by a similar mechanism: perhydrohistrionicotoxin (Albuquerque *et al.*, 1973), hexafluorenium (Nastuk and Karis, 1964) and thiopentone (Adams, 1974).

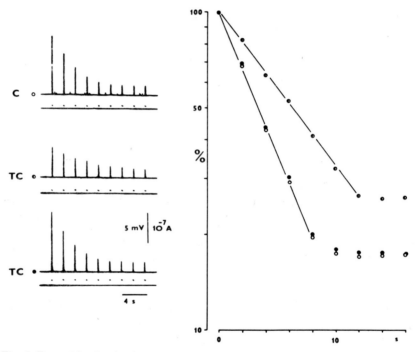

Fig. 4. Desensitization in the presence of 5 × 10⁻⁷M chlorpromazine in the muscle bath. On the left, oscillographic records from the same sartorius end-plate, C, ○ = control, TC, ◐ = after adding 6 × 10⁻⁷M *d*-tubocurarine into the bath; TC, ● = after increasing the amount of ACh in each pulse to overcome the lowered sensitivity. On the right, desensitization curves on a semilogarithmic scale (100% = amplitude of the first ACh potential in each train), showing the exponential character of DS onset and identical rate of C, ○ and TC, ● desensitization. Frog sartorius, 22°C.

Some other compounds apparently possess cholinolytic properties together with the ability to potentiate DS: arpenal (diethylaminopropyl diphenylacetate) and its quaternary analogue mesphenal, adiphenin (2-diethylaminoethyl diphenylacetate), karamiphen (2-

diethylaminoethyl phenylcyclopentanacetate) and its quaternary analogue merpanit, cypenam (2-diethylaminopropyl phenylcyclopentanacetat) (Magazanik, 1971b) and derivatives of benzylic acid-pipenzolate and oxyphenonium (Terrar, 1974). Qualitatively, the effect of such drugs is similar to SKF–525A or chlorpromazine, i.e. they strongly increase the rate of DS onset without affecting recovery, but they also affect the amplitude of the end-plate potential when their concentration is raised 10–30 times. Moreover, many of them are known as rather potent cholinolytics.

It is noteworthy that compounds potentiating DS can also be effective when applied intracellularly, from an ionophoretic pipette inserted into the muscle fibre near the end-plate (Vyskočil and Magazanik, 1972; Magazanik and Vyskočil, 1973, 1975). Rang and Ritter (1969, 1970b) described several compounds which enhance DS, induced by carbachol and other angonists. Some of them were weak antagonists, while some were found completely without cholinolytic action. They have been called metaphilic agents. Unfortunately, there is no information about their effect on ACh evoked end-plate responses.

It is typical for almost all the enumerated DS-potentiating compounds that they have large hydrophobic radicals such as phenyl, cyclopentane, naphtyl or long aliphatic chains. On the other hand, many of them also carry charged groups such as substituted amines and cation ammonium heads. Similar structures can provide a combination of effective hydrophobic and ionic interactions which in turn gives them great affinity for molecular components of the cell membrane (Seeman, 1972). Actually, some of the above mentioned DS potentiating compounds (Magazanik, 1971a), are known as stabilizers of the erythrocyte membrane. It is also very probable, that many detergents possess the ability to influence the kinetics of DS.

Antagonists of ACh, which presumably compete with it for the receptor site (such as d-tubocurarine, d-TC) only reduce DS because they depress the sensitivity of the PJM to applied ACh. When the concentration of ACh is raised to overcome the lowered sensitivity, the rate of DS recovers towards its initial level (Magazanik, 1968; Magazanik and Vyskočil, 1973; Terrar, 1974). The rate of recovery from DS, however, remains the same. Analogous data have been obtained with atropine (Magazanik and Vyskočil, 1973). Furthermore, this effect can be demonstrated in the presence of DS-potentiating drugs such as chloropromazine (Fig. 4).

VII   Desensitization in fragments of cholinoreceptive membranes

Considerable progress has recently been made in isolating cholinergic membranes or even cholinoreceptive proteins which makes it possible to study many receptor properties and also DS under *in vitro* conditions.

The fragments of the isolated electroplax from *Torpedo marmorata* can form small vesicels, microsacks, and when loaded with ^{22}Na, the release of this radioactive ion serves as a measure of membrane permeability. The release is stimulated by cholinomimetics (carbachol, decamethonium) and can be blocked by classical antagonists (*d*-TC) and also by the α-toxin from *Naja nigricollis*. Preincubation of a suspension of microsacks in the presence of cholinomimetics diminishes or completely blocks the effect of the next dose of the same cholinomimetic, even if given in a higher concentration. The longer the preincubation, the more complete was the block. After washing out the cholinomimetics the sensitivity to it progressively recovers. Higher temperature and [Ca]$_o$ or the presence of local anaesthetics (procaine, lidocaine, etc.), and SKF–525A enhance DS. It is interesting that DS takes place even in a Ca-free solution in the presence of 1 mM EGTA (Hazelbauer and Changeux, 1974; Popot *et al.*, 1974; Popot *et al.*, in prep.), as was also observed at the frog neuromuscular junction (Vyskočil, unpublished observation).

Local anaesthetics and Ca^{2+} in concentrations potentiating DS are also able to increase 2–3 times the binding of labelled ACh to membrane fragments. The number of ACh binding sites is not increased in this case, but these agents increase the affinity of ACh to the membranes. According to the data obtained with the fluorescent ligand DNS-chol the sites of local anaesthetic and ACh fixation are not identical. Solubilization of the fragments by detergent leads to the appearance of several classes of ACh-binding sites, which differ in their dissociation constants. It is noteworthy, that local anaesthetics as well as Ca^{2+} lose their DS-potentiating effect on the solubilized receptor protein, but their effect is regained after washing out of detergent and reaggregation of the membrane complexes. Also careful purification of the receptor protein leads to a loss of the DS-potentiating effect of local anaesthetics and of Ca^{2+} (Cohen *et al.*, 1974; Sugiyama *et al.*, 1975; Sugiyama and Changeux, in press) and also to a loss of the increased affinity to ACh. Many features of DS, observed in electrophysiological experiments, are therefore present in preparations of membrane fragments. But it is obvious from these experiments that for adequate reproduction of the phenomenon of DS under these conditions a maintenance of the original organization of membrane appears essential.

VIII Molecular background of desensitization

Despite the great factography accumulated about different aspects of DS (see Thesleff, 1966; Nastuk, 1967; Magazanik and Vyskočil, 1973; Magazanik, 1976) no clear-cut idea has yet emerged about its molecular nature.

It has been presumed, in order to bring the cyclic scheme (1) and experimental observations into agreement (Katz and Thesleff, 1957) that the processes of PJM inactivation and recovery of sensitivity are accompanied by changes in receptor affinity. It does not follow, however, from the cyclic scheme that these changes are the primary cause of DS. The loss of ability of the PJM to be activated can formally be ascribed to a functional change of any membrane component, participating in activation. This is the reason why the cyclic scheme has recently been transcribed into a more complex form (Magazanik and Vyskočil, 1973):

$$A + R - S \underset{K_{-1}}{\overset{K_1}{\rightleftharpoons}} AR - S^A$$
$$K_R \Big\updownarrow K_{-R} \qquad K_{-D} \Big\updownarrow K_D \qquad (6)$$
$$A + R' - S^I \underset{K_{-2}}{\overset{K_2}{\rightleftharpoons}} AR' - S^I$$

where $R - S$ represents the complex receptor moiety consisting of a component with a recognition site for ACh (R) and membrane components, which are responsible for changes in ionic permeability (S).

Correspondingly, the complex can exist in one of three states: (a) resting $R - S$ state, when ion channels are closed with zero probability that they will be spontaneously opened (b) agonist complex $AR - S^A$, with great probability of opened channels and (c) inactive state $AR - S^I$ in which the low probability of opening of channels exists despite the presence of agonist. As has already been shown, it is necessary to presume that $K_2/K_{-2} > K_1/K_{-1}$, i.e. that the change $AR - S^A \rightarrow AR' - S^1$ is accompanied by an increase of affinity to A. In this case the properties of R should also be changed: $R - S \rightarrow R' - S^I$. The scheme (6) differs from (1) only by substitution of symbol R by $R - S$ indicating that activation as well as inactivation should not only include changes in the recognition part of the molecule but also in some other molecular components, controlling the ionic permeability.

The cyclic scheme closely describes the kinetics of the action of ACh and other agonists, but it is rather difficult to use for explaining the effect of many DS-modifying influences. According to the theoretical analysis by Rang and Ritter (1970a), all compounds which compete with ACh for receptor (it is irrelevant whether the receptor is in state R or the "desensitized" state R') have to slow the rate of equilibration. In the case of preferential affinity of the compound to R' (metaphilic drugs), the equilibrium is established on a lower level, i.e. a majority of the receptors will be in the DS state. Moreover, the metaphilic drugs should slow the rate of recovery. The analysis showed, that this is theoretically true for equilibrium scheme (1) and (6). But these theoretical predictions encounter direct experimental objections: d-TC and atropine do not change the rate of equilibration, when the concentration of agonist is kept high to overcome the blocking effect of the drugs. Many other agonists such as adiphenine or mesphenal *increase* the rate of DS onset, having no effect on the recovery rate. Except for this, compounds were also found, which have no cholinolytic effect at the given concentrations (SKF–525A, chlorpromazine and others) on short-lasting ACh responses (e.p.p.) but which strongly increase the rate of onset of DS without changing the recovery. There is direct evidence that these compounds do not react with the recognition site of the receptor.

1. The effect on DS can be obtained after intracellular application of the drugs (Vyskočil and Magazanik, 1972; Magazanik and Vyskočil, 1973, 1975).

2. The site of their fixation on the membrane fragments (experiments with DNS-chol) differs from the recognition site (Cohen *et al.*, 1974). These compounds cannot therefore be classified as metaphilic according to the hypothesis of Rang and Ritter. Generally, their concept that DS is affected by a preferential combination of metaphilic drugs with the desensitized receptor (R') needs more direct experimental proof, since these compounds are structurally very similar to membrane stabilizers.

Let us try to evaluate the effect of DS-potentiating drugs from the point of view of the cyclic scheme. The equilibrium constant during DS can be derived from the following equation (Katz and Thesleff, 1957):

$$\frac{1}{\tau_D} = \frac{K_D aA + K_{-R}}{1 + aA} + \frac{K_R bA + K_{-D}}{1 + bA}$$

where A = concentration of agonist, a = affinity constant of the agonist to R(a = K_1/K_{-1}), b = affinity constant to R'(b = K_2/K_{-2}); K_D, K_{-D}, K_R, K_{-R} are the rate constants of the reversible reactions according to schemes 1 and 6.

One can postulate that the potentiating compound can change the affinity of R or R' by an allosteric mechanism, i.e. by selectively changing the constants a or b respectively.

The selective effect on the affinity to R (a) is very improbable, because it should be reflected in the altered probability of AR complex formation even after short-lasting activation (e.p.p.'s) and this is not the case.

Higher affinity to R' (increase of b) would lead to the slowing the DS (as has been discussed above), and this possibility has not been confirmed experimentally. Lowering of b should increase the rate of DS, but this would be contrary to the findings of Cohen et al. (1974) concerning the fact that potentiation of DS by drugs and Ca^{2+} is accompanied by increased binding of labelled ACh.

Similarly, the next possible site of action of DS-potentiating drugs, namely the change of K_R seems highly improbable, because the rate of recovery can be expressed according to scheme (1) and (6) as

$$\frac{1}{\tau_D} = \frac{K_R}{K_{-R}} \quad \text{and} \quad K_R > K_{-R},$$

and this rate is not influenced by drugs at all.

The most acceptable remaining possibility is that the process $AR - S^A \rightarrow AR' - S^I$, i.e. the ratio K_D/K_{-D}, is affected by these drugs. Because it is reasonable to assume that K_{-D} should be smaller than K_D, one can postulate that K_D is increased during DS potentiation (Magazanik and Vyskočil, 1973).

The question naturally arises, how this change is structurally materialized. The crucial step in DS development is apparently the formation of the complex $AR' - S^I$ which cannot be activated. It may be accompanied by changes in affinity to the agonist, but this change would slow down the establishment of equilibrium.

It is known, that Ca^{2+} ions potentiate DS. Nastuk and co-workers suggested that these ions affect the inner surface of the PJM. The following experimental data support this view: DS is increased when the intracellular concentration of Ca^{2+} is increased by caffeine (Cochrane and Parsons, 1972), during the substitution of external Na^+ by Li^+ (Parsons et al., 1973), or during the action of some metabolic inhibitors which may block the internal Ca-pump (Vyskočil, unpublished). On the other hand, blockade of the Ca^{2+} channels in the membrane by Mn^{2+} ions leads to an increased instead of a decreased rate of DS onset (Nastuk and Parsons, 1970). Desensitization can be demonstrated in electrophysiological experiments on the end-plate (Vyskočil, unpublished) as well as in experiments on electroplaxs (Cohen et al., 1974) under Ca^{2+} free conditions and in the presence of EGTA. But it is not yet known whether EGTA affects the rate of DS by itself.

It cannot be excluded, that the mechanism of action of Ca^{2+} and potentiating drugs differs. According to Cohen et al. (1974) local anaesthetics lower Hill's coefficient for the binding of labelled ACh to membrane fragments, but higher $[Ca^{2+}]_o$ enhances binding without changing Hill's coefficient. Membrane stabilizers may affect the character of the hydrophobic interactions between membrane proteins and phospholipids, which may in turn change the energetic barriers for protein conformation from one state into another.

We do not yet know whether the protein molecule carrying the recognition site for ACh also possesses the properties of an ionic gate and/or ionic "filter". If it is so, and the receptor molecule is really polyfunctional, then DS is apparently controlled by an "internal affairs' cabinet" of the whole protein complex which regulates the actual conformational situation in the PJM.

In any case, it is necessary to continue the analysis of DS kinetics and its changes during different influences and also to overcome the contradictions, which appear when the cyclic scheme is applied for explaining actual experimental data.

IX Functional role of desensitization

Rhythmic stimulation of the motor nerve causes a gradual reduction of e.p.p. amplitude. It is now clear that this phenomenon of Wedensky inhibition is caused mainly by a decrease in the number of ACh quanta, released by each stimulus. But it was also quite reasonable to assume, that inhibition is caused, at least partly, by DS which may result from the accumulation of ACh at the PJM during repetitive stimulation

(Thesleff, 1959). In this case the DS would play a specific role as a feed-back mechanism, regulating the action of ACh under conditions of transmitter excess. But the experimental proof of this idea has met serious difficulties. It would be desirable to perform such experiments on the neuromuscular preparation without initially lowering the sensitivity of the PJM to ACh (without d-TC) and also without lowering the quantum content of e.p.p. However, under these conditions intracellular recording is almost impossible due to the presence of action potentials and muscle fibre contraction.

Thesleff (1959) tried to overcome these difficulties by raising 2·5 times the NaCl concentration in the perfusion solution. He observed on the phrenic diaphragm preparation of the rat that the response to iono-phoretic ACh application decreased when the pulse was applied not later than 0·2 s after a train of e.p.p.'s at a frequency of 40–60 Hz. If the lowering of the ACh-potential is caused by DS, this means that "end-plate potential DS" develops and disappears at a rate of several hundred milliseconds, which is 10–50 times greater than the rate during artificial application of ACh to the PJM. Attempts to reproduce DS, evoked by repetitive nerve stimulation in the frog neuromuscular preparation were not successful (Otsuka et al., 1962; Vladimirova, 1964). But there exist, in the case of the frog muscle, some complications: (a) the rate of DS onset in frog muscle is lower than in rat muscle (Axelsson and Thesleff, 1958; Vyskočil, 1976) and (b) the d-TC, used in the experiments with frog muscle actually lowered the effective concentration of ACh. DS would not develop under these conditions even if this phenomenon really existed during stimulation of the intact frog neuromuscular preparation.

We tried to demonstrate DS at the frog neuromuscular junction treated with Mg^{2+} in the presence of DS potentiating drugs such as chlorpromazine or SKF–525A. Figure 5 shows that the pattern of e.p.p. amplitudes in the train is the same in the absence and in the presence of chlorpromazine. This indicates that DS failed to develop during repetitive stimulation of the nerve.

Our knowledge about the sites of release for transmitter quanta has recently been considerably extended. It is known that ACh quanta are released from specific "active zones" in the nerve terminals, which are located above the subsynaptic membrane infoldings and are spatially separated. It is, therefore, very improbable that the areas of the sub-junctional membrane which are irrigated by synchronously released

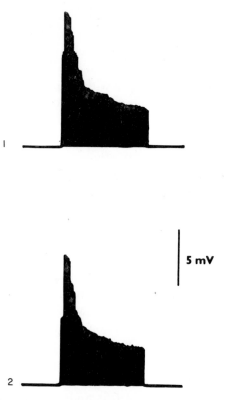

Fig. 5. Trains of 50 end-plate potentials evoked by nerve stimulation of a single sartorius muscle fibre at a frequency of 40 Hz. 1 = control, 2 = after 30 min in the presence of 5×10^{-7}M SKF 525A in the muscle bath. Resting potential = 86 mV. The muscle was pretreated with Ringer solution containing 10^{-6}g/ml of prostigmine and 11 mM Mg^{2+} and 2 mM Ca^{2+}. Stimulus artefacts are retouched.

ACh from these zones overlap each other. The number of active zones in the frog end-plate varies between 1200–2000 in dependence on the total length of the individual junction (Peper *et al.*, 1974). If it is assumed that the mean number of quanta in one e.p.p. is about 100 (and it is subsequently considerably decreased during repetitive stimulation) then it is probable that the next quantum will be released after 300–500 ms) at the same site of PJM at the stimulus frequency of 40 Hz. After such a time interval, no trace of the previously released ACh is present in this area. This simple calculation therefore shows that conditions for DS during repetitive stimulation are very unfavourable. This, however, does not mean that DS could not exist at all under physiological conditions, but apparently it can only occur during long-lasting accumulation of the transmitter in the synaptic zone.

X   Final remarks

It would presumably be an overexaggeration to say that DS, particularly at the motor end-plate, is a physiologically essential mechanism similar to the synthesis and release of ACh, activation of the subsynaptic membrane or the enzymatic hydrolysis of the transmitter. In a typical excitatory synapse, such as the motor end-plate, where successful transmission of nerve impulses at physiological frequencies is guaranteed and ensured very precisely on both the prejunctional as well as on the postjunctional level, the existence of the DS phenomenon seems to be rather superfluous and illogical. But in the central nervous system, where processes of activation and inhibition of individual cells in the neuronal network must be balanced precisely, DS-like phenomena may occur as a supplementary inhibitory mechanism, which can effectively lower the stimulating action of the transmitter for a relatively long period of time. The small extracellular space between cells in the central nervous system gives a better opportunity for accumulation of the transmitter than in the case of the motor end-plate, where a large proportion of ACh diffuses out of the subsynaptic zone. However, up to now, we have no direct experimental evidence about such an "inhibitory" role of DS in the central nervous system.

The aim of this article was not to supply the reader with an exhaustive description of all the facts and literature concerning DS. We have only tried to point out some of the interesting observations and ideas which can, at least partly, help in understanding this phenomenon. For example, the comparison of DS at the muscle end-plate and at its derivative electroplax is very informative. It shows in particular, that DS cannot simply be explained by the presence or absence of some substance or ion at the receptor recognition site or ion channels. We can rather say that DS, which always follows activation, takes place when structural units (or subunits) on PJM pass into a specific conformational stage during activation. Many factors and chemical compounds may modify this process. It is also possible that the mutual arrangement or orientation of neighbouring receptor molecules is important for the induction of DS and this may explain, why DS on electroplax fragments cannot be affected by DS-potentiating factors when the receptor protein is solubilized or purified.

XI References

Adams, P. R. (1974). Drug concentration—conductance curves at frog-end-plates determined by voltage-clamp. *J. Physiol.* **241**, 41–42P.

Albuquerque, E. X., Barnard, E. A., Chiu, T. H., Lapa, A. J., Dolly, J. O., Jansson, S.-E., Daly, J. and Witkop, B. (1973). Acetylcholine receptor and ion conductance modulator sites at the murine neuromuscular junction: Evidence from specific toxin reactions. *Proc. Nat. Acad. Sci.* **70**, 949–953.

Anderson, C. R. and Stevens, C. F. (1973). Voltage clamp analysis of acetylcholine produced end-plate current fluctuations at frog neuromuscular junction. *J. Physiol.* **235**, 655–691.

Axelsson, J. and Thesleff, S. (1958). The "desensitizing" effect of acetylcholine on the mammalian motor end-plate. *Acta Physiol. Scand.* **43**, 15–26.

Bregestovski, P. D., Vulfius, E. A. and Veprintsev, B. N. (1975). *In* "Nature of Cholinoreceptors and Structure of its Active Site," pp. 113–139, Puschino (in Russian).

Cochrane, D. E. and Parsons, R. L. (1972). The interaction between caffeine and calcium in the desensitization of muscle post-junctional membrane receptors. *J. Gen. Physiol.* **59**, 437–461.

Cohen, J. B., Weber, M. and Changeux, J.-P. (1974). Effects of local anesthetics and calcium on the interaction of cholinergic ligand with the nicotinic receptor protein from *Torpedo marmorata*. *Molec. Pharmacol.* **10**, 904–932.

Colquhoun, D., Dionne, V. E., Steinbach, J. H. and Stevens, C. F. (1975). Conductance of channels opened by acetylcholine-like drugs in muscle end-plates. *Nature Lond.* **253**, 204–206.

Del Castillo, J. and Katz, B. (1957). Interaction at end-plate receptors between different choline derivatives. *Proc. Roy. Soc. B.* **146**, 369–381.

Gissen, A. J. and Nastuk, W. L. (1966). The mechanisms underlying neuromuscular block following prolonged exposure to depolarizing agents. *Ann. N.Y. Acad. Sci.* **135**, 184–194.

Hazelbauer, G. L. and Changeux, J.-P. (1974). Reconstitution of a chemically excitable membrane. *Proc. Nat. Acad. Sci.* **71**, 1479–1483.

Jenkinson, D. H. and Terrar, D. A. (1973). Influence of chloride ions on changes in membrane potential during prolonged application of carbachol to frog skeletal muscle. *Br. J. Pharmac.* **47**, 363–376.

Katz, B. and Miledi, R. (1972). The statistical nature of the acetylcholine potential and its molecular components. *J. Physiol.* **224**, 665–699.

Katz, B. and Miledi, R. (1973). The characteristics of "end-plate noise" produced by different depolarizing drugs. *J. Physiol.* **230**, 707–717.

Katz, B. and Thesleff, S. (1957). A study of the "desensitization" produced by acetylcholine at the motor end-plate. *J. Physiol.* **138**, 63–80.

Kordaš, M. (1969). The effect of membrane polarization on the time course of the end-plate current in frog sartorius muscle. *J. Physiol.* **204**, 493–502.

Kordaš, M. (1972). An attempt at an analysis of the factors determining the time course of the end-plate current. I. The effects of prostigmine and of the ratio of Mg^{2+} to Ca^{2+}. *J. Physiol.* **224**, 317–332.

Lambert, D. H. and Parsons, R. L. (1970). Influence of polyvalent cations on the activation of muscle end plate receptors. *J. Gen. Physiol.* **56**, 309–321.

Magazanik, L. G. (1968). Mechanism of desensitization of the postsynaptic muscle membrane. *Biofizika* **13**, 199–203.

Magazanik, L. G. (1969). Effect of sympathomimetic amines on the desensitization of frog motor end-plate to acetylcholine. *Sechenov J. Physiol.* **55**, 1147–1155 (in Russian).

Magazanik L. G. (1970). About the mechanism of the SKF-525A action on the neuromuscular synapse. *Bull. Exp. Biol. Med.* **3**, 10 (in Russian).

Magazanik, L. G. (1971a). Influence of certain membrane stabilizers on the function of a neuromuscular synapse. *Sechenov J. Physiol.* **57**, 1313–1321 (in Russian).

Magazanik, L. G. (1971b). On the mechanism of antiacetylcholine effects of some mononitrogen anticholinergics in the neuromuscular synapse. *Pharmacol. Toxicol.* **3**, 292–297.

Magazanik, L. G. (1976). Functional properties of the post-synaptic membrane. *Ann. Rev. Pharmacol.* **16**, 161–175.

Magazanik, L. G. and Shekhirev, N. N. (1970). Desensitization to acetylcholine in various frog muscles. *Sechenov. J. Physiol.* **56**, 582–588 (in Russian).

Magazanik, L. G. and Vyskočil, F. (1970). Dependence of acetylcholine desensitization on the membrane potential of frog muscle fibre and on the ionic changes in the medium. *J. Physiol.* **210**, 507–518.

Magazanik, L. G. and Vyskočil, F. (1973). Desensitization at the motor end-plate. *In* "Drug Receptors" (H. P. Rang, ed.), pp. 105–119. Macmillan, London.

Magazanik, L. G. and Vyskočil, F. (1975). The effect of temperature on desensitization kinetics at the post-synaptic membrane of the frog muscle fibre. *J. Physiol.* **249**, 285–300.

Magleby, K. L. and Stevens, C. F. (1972a). The effect of voltage on the time course of end-plate currents. *J. Physiol.* **223**, 151–171.

Magleby, K. L. and Stevens, C. F. (1972b). A quantitative description of end-plate currents. *J. Physiol.* **223**, 173–197.

Manthey, A. A. (1966). The effect of calcium on the desensitization of membrane receptors at the neuromuscular junction. *J. Gen. Physiol.* **49**, 963–976.

Manthey, A. A. (1970). Further studies of the effect of calcium on the time course of action of carbamylcholine at the neuromuscular junction. *J. Gen. Physiol.* **56**, 407–419.

Nastuk, W. L. (1967). Activation and inactivation of muscle postjunctional receptors. *Fed. Proc.* **26**, 1639–1646.

Nastuk, W. L. and Karis, J. H. (1964). The blocking action of hexafluorenium on neuromuscular transmission and its interaction with succinylcholine. *J. Pharmacol. Exp. Ther.* **144**, 236–252.

Nastuk, W. L. and Parsons, R. L. (1970). Factors in the inactivation of postjunctional membrane receptors of frog skeletal muscle. *J. Gen. Physiol.* **56**, 218–249.

Otsuka, M., Endo, M. and Nanomura, Y. (1962). Presynaptic nature of neuro-muscular depression. *Jap. J. Physiol.* **12**, 573–584.

Parsons, R. L., Cochrane, D. E. and Schnitzel, R. M. (1973). Desensitization of end plate receptors produced by carbachole in lithium Ringer solutions. *J. Gen. Physiol.* **61**, 263.

Popot, J. L., Sugiyama, H. and Changeux, J.-P. (1974). Neurobiologie molèculaire. *C.R. Acad. Sci. Ser.* D **279**, 1721–1724.

Popot, J. L., Sugiyama, H. and Changeux, J.-P. (in preparation).

Rang, H. P. and Ritter, J. M. (1969). A new-kind of drug antagonism: evidence that agonists cause a molecular change in acetylcholine receptor. *Molec. Pharmacol.* **5**, 394–411.

Rang, H. P. and Ritter, J. M. (1970). On the mechanism of desensitization at cholinergic receptors. *Molec. Pharmacol.* **6**, 357–382.

Rang, H. P. and Ritter, J. M. (1970). The relationship between desensitization and the metaphilic effect at cholinergic receptors. *Molec. Pharmacol.* **6**, 383–390.

Seeman, P. (1972). The membrane actions of anesthetics and tranquilizers. *Pharmacol. Revs.* **24**, 583–655.

Sugiyama, H. and Changeux, J.-P. (in press).

Sugiyama, H., Popot, J. L., Cohen, J. B., Weber, M. and Changeux, J. P. (1975). *In* "Protein-ligand Interactions", pp. 289–305.

Terrar, D. A. (1974). Influence of SKF–525A congeners, strophanthidin and tissue-culture media on desensitization in frog skeletal muscle. *Br. J. Pharmac.* **51**, 259–268.

Thesleff, S. (1955). The mode of neuromuscular block caused by acetylcholine, nicotine, decamethonium and succinylcholine. *Acta Physiol. Scand.* **34**, 218–231.

Thesleff, S. (1959). Motor end-plate "desensitization" by repetitive nerve stimuli. *J. Physiol.* **148**, 659–664.

Vladimirova, I. A. (1964). Investigation of end-plate chemosensitivity in striated muscle fiber of the frog during its rhytmical activity. *Fiziol. Ž. U.S.S.R.* **50**, 1358–1363 (in Russian).

Vyskočil, F. (1975). Recovery of sensitivity to acetylcholine following desensitization in muscle of different vertebrate species. *Pflüg. Arch.* **361**, 83–87.

Vyskočil, F. and Magazanik, L. G. (1972). The desensitization of post-junctional muscle membrane after intracellular application of membrane stabilizers and snake venom polypeptides. *Brain Res.* **48**, 417–419.

6

Neuromuscular Transmission after Immunization with Nicotinic Acetylcholine Receptor

Edith Heilbronn

Dept. of Biochemistry, Research Institute of National Defence, Dept. 4, Sundbyberg, Sweden

I Introduction

The nicotinic acetylcholine receptor (nAChR) is an integral protein component of the postsynaptic membrane at neuromuscular junctions and at synapses in the electric organ of certain fishes. It recognizes the cholinergic neurotransmitter acetylcholine (ACh) and triggers, in a series of events not yet entirely understood, a response in the muscle fibre or electroplaque. During the years 1971–1973, several laboratories succeeded in the isolation and purification of this receptor from the electric organ of fishes belonging to the group of Teleosts (*Electrophorus electricus*) and Elasmobranchs (*Torpedo marmorata*, *T. californica* and others). Solubilization of nAChR was generally achieved with the aid of neutral detergents, in one case (Aharonov *et al.*, 1974) by dialysis against low ionic strength buffer followed by controlled tryptic digestion. The receptor was purified by affinity chromatography; the matrix bound ligand of the affinity columns being either native or modified snake α-toxin (Karlsson *et al.*, 1972; Klett *et al.*, 1973), quaternary ammonium compound (Schmidt and Raftery, 1973), flaxedil analogue (Olsen *et al.*, 1972) of *d*-tubocurarine (*d*-tc) (Green *et al.*, 1975. A further purification step, ion-exchange (Biesecker, 1973; Heilbronn and Mattsson, 1974) or hydroxylapatite (Klett *et al.*, 1973) chromatography, centrifugation (Meunier *et al.*, 1974) or chromatography on matrix bound antibody (Aharonov *et al.*, 1975b) had to be used. The purified receptor was shown to be a protein oligomer consisting of several subunits. The exact number of subunits is not yet known and may differ for nAChR from *Torpedo* (Karlsson *et al.*, 1972; Gordon *et al.*, 1974; Weill *et al.*, 1974; Raftery *et al.*, 1975; Mattsson and Heilbronn, 1975) and *Electrophorus* (Klett *et al.*, 1973; Biesecker, 1973; Meunier *et al.*, 1974; Patrick *et al.*, 1975; Chang, 1974). A subunit of about 42 000 daltons has been suggested to be the ACh-binding unit (Karlin and Cowburn, 1973) and is generally found after treatment with SDS and reduction. It has, however, also been suggested that this subunit could be an artefact due to proteolysis (Patrick *et al.*, 1975). Certainly, caution is appropriate; inhibitors of proteolytic enzymes have seldom been used in receptor studies. The amino acid composition of nAChR has been described (Eldefrawi and Eldefrawi, 1973; Klett *et al.*, 1973; Meunier *et al.*, 1974; Michaelsson *et al.*, 1974; Mattsson and Heilbronn, 1975) and revealed also the presence of N-acetyl D-glucoseamine (Heilbronn *et al.*, 1973). The glycoprotein nature of nAChR was established (Meunier *et al.*, 1974; Moore *et al.*, 1974) and the carbohydrates present identified as mannose, glucose, galactose and N-acetyl D-glucoseamine (Heilbronn *et al.*, 1974; Raftery, 1975; Mattsson and Heilbronn, 1975). Isolated subunits carry carbohydrate as indicated by staining. The purified glycoprotein material has been characterized as the nicotinic acetylcholine recognition site by binding studies in many laboratories (see Chapter I). A further function as an ion translocation unit has been discussed on the basis of its structure as revealed by electron microscopy and on some results obtained in membrane reconstitution experiments (Michaelson and Raftery, 1974). Immunological, gelelectrophoretic and kinetic analysis of purified nAChR have in the *Torpedo* preparation revealed two molecules, not differing in amino acid composition and binding specificity (Mattsson and Heilbronn, 1975), but in immunological properties, molecular weight and rates of binding (Mattsson *et al.*, 1976). The two molecules have been named by us nAChRI and nAChRII.

   Immunization of rabbit, goat, guinea pig, rat, mouse and monkey with purified (or crude) nAChR from either *Torpedo* or *Electrophorus* induced the formation of nAChR antibodies. In all animals mentioned, it caused a disease characterized by progressive muscle fatiguability which led, in some animals, to paralysis and death. Symptoms observed point to a neuromuscular transmission deficiency related to human myasthenia gravis (MG). The experimental disease, generally termed experimental autoimmune myasthenia gravis (EAM) will be described and discussed in this chapter. A comparison with MG will be attempted.

II Immunization

A PRODUCTION OF RECEPTOR ANTIBODIES

Antibodies directed against nAChR isolated from either the electric organ of *Electrophorus electricus* (Patrick and Lindstrom, 1973; Sugiyama *et al.*, 1973) or from that of *Torpedo marmorata* (Heilbronn and Mattsson, 1974) were originally produced in rabbit, later in rat and guinea pig (Lennon *et al.*, 1975; Foldes, *et al.*, unpublished), goat (Lindstrom, 1975), *Rhesus* monkey (Tarrab-Hazdai *et al.*, 1975a) and mouse (Smith *et al.*, unpublished).

Immunization was generally performed with μg quantities of nAChR carrying some non-ionic detergent and emulsified in Freunds adjuvant. The injections were either *s.c.* at several sites in complete (Green *et al.*, 1975) or *i.d.* and *i.m.* in incomplete adjuvant (e.g. in the hindlegs or hindfoot pads) (Patrick and Lindstrom, 1973; Heilbronn and Mattsson, 1974). *B. pertussis* vaccine has sometimes been added. Injections were either single ones or repeated. Occasionally, crude extracts of *Torpedo* membranes have been used as antigen (Clementi *et al.*, 1975). Immunization with sterile filtered receptor in Freunds adjuvant resulted in rather low antibody titres and a disease without transmission decrement (see below; Heilbronn *et al.*, 1976a).

The purified receptor preparations, usually but not always (Aharonov *et al.*, 1975), contained traces of active acetylcholinesterase (AChE). Formed antibodies, however, did not react with AChE. This enzyme seems to be considerably less antigenic than nAChR. Immunization with purified AChE caused no disease. In one case a disease without decrement of neuromuscular transmission was observed (authors lab.). The presence of some ligand-receptor complexes in purified receptor preparations was found by Patrick *et al.* (1973). Mattsson and Heilbronn found such complexes only after use of affinity columns older than two months. Receptor antibodies were not directed against α-neurotoxin, nor were α-neurotoxin-antibodies directed against receptor used for immunization (Heilbronn *et al.*, 1975a, 1976a).

Antibodies directed against subunits from nAChR treated with SDS and reducing agents have been produced in rabbits (Mattsson *et al.*, 1976; Lindstrom *et al.*, 1976) and rats (Lindstrom *et al.*, 1976) but much higher doses of antigen than with native receptor were required. Generally such immunized animals did not develop clinical symptoms. Antibodies directed against the subunits reacted with nAChR macromolecules but obtained sera had a low titre. Antibodies directed against the nAChR macromolecule gave a very weak reaction with the subunits.

Antibody titres in animal sera were determined using radioimmunoassay (Patrick *et al.*, 1973), nAChR trace-labelled with ^{125}I-α-bungarotoxin (αbgt) (Green *et al.*, 1975a), quantitative microscale complement fixation assay (Aharonov *et al.*, 1975a), rocket immunoelectrophoresis and haemagglutination tests (Mattsson and Heilbronn, 1975).

B. ANTIBODY NATURE, ANTIBODY SPECIFICITY AND SITE OF ACTION

1. *Globuline type*
By ammonium sulphate precipitation and gel filtration a small part of rabbit antibodies in the late stages of EAM was characterized as IgM; the main fraction was of IgG type (Mattsson and Heilbronn, 1976). Lennon *et al.* (1975; 1976) have pointed out that, in chronic stages of EAM rats, a shift has occurred from mainly antibodies with sedimentation coefficients of 19 S (IgM) to mainly 7 S (IgG).

2 *Antibody specificity*
Immunoelectrophoresis of nAChR from either *Torpedo* or *Electrophorus* revealed only a single precipitation line, suggesting immunological homogeneity of the receptor preparations. Recently, it was found that the two receptor molecules nAChRI and nAChRII obtained from *Torpedo* are only partially immunoidentical (Mattsson *et al.*, 1976).

nAChR from different animal species are not necessarily immunoidentical. *Electrophorus* nAChR antibodies crossreacted with *Torpedo* nAChR (Valderrama *et al.*, 1976). Antibodies to crude skeletal rat muscle nAChR, raised in rabbits, crossreacted with *Torpedo* nAChR, while those to crude human muscle receptor reacted very weakly (Uddgård and Heilbronn, 1976). An anti-AChR factor, evidently of antibody nature, was detected in serum from MG patients using denervated rat muscle receptor (Almon *et al.*, 1974; Bender *et al.*, 1975). Several investigators (Lindstrom *et al.*, 1976; Fischer *et al.*, unpublished) found negligible crossactivity of this antibody with *Electrophorus* or *Torpedo* nAChR. In contrast to this, Aharonov *et al.* (1975b), who used a quantitative microscale complement fixation assay for analysis, found that 80% of MG patients in various stages of disease, responding positively to Tensilon tests and exhibiting characteristic electromyographic patterns (see below) had humoral receptor antibodies directed against *T. californica* nAChR. Lindstrom *et al.* (1976) stated that about 10% of human nAChR antibody crossreact with rat muscle nAChR and suggested that a small fraction of nAChR antibodies might be directed against the ACh binding site. Almon *et al.* (1974, 1975) found that human nAChR antibody cross-reacted to about 40% with denervated rat skeletal muscle nAChR. It has to be kept in mind that these inconsistent results have been obtained with rather different methods.

Receptor antibodies from EAM animals are not all directed against the injected antigen; rats in early stages of the disease had a small amount of autoantibodies besides *Electrophorus* nAChR antibody. The autoantibody titre increased just before the onset of chronic EAM (Lindstrom *et al.*, 1976).

Sera from myasthenic patients contain, besides human muscle nAChR antibodies, a variety of antibodies directed, e.g. against skeletal muscle, thymus and thyroid tissue (Strauss *et al.*, 1960; van der Geld *et al.*, 1963). A correlation in the occurrence of nAChR antibodies and muscle antibody was observed (Ringel *et al.*, 1975). As shown by Aarli *et al.* (1975) in rabbits, antibodies to purified *Torpedo* nAChR and to human citric acid extractable muscle antigen do not crossreact with the antigens indicating that the latter are different.

3 Site of action

Rabbit nAChR antisera precipitated their antigen from a solution also containing detergent, as measured by the α-neurotoxin binding capacity of the supernatant (Patrick and Lindstrom, 1973; Heilbronn *et al.*, 1975b).

The important question if the nAChR antibodies induced in EAM-animals and the human muscle receptor antibodies found in sera from nearly all myasthenic patients are directed against determinants at the ACh recognition site of the receptor molecule is not yet clearly answered. Binding and precipitation studies suggested that antibodies are partly or not at all directed (Patrick *et al.*, 1973; Lindstrom, 1975; Lindstrom *et al.*, 1976) against the ACh binding site. nAChR antibodies in MG sera caused partial reduction of αbgt binding to nAChR extracted from denervated rat muscle but not of that to junctional receptors (Almon *et al.*, 1974; Bender *et al.*, 1975). Crude preparations of human muscle nAChR interacted with myasthenic serum IgG as measured by radioimmunological techniques (Almon and Appel, 1976). Such receptor preparations may contain both junctional and extrajunctional nAChR and also other components. An αbgt binding site need not be identical with the ACh binding site. Also, αbgt binding proteins, not identical with nAChR are present in membranes.

Torpedo or *Electrophorus* nAChR antisera blocked the response of electroplaque to carbamylcholine (Patrick *et al.*, 1973) and that of cultured myofibres to iontophoretic application of ACh (Patrick and Lindstrom, 1973). Antibodies from immunized animals decreased m.e.p.p. size in phrenic nerve diaphragm preparations from frog, rat and rabbit and changed the amplitude histograms to resemble those of immunized animals (Green *et al.*, 1975). A reducing effect on the ACh-response of a leech muscle preparation was observed by Heilbronn *et al.* (1974), particularly in the presence of complement. Serum complement which destroys muscle fibres and causes loss of resting potential (Ito *et al.*, 1974) was, however, not necessary for blockage (Green *et al.*, 1975) in frog end-plates, though deep muscle fibres were more slowly affected in its absence. Blocking with nAChR antisera reduced the delay of ACh-diffusion out of the synaptic gap. Such delay is thought to be due to binding of ACh to nAChR (Katz and Miledi, 1973) and is reduced with αbgt. Age and species variations observed were assumed to be due to differences in the immunological sensitivity to nAChR and perhaps to variations in the safety margin of neuromuscular transmission.

Thus, both studies on isolated biological preparations and *in vivo* studies showed that nAChR antibodies block neuromuscular transmission. This might indicate that ACh molecules are prevented from reacting with their receptor either because of a direct, possibly competitive block or because of steric hindrance after a reaction close by. Another possibility is that antibodies are directed against another functionally important site of nAChR than the ACh recognition site.

Albuquerque *et al.* (1976) analysed the point of antibody attack on mouse phrenic nerve-diaphragm muscle preparations. IgG from the sera of myasthenic patients or from controls were not able to prevent αbgt induced block of neuromuscular transmission, as e.g. the competitive antagonist *d*-tc might have done. Junctional and extrajunctional sensitivity to iontophoretically applied ACh in soleus and extensor rat skeletal muscles and in frog sciatic nerve sartorius muscle preparations and spontaneous or evoked ACh release were not affected by MG serum. Albuquerque *et al.* concluded that myasthenic serum does not affect the ACh recognition site. There may, of course, be penetration barriers.

III Analysis of immunized animals

A CLINICAL STATUS

Injections of nAChR from either *Electrophorus* or *Torpedo* induced antibody production and resulted, in rabbits, in a neostigmine-sensitive, general, flaccid paralysis (Patrick and Lindstrom, 1973; Sugiyama *et al.*, 1973; Heilbronn and Mattsson, 1974). In animals treated in our laboratory, paralysis started in the hindlegs (point of injection), moved to the forelegs and finally to the whole body. Limb, neck, head, laryngeal and respiratory muscles of the animals were finally involved (Fig. 1.1a). Observed symptoms remind of myasthenia gravis. One injection of nAChR into rabbits, even if distributed over many sites, resulted in low levels of nAChR antibody for many weeks and no clinical symptoms. Three to six days after a second or third injection rabbits generally developed paralysis. The rabbits died unless intensive care was given, including control of water balance, etc. Anticholinesterase treatment produced (Fig. 1.1b) usually a transient relief, but some severely ill animals did not respond visibly to such treatment. Onset and severity of paralysis was correlated to the level of nAChR antibodies in the serum as measured by the precipitin reaction (Green *et al.*, 1975) or the rocket immunoelectrophoresis test (Mattsson and Heilbronn, 1976). Paralysis became obvious with nAChR antibody levels of more than 150–2000 μg/ml (Patrick *et al.*, 1973; Sugiyama *et al.*, 1973; Green *et al.*, 1975a; Mattsson and Heilbronn, 1975b).

Observations of the same kind as in rabbits were also made on rat, guinea pig, goat and monkey. Mice developed nAChR antibodies; recently symptoms of muscular weakness and decrement (see below) could be demonstrated (Fuchs, personal communication). Early signs of weakness in rat and guinea pig (Lennon *et al.*, 1975) were loss of cry and hypotonia associated with weakness of the neck and forelimbs. Rats and guinea pigs showed incontinence of urine and inability to swallow. In rats chronic respiratory distress often became manifest as porphyrin staining of nasal secretion and tears. Signs of chronic respiratory distress disappeared in cases where muscle strength was regained. Rechallenging of animals several weeks after the first symptoms appeared generally led to severe respiratory difficulties and death.

Fig. 1.1. EAM rabbit showing paralysis of legs, neck and trunk three weeks after immunization with nAChR (a) before and (b) 10 min after injection of Tensilon (from the author's laboratory).

Several rabbits (Thornell *et al.*, 1976) and rats, in general, exhibited two distinct periods of weakness, accompanied by weight loss (Lennon *et al.*, 1976). Rats had an acute period between days 7–12. Apparently full strength was recovered by days 12–15. The second episode, one of

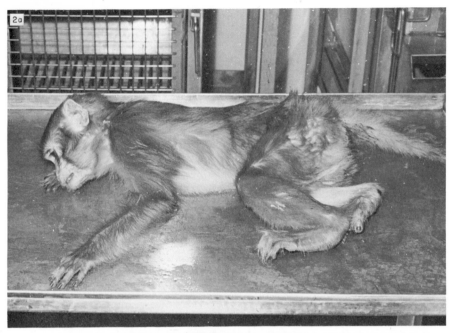

Fig. 1.2. EAM Rhesus monkey (a) one week after 4th injection of nAChR and (b) 10 min after injection of Prostigmine (courtesy of Sara Fuchs).

progressive weakness, was observed days 26–35 by signs of worsening chronic respiratory difficulties. Guinea pigs showed numerous relapses of weakness, particularly if some form of stress was applied. The disease became progressive and severe after day 42.

Rhesus monkeys immunized with purified *T. californica* nAChR (Tarrab-Hazdai *et al.*, 1975a) developed, after three injections, signs of fatigue, hypoactivity, anorexia and weight loss. One week after a fourth injection an acute, extreme flaccid paralysis of the limbs, neck and trunk developed (Fig. 1.2a) with severe breathing difficulties. Bulbar signs included ptosis, external ophthalmoplegia, facial diplegia and paralysis of the jaw muscles, with uncontrolled salivation and severe dysphagia. The disease was prostigmine-sensitive with recovery lasting about 4 h (Fig. 1.2b).

B NEUROPHYSIOLOGY

1 *Decremental response*

Most immunized rabbits showed a decremental response to repetitive nerve stimulation (Fig. 2) not seen in controls or in animals injected only once (Heilbronn *et al.*, 1975b; c, 1976; Elmqvist *et al.*, 1976). This was, to a certain level, correlated to the presence and titre of nAChR antibodies (Fig. 3). A decremental response to repetitive supramaximal nerve stimulation is a sign of disturbed neuromuscular transmission and is found in patients with myasthenia gravis. Animals with decrement often had little post-tetanic potentiation as found in myasthenia gravis. Injection of neostigmine or Tensilon increased the initial low amplitude, but not always to the level of control animals. No increment in response amplitude was seen on repetitive stimulation. The conduction time from the stimulus site to the recording site was the same (1·0–1·7 ms) in control and immunized animals.

Green *et al.* (1975a) found that in immunized and heavily affected rabbits the twitch of the diaphragm muscle evoked by a single stimulus of the phrenic nerve was weaker than normal. Neither resting nor conducted action potentials of the muscle fibres were different from normal suggesting that the main effect of immunization against nAChR was at the neuromuscular junction. Injection of an anticholinesterase temporarily reversed the decrement.

Guinea pigs frequently exhibited severe clinical weakness without demonstrable EMG abnormalities (Lennon *et al.*, 1975), however, de-

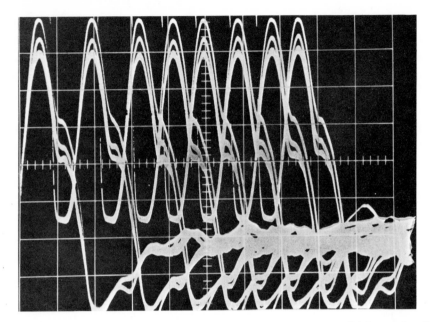

Fig. 2. The compound muscle action potential of the anterior tibial muscle of immunized rabbit. The peroneal nerve was stimulated with 4 imp at 4 Hz every 30 s, to show the constancy of decrement.

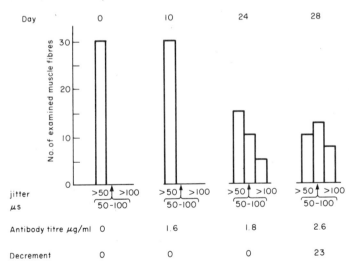

Fig. 3. Variability in neuromuscular transmission time in relation to antibody titre and decrement. Rabbit immunized with nAChR on days 0, 10, 24 and 28. (From the author's laboratory.)

cremental response and post-tetanic facilitation, not observed in controls, was demonstrated in some cases (Foldes *et al.*, unpublished). Rats exhibited decremental EMG patterns at the peak of clinical weakness in about 80% of the cases and the decrement was reduced with clinical signs of improvement. Reproducible decrements of action potential were sometimes seen in forelimbs only and were transiently normalized by *i.p.* injections of neostigmine. Postactivation facilitation and exhaustion occurred. When clinical weakness disappeared only one-tenth of the rats had a significant decrement and this did not reappear when the period of weight loss and progressive weakness came (Lennon *et al.*, 1975).

Immunized *Rhesus* monkey showed a decrease in action potential amplitude during repetitive nerve stimulation (Tarrab-Hazdai *et al.*, 1975a).

2 Single fibre EMG

Single fibre EMG measures the variability, the jitter, in consecutive stimulus-response latency values and reflects the variability of the neuromuscular transmission time. Such jitter is increased already when neuromuscular transmission is slightly disturbed, before impulse blockade occurs (Stålberg and Ekstedt, 1973). Single fibre EMG is a sensitive test for myasthenia gravis (Stålberg *et al.*, 1974). Single muscle fibre recordings made during stimulation of the peroneal nerve of EAM rabbits revealed abnormal values three days after a third challenge with nAChR (Heilbronn *et al.*, 1976). Before immunization and after the first challenge, the latency variability was below 50 μseconds. Abnormal values were found also in immunized rabbits who showed no decremental response (Fig. 3). Three days later more abnormal recordings were obtained including occasional blocking in rabbits with decremental response.

3 Microelectrode studies

The resting membrane potential of intercostal muscle fibres from control and immunized rabbits were equal. The amplitude of the miniature end-plate potentials (m.e.p.p.) were about 1 mV in most muscle fibres from control rabbits. In immunized animals the m.e.p.p. amplitude was variable, usually around 0·2 mV, but often so low that the m.e.p.p.s could not be clearly separated from baseline noise (0·1 mV). The frequency of the m.e.p.p. at motor end-plates in muscles from immunized rabbits was slightly lower than at end-plates from control animals, probably due to disappearance of the smallest m.e.p.p.s in the baseline noise (Heilbronn *et al.*, 1976a; Elmqvist *et al.*, 1976).

The relative fall in the amplitude of successive end-plate potentials e.p.p. with higher frequency of nerve stimulation was similar in immunized and control rabbits. Green *et al.* (1975a) found smaller e.p.p. in superficial and deep fibres in the diaphragms of immunized animals than in those of normal rabbits. The e.p.p. rose more slowly and had a clear step before the muscle fibre action potential was triggered (Green *et al.*, 1975). In very heavily affected animals, some fibres responded with only subthreshold e.p.p. which probably accounts for the weaker twitch observed in these muscles. When repetitive stimulation was used, the e.p.p. amplitude declined further and more fibres failed to contract. However, in animals which were unable to move, most muscle fibres in the diaphragm still gave impulses to single nerve stimuli. Probably junctional paralysis progresses during repetitive nerve activity *in vivo*. Green *et al.* suggest that the safety margin for transmission may be lower in limb muscles than in the diaphragm.

Estimation of the number of acetylcholine quanta released by a nerve impulse (quantum content) (Lambert and Elmqvist, 1971) gave similar results in immunized and control rabbits (Heilbronn *et al.*, 1976a).

For total block of neuromuscular transmission with *d*-tc lower doses (10^{-7} g/ml) were sufficient in immunized rabbits than in controls (2×10^{-6} g/ml). Guinea pigs as well showed increased curare sensitivity (Foldes *et al.*, unpublished).

Lambert (quoted in Lindstrom, 1975) found greatly reduced m.e.p.p. amplitudes in immunized rats, who also showed an increased curare sensitivity and an unaltered number of ACh quanta available for release.

4   *EMG*

Normal rabbits exhibited a burst of EMG activity at the insertion of an EMG electrode into a muscle. This occurred usually as irregular discharges and occasionally as regular high frequency bursts, which ceased almost completely after 5–10 seconds. Only exceptionally spontaneously firing muscle fibres were observed. No afterdischarges were seen in normal animals following nerve stimulation.

EAM rabbits had a more pronounced insertion activity, composed both of irregular activity and high frequency discharges similar to "pseudo myotonic" bursts. This activity continued over a longer time than in normal muscle. There was also a varying degree of spontaneous activity of the same kind as fibrillation potentials in denervated muscles. The compound evoked action potential of experimental rabbits was followed by low amplitude afterdischarges for up to 50 ms (Stålberg *et al.*, unpublished).

There was a quantitative difference between normal and the EAM rabbits concerning both insertion and continuous EMG activity and it could not be excluded that this resulted from denervation. In myasthenia gravis patients, EMG signs of denervation have been reported in about 25% of the cases (Stålberg *et al.*, 1976).

C NUMBER OF RECEPTOR BINDING SITES

Animals immunized with nAChR showed a reduction in specific endplate binding of α-neurotoxin (Heilbronn *et al.*, 1976a), according to Green *et al.* (1975a) correlating well with the severity of the clinical symptoms and with the titre of nAChR antibody found in the serum. Extrajunctional α-neurotoxin binding increased in immunized animals. These extrajunctional sites may differ from the junctional ones, as they seemed not to be ACh-sensitive. Green *et al.* found that ACh reduced end-plate binding of α-toxin by 70–80% while extrajunctional binding was, if at all, only slightly reduced. In this connection it is of interest that Mattsson *et al.* (1976) recently characterized an α-neurotoxin binding membrane protein from *Torpedo* electroplaques which was not identical with nAChR.

D HISTOPATHOLOGY

1 *Occurrence of inflammatory cells*

Infiltrations with predominantly mononuclear inflammatory cells or macrophages were seen in muscles of rats (Lennon *et al.*, 1976) and rabbits (Fig. 4) (Thornell *et al.*, 1976) in the acute phase of EAM. Mild lymphorrhage foci in buccal muscle and lymphocytes between muscle cells were observed in EAM monkey (Tarrab-Hazdai *et al.*, 1975a).

2 *Motor end-plate*

Motor end-plates of the internal intercostal and the anterior tibial muscles of rabbits, immunized with purified *Torpedo* nAChR, were

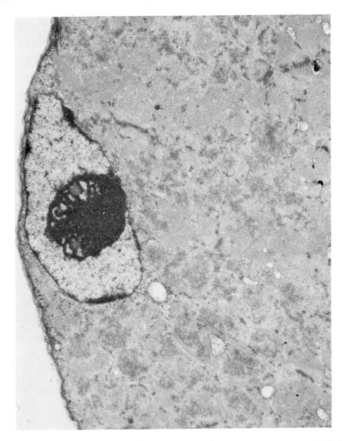

Fig. 4. Invasion of degenerated intercostal muscle fibre from EAM rabbit by macrophage. Note prominent nucleolus. × 19 000.

studied during the first paralytic EAM state and after recovery, shortly after a renewed challenge with receptor, resulting in a relapse into the paralytic state (Thornell *et al.*, 1976; Heilbronn *et al.*, 1976a).

In light microscopy axons, terminal arborizations and terminal swellings appeared equal in immunized and control animals. No terminal sprouts were observed. The subneural apparatus of controls and end-plates of rabbits with one period of paralysis showed no obvious differences while rabbits with two periods of paralysis had marked changes in the sub-neural apparatus of some motor end-plates. Staining could be irregular in density or form rounded substructures or extensions and minor extra parts could be seen.

Electron microscopy on crosscut fibres from controls showed well-defined end-plate regions with nerve terminals in close relation to the folded postjunctional area of the muscle fibre. The primary clefts were of uniform size and they, as well as the secondary clefts, contained a basement membrane. Striations about 10 nm in height and projecting out from the membrane on the top of the folds in the denser area of the membrane were interpreted as nAChR sites. Size of nerve terminals, vesicle and mitochondria content, postsynaptic foldings and amount of junctional sarcoplasm varied. Some end-plates of immunized rabbits (both one and two periods of paralysis) could not be distinguished from controls, but most end-plates contained a wide range of ultrastructural abnormalities. This variability confirmed single fibre EMG investigations. Nerve terminals were sometimes small, with no or few synaptic vesicles, occasionally they were lacking. The primary synaptic clefts were irregular and wide and often filled with electron dense material, seen also in secondary clefts if those remained (Fig. 5). Processes of Schwann cells were seen between nerve element and muscle fibres.

Rabbits with two periods of paralysis had shallow secondary clefts with few and broad or with no foldings (Fig. 6). In areas with simplified or with no nerve terminals the folds were thin or of short length. Empty profiles or folds of basement membrane were seen as well as granules or vesicles outside the folds, enclosed by basement membrane. Membrane thickenings with striations were seen at the top of the postsynaptic foldings and on foldings lying without proximity to the nerve endings (Fig. 7). Thornell *et al.* suggested that this could account for a lower number of receptors available at junctions of immunized animals. The amount of junctional sarcoplasm of the end-plate in immunized rabbits varied and consisted mainly of ribosomes, filaments, microtubuli or rough and smooth endoplasmic reticulum, mitochondria and nuclei. Vacuoles, with or without myelin figures occurred, as well as large amounts of membranes, with empty gaps between them, seemingly in contact with the irregular synaptic clefts. In conclusion, and in comparison with descriptions of the ultrastructure of end-plates from myasthenic patients (Zacks *et al.*, 1962; Woolf, 1964; Engel and Santa, 1971; Santa *et al.*, 1972), particularly end-plates from rabbits in later stages of EAM have features in common with MG end-plates. However, considerable changes in terms of degeneration and repair of the neuromuscular junction seem to have occurred in animals at a chronic stage. Similar results were recently described by Engel *et al*, (1976).

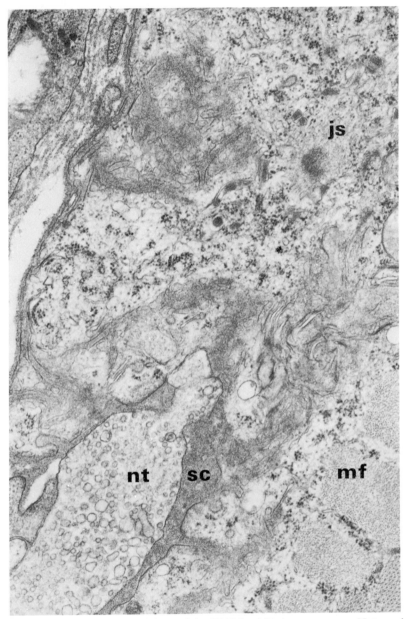

Fig. 5. End-plate (anterior tibial muscle) of EAM rabbit in acute stage. Note variations of width and deposited dense material in the synaptic cleft and elements of the sarcotubular system in the junctional sarcoplasm. (Thornell *et al.*, 1976). nt = nerve terminal, sc = synaptic cleft, mf = microfilaments, js = junctional sarcoplasm. × 36 000.

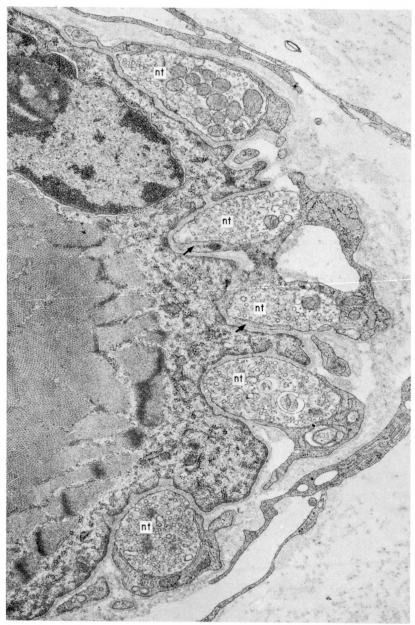

Fig. 6. Cross-sectioned motor end-plate from immunized rabbit with two periods of paralysis. Postsynaptic folds and secondary synaptic clefts are few. Schwann cell processes occur between the nerve terminals (nt) and the muscle fibre in the primary synaptic gap. × 21 600. (Thornell *et al.*, 1976.)

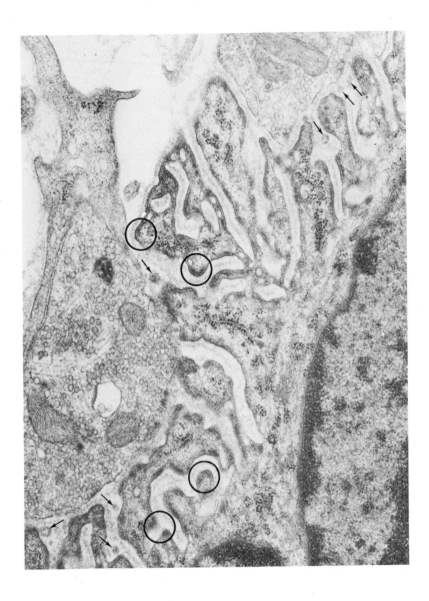

Fig. 7. 1. Endplate of EAM rabbit showing dense particles in synaptic cleft (arrows) and tips of postsynaptic foldings (rings) without contact with the presynaptic nerve terminal. × 48 000. (Thornell *et al.*, 1976.)

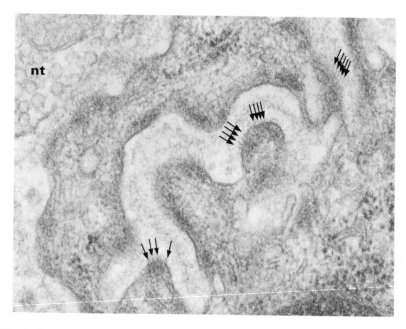

Fig. 7. 2. Enlargement of tips of postsynaptic foldings carrying striations interpreted as synaptic receptor areas (arrows) nt = nerve terminal. × 80 000. (Thornell *et al.* 1976.)

IV   Humoral and cellular response, passive transfer of EAM and of MG

A   CIRCULATING ANTIBODIES

As previously discussed all nAChR immunized animal species have produced circulating antibody to the antigen, demonstrable before signs of disease occurred. In a recent publication describing passive transfer of EAM with lymph node cells from immunized guinea pigs (Tarrab-Hazdai *et al.*, 1975a), no circulating antibodies were detectable at the time of transfer.

B   LYMPHOCYTE STIMULATION

Cellular immune response to nAChR has been reported from several laboratories. Lymphocytes from MG patients but not from controls were stimulated when cultured *in vitro* with a water-soluble fraction of nAChR from *Electrophorus electricus* (Abramsky *et al.*, 1974; 1975 b, c). Stimulation was reduced in lymphocytes from patients improving due to steroid treatment. Other antigens of muscle or nerve origin did not stimulate. Lymph node cells from rabbits preimmunized with *Electrophorus* (eel) nAChR and incubated with calf thymus or eel nAChR for 24 h followed by ^{14}C) thymidine showed a small but significant stimulation with both antigens (Aharonov *et al.*, 1975c). Lymphocytes from animals preimmunized with purified nAChE (a possible impurity in nAChR preparations) were not stimulated *in vitro* with calf thymus extracts indicating that AChE and thymus do not cross-react. Lymphocytes from animals preimmunized with calf thymus extract, however, were not stimulated by eel nAChR, possibly due to low immunogenicity of the relevant determinants in the thymus. Non-related proteins, among them human muscle extracts, caused no stimulation either. Antiserum to eel nAChR crossreacted with several calf thymus fractions, as tested by microcomplement fixation. No complement fixation was obtained with AChE, with human aqueous muscle extract or with *Naja naja siamensis* α-neurotoxin. Rabbit thymus fraction antiserum gave complement fixation with its antigen and crossreacted with eel nAChR. The authors concluded that the electroplaque and the thymus may share an antigen. Using lymphocyte transformation tests cellular sensitivity towards nAChR was demonstrated in lymph node cells of EAM monkey (Tarrab-Hazdai *et al.*, 1975a). A specific cellular immune (peripheral blood lymphocytes) response towards nAChR was also demonstrated in MG patients (Abramsky *et al.*, 1974; 1975a). In cases showing clinical improvement due to steroid treatment this sensitization was suppressed. On the basis of these experiments and the passive transfer of EAM (see below), the Israelian authors suggested that cell-mediated autoimmune mechanisms involving T-cell reactions may be an important factor in the pathogenesis of the neuromuscular block in MG.

C DELAYED TYPE CUTANEOUS REACTIVITY

Delayed cutaneous reactivity to eel nAChR was observed four days after inoculation in rats (Lennon *et al.*, 1976) and preceded the clinical and histological onset of EAM. It disappeared in the chronic phase.

D PASSIVE TRANSFER OF EAM OR MG

Early attempts to transfer MG from humans to animals were mostly negative (Bergh, 1953; Nastuk *et al.*, 1959). Tarrab-Hazdai *et al.* (1975c) and Lennon *et al.* (1976) demonstrated passive transfer of EAM in guinea pigs and rats using lymph node cells 10 (7) days after sensitizing animals with purified *T. californica* (*Electrophorus*) nAChR. Recipient guinea pigs but no controls showed positive skin tests (erythema after 24 h) when challenged with nAChR five days after cell transfer. Positive skin tests were also seen in actively challenged animals 10 days after immunization. Donor animals had no circulating nAChR antibodies when used for transfer. 12–22 days after active challenging, 60% of animals showed Tensilon sensitive clinical signs of EAM such as mild muscle weakness, sinking of head, hypoactivity, anorexia and weight loss. The same was observed in 57% of recipient animals 7–14 days after cell transfer. Symptoms were transient, usually lasting 3–4 days. No response was seen with nonrelated antigen (lysozyme). Repetitive nerve stimulation caused a decremental decrease in EMG amplitude in all recipient guinea pigs and in all nAChR injected guinea pigs. Recipient rats developed intermittent signs of EAM 6–20 days after cell transfer and in some "weakness developed after bleeding" on the 19th day. No antibody to rat muscle nAChR was seen, and no abnormal sensitivity was observed to curare.

Toyka *et al.* (1975) described passive transfer from man to mouse after daily *i.p.* injections for 9–14 days of 10–11 mg patient IgG. Cyclophosamide was given in order to induce tolerance to the human serum proteins. Diaphragm m.e.p.p. amplitudes were decreased, the mean m.e.p.p. amplitude being 0·10 mV in myasthenic IgG recipients as compared with 0·62 mV in control animals. Resting membrane potential and input resistance of muscle fibres were unchanged. The mean number of junctional receptors (^{125}I-αbgt labelling) in the experimental animals was 42% lower than in control animals. Repetitive nerve stimulation resulted in decremental response in some mice, one mouse improved by neostigmine.

Neonatal MG, i.e. transient myasthenic symptoms observed in newborn babies of myasthenic mothers (Namba *et al.*, 1970), has been interpreted as transfer of circulating antibody from mother to child.

V Model experiments with snake α-toxin

The α-neurotoxins from snake venoms are known to block nAChR. Various toxins act more or less reversibly. Within about 10 min after *i.v.* injection into rats inability to raise the head and to extend the forelimb digits, hunched posture and forelimb weakness were seen. The rats died after about 30 min from respiratory paralysis. A progressive decline in amplitude of the response to stimulation was noted. Decremental responses were found but did not progress with time (Lennon *et al.*, 1975). Drachman *et al.* (1976), using more reversibly acting α-toxin from venom of *Naja naja atra i.v.* (12–20 μg) found single evoked potentials progressively decreasing in amplitude for several hours, subsequently recovering to normal or nearly normal levels. A decremental response was present in all animals. When the amplitude of the initial potential had returned to near normal the mean decrement was 46%. Most α-toxin treated rats showed abnormal curare sensitivity and all rats responded to neostigmine treatment. Toxin treated rats also showed post-tetanic potentiation followed by postactivation exhaustion. Chang and Lee (1966) have demonstrated a reduction in m.e.p.p. amplitude in rat diaphragm after α-toxin treatment, dependent on toxin concentration. Duchen *et al.* (1975) injected a single dose of 3–5 μg of α-neurotoxin from *Naja naja siamensis* into the soleus of albino mice, which caused total paralysis of the muscle response to nerve stimulation for up to three days and impaired neuromuscular transmission for up to nine days. Within 3–5 days extrajunctional sensitivity to ACh became detectable and muscles remained supersensitive for up to 10 days. The degree of extrajunctional supersensitivity was comparable to that caused by structural denervation. Fibrillation was observed. Histological examination showed little damage to muscle fibres, not more than caused by injections of saline. Preterminal and terminal axons appeared normal. The model experiments suggest that a reduction of available acetylcholine receptors *per se* could account for the defects of neuromuscular transmission in EAM and in myasthenia gravis. They also suggest that block of junctional receptor may cause formation of extrajunctional ones. Further they clearly show that EAM could not have been caused by traces of neurotoxin bound to injected nAChR.

VI The thymus

The thymus of EAM rabbits, guinea pigs and monkeys was found to be hyperplastic (Tarrab-Hazdai *et al.*, 1975a).

Animal thymus has been analysed for the presence of nAChR, which might crossreact with humoral nAChR antibodies. Small amounts of nAChR were found in rat thymus, about 0·5% of that of rat muscle or 0·005% of that of eel electric organ, as shown with immune sera from nAChR immunized rats and with α-toxin labelling (Lindstrom et al., 1976). Some nAChR was also found in calf thymus (Aharonov et al., 1975d) as shown with lymphocyte transformation test and with sera from EAM rabbits. Recently, α-neurotoxin binding to thymus from normal and MG patients was demonstrated (Uddgård et al., to be published): neurotoxin binding could partly be prevented with d-tc, and crossreaction with antibodies to human skeletal muscle receptors and with circulating antibodies from MG patients were observed indicating that binding occurred to nAChR.

Thymectomy of rats early in the chronic phase of EAM did not alter the clinical cause of the disease (Lennon et al., 1976). However, rats depleted of T-cells by thymectomy and X-radiation before EAM induction produced no nAChR antibodies after challenge. Administration of B- and T-cells before challenge with nAChR gave rise to antibodies, seldom seen after administration of B-cells only. No EMG decrements were seen but could be induced by low doses of d-tc, except in "B-cell only" animals. Administration of antithymocyte serum shortly before challenge with nAChR suppressed the acute phase of EAM. According to Lennon et al., T-lymphocytes were responsible for delayed type hypersensitivity reactions and they were major components of the rat lymph node cell population. The authors suggested that these T-cells might be involved as helpers to B-cells in the production of antibody to nAChR.

The role of thymus in MG has been discussed for many years. Most MG patients have either neoplastic or hyperplastic changes in the thymus (Castleman, 1966). In 10–15% of cases, thymomas are seen. Repeated lymph drainage (Bergström et al., 1973) and thymectomy are beneficial in some patients (Keynes, 1954; Perlo et al., 1966; Papatestas et al., 1971). This may be due to general immunodepression (Miller et al., 1967), but could also indicate that the thymus is a primary cause of the disease. A humoral as well as a cellular immune response towards thymic tissue has been demonstrated (van der Geld and Strauss, 1966; Field et al., 1973) and a cytotoxic effect of MG lymphocytes on cultured thymus cells has been shown (Mori and Kawanami, 1973). Profound depression of lymphocyte reactivity and diminution of T-cells in blood of MG patients was observed after thymectomy (Taylor, 1965; Metcalf, 1965; Schechter et al., 1974). A six times higher synthesis of IgG was observed in MG thymus as compared to normal (Smisley *et al.*, 1969) and was suggested to depend upon an abnormal immune reaction in the dysplastic MG tymus. Sera of MG patients reacted with both muscle (Aarli and Tönder, 1970) and thymus and contain antibodies to both organs or shared antibodies (van der Geld and Strauss, 1966). Crossreaction seemed to be caused by myoid cells (van der Velde and Friedman, 1966). Eel nAChR antibodies may cross-react with nAChR on these myoid cells. Patients with thymoma but no MG have insignificant receptor antibody titres (Lindstrom and Lennon, 1976).

Several workers (Goldstein and Whittingham, 1966; Goldstein, 1968; Kalden *et al.*, 1969; Goldstein and Hofmann, 1971; Kawanami and Mori, 1972) have produced thymitis and partial neuromuscular block in guinea pigs by immunizing them with extracts of calf thymus or calf skeletal muscle in Freunds complete adjuvant. Extracts of calf lymph nodes had no effect. Thymectomized animals developed no neuromuscular block. It was suggested that a primary involvement of the end-plate may initiate both cellular and humoral autoimmune reactions against a neuromuscular junction component and that sensitized lymphocytes and antibodies coud also crossreact with a thymic component. The tissues may be affected concordantly or one of them may be primarily affected, and the other secondarily, due to immunological crossreaction between them (Rule and Kornfeld, 1971). Alternatively, it was suggested that a thymus factor may be released and cause neuromuscular block. Goldstein and collaborators (Goldstein, 1974; Goldstein and Schlesinger, 1975) isolated a bovine thymic hormone thymopoietin ("thymin"), a polypeptide (Schlesinger and Goldstein, 1975) that induced differentiation of lymphopoietic stem cells and alterations in neuromuscular transmission. A synthetic peptide, corresponding to part of thymopoietin, also induced the differentiation of thymocytes *in vitro* (Schlesinger *et al.*, 1975) and caused a neostigmine-responsive neuromuscular block in mice as shown by EMG tests (Goldstein and Schlesinger, 1975). At 50 Hz stimulation some decrement was seen. The effect of these substances on the end-plate was suggested to be due to the presence of a synaptic receptor similar to some receptor of the stem cell. Differentiation within the thymus could lead to subclasses of T-cells with discrete functions (killer, helper and suppressor) after reaction with antigen (Canton and Boyse, 1975), a process where thymopoietin might be involved. It was suggested that the neuromuscular lesion in MG might be the result of systemic release of thymopoietin resulting from autoimmune thymitis.

VII Comparison between MG and EAM

A MYASTHENIA GRAVIS (MG)

Before a comparison between MG and EAM is made, it may be appropriate to summarize some of the information available on MG.

Myasthenia gravis is a neuromuscular disease characterized by striated muscle fatigue, aggravated by exercise and improved by rest. The disease is treated with anticholinesterases to raise the level of ACh, and with immunosuppressive steroids. Thymectomy is applied. Delayed remission after thymectomy may reflect the time required for depletion of lymphocytes (T-cells) which are sensitized to unknown components of the neuromuscular junction. Circulating antibodies to nAChR and various tissues are present. Lymphorrhages in skeletal muscles suggest lymphocyte mediated immunological reactions. Peripheral and thymic lymphocytes from patients with MG have cytotoxic effects on muscle cells grown in tissue culture and their migration from capillary tubes is sometimes inhibited in the presence of muscle, thymus or brain antigens (Rule and Kornfeld, 1971; Field et al., 1973). In MG muscle fatigue usually begins in the face and neck and spreads to the rest of the body, causing death in severe cases.

Neuromuscular transmission in patients is characterized by a decremental response to slow repetitive stimulation and by post-tetanic facilitation and exhaustion (Desmedt, 1973). The amplitude of the decremental responses is changed toward normal with anti-AChE drugs. The jitter (Stålberg et al., 1976) when recording from two muscle fibres belonging to one and the same motor unit is in patients with MG normal in some muscles and up to 10-fold increased in others, often with frequent blocking. Thus in a myasthenic muscle the different motor end-plates are affected to a different degree. Some are entirely normal. Spontaneous remissions occur frequently.

Faulty neuromuscular transmission seems to be the physiological cause of MG, but it is not known if the abnormal events at the motor end-plate are primary etiological factors. Neither the nature of the defect nor the exact site of the neuromuscular junction lesion are yet understood. A possible presynaptic defect in the ACh mechanism has been suggested by Elmqvist et al. (1964), who observed a reduction in the amplitude of m.e.p.p. in MG. The observation was interpreted as a reduction in the size of the ACh quantum. Another possibility would be a partially inactive neurotransmitter or perhaps a false transmitter.

Other authors suggested that atrophied nerve terminals and a 40% reduction in surface area of the presynapse might be the primary cause for decreased ACh contents. It has also been suggested that a widened synaptic cleft may reduce the amount of ACh reaching the postsynaptic receptors. Finally, a postsynaptic defect caused by a curare-like substance has been proposed—and is substantiated by model experiments with α-neurotoxins—and a reduced number of functional receptor sites was demonstrated at neuromuscular junctions from motor point muscle biopsies of MG patients in various clinical stages of the disease (Hartzell *et al.*, 1972; Fambrough *et al.*, 1973; Fambrough, 1974). Increase in extrajunctional receptors, suggesting denervation was occasionally observed (Drachman *et al.*, 1973).

Simpson (1960) and Nastuk *et al.* (1959) first suggested that an autoimmune process might be the primary cause in MG and various features of autoimmune disease are often found to accompany MG (Simpson, 1966, see also articles on MG and on EAM in Proc. of Myasthenia meeting, May 1975, *N.Y. Acad. Sci.*, 1976. The occurrence of transient neonatal MG (Namba *et al.*, 1970), of human skeletal nAChR and other antibodies, the positive effects of thymectomy and of lymph drainage, the occurrence of lymphorrhages in skeletal muscle, cytotoxic effects of MG lymphocytes and the occurrence of thymic abnormalities in many MG cases belong to this category. MG is also often associated with other autoimmune diseases.

B   MG AND EAM

The *acute* stage of EAM has features which differ from MG. These are particularly revealed by histological studies. The signs of inflammation and those of violent tissue destruction may have been aggravated by adjuvant present during immunization. Also, MG in early stages of the disease may not have been histologically analysed. Acute EAM may well mirror events during early stages of MG but be compressed in time. The *late* stages of EAM resemble closer MG both histologically, electrophysiologically and clinically. Heilbronn *et al.* (1976b) found electrophysiologic incidents suggesting denervation perhaps more often in EAM than in MG. These too may be the result of the violent processes during the first stage of EAM. Histological findings suggest denervation to occur in MG and electrophysiological signs of reinnervation are found in about 25% of MG cases. Electrophysiological studies of

EAM and MG confirm lack of postsynaptic response to the release of ACh and reveal similarities such as decremental electromyograms, increased jitter, increased curare sensitivity and decreased m.e.p.p. amplitude. Some immunized animals (mice, most rats, some guinea pigs) do not show decrement.

A reduced number of α-toxin binding receptor sites has been observed both in EAM and in MG. A block caused by circulating nAChR antibodies present at the postsynaptic membrane might explain this but so far no antibodies have been demonstrated in the synaptic cleft. What has been observed in EAM is a removal of receptor sites from the proximity of the presynaptic nerve terminal in the acute stage and a reduction and simplification (loss of secondary folds) of the postsynaptic membrane at the muscle end-plate in the late stage of EAM and in MG. This could be the cause of the observed decrease in the number of receptor sites. Further, during acute EAM a barrier to ACh may be provided by the unidentified electron dense material found in the wide and irregular synaptic cleft in early stages of EAM. The debris could have been caused by antibody attachment to the postsynaptic membrane, which may cause atrophy of the postsynaptic foldings and other surrounding tissue, particularly after complement binding. Several results suggest a complement mediated nAChR antibody attack at the motor end-plate resulting in tissue damage and possibly followed by release of junctional receptors. On the other hand, however, traumatic release of tissue constituents need not elicit antibody formation. Increased concentrations of ACh due to treatment with anti-AChE may allow some transmitter to reach receptors further away or blocked by debris.

In EAM as in MG, circulating antibodies directed against skeletal muscle nAChR were found and may be involved in the impairment of neuromuscular transmission. This theory is supported by the occurrence of transient neonatal MG—which however could also be caused by non-IgG compounds—and by the successful passive transfer, in one laboratory, of MG to animals after injection of patient IgG. Neurotransmission in tissue preparations was blocked by nAChR antibody from EAM or MG sera and decremental EMG responses after application of EAM sera have been described. Antibodies, however, seem not to be directed towards the ACh binding site of nAChR. Yet, *in situ*, in order to be accessible to the transmitter, this part of nAChR should be the one projecting into the synaptic cleft, while other parts of the receptor may be buried within the postsynaptic membrane. The conclusion would be that nAChR antibodies, if acting in the synaptic cleft, block ACh passage but do not react with the ACh recognition site.

A primary attack of circulating antibodies in the synaptic cleft or nearby, however, need not be the cause of block of neuromuscular transmission in MG or EAM. There is considerable evidence indicating the involvment of thymus and cellular immune response in EAM as well as in MG. Passive transfer of EAM with lymph node cells supports this. The presence of small amounts of α-toxin binding sites in thymus with which fish electric organ nAChR antibodies crossreact suggests a way for nAChR antibodies to attack here. Such attack may cause a defective T-lymphocyte function resulting in loss of control of autoantibody producing B-cells. Cytotoxic lymphocytes were shown to exist in MG. Lymphorrhages and macrophage invasion of muscle fibre have been observed in EAM and in MG. Thymectomy as well as lymph drainage are beneficial in MG. Antibody induced changes in the thymus may cause the release of thymopoietin, a thymic hormone able to block neuromuscular transmission. Most workers agree that cellular immune response is indicated as a main factor in both EAM and MG.

VIII Concluding remarks

So far, the study of EAM has not solved the riddle of MG etiology, but it has provided a number of clues. The involvment of humoral and cellular immune response and that of the thymus and its T-lymphocytes seems obvious, but the primary cause of MG and the order of events leading to an autoimmune response in MG is still unknown. The role of circulating antibodies, which could be blocking and/or destructive as well as nonrelevant or even protective is not clear, nor is it clear if and where they react with nAChR *in situ*. Conclusive experiments concerning transfer of MG or EAM as well as conclusive lymphocyte stimulation experiments are still lacking. Neither presence nor absence of antibody within the synaptic cleft has yet been demonstrated. The relation between thymus, T-lymphocytes and antibodies is not clear. Modifications of nAChR, including viral modifications are needed to elucidate possible transformation of self-receptor into antigen. The effect of antibody binding to thymus as well as possible links between an immune response to nAChR and the release of thymopoietin remain to be studied. Phylogenetic studies on nAChR are needed. Continued and careful analysis of ways to produce EAM and of the nature of the disease may provide further information as a basis for the understanding of myasthenia gravis.

IX   References

Aarli, J. A. and Tönder, O. (1970). Antiglobulin consumption test with sera from patients with myasthenia gravis. *Clin. Exp. Immunol.* **7.** 11–21.

Aarli, J. (1972). Myasthenia gravis. Antibodies to an acid-soluble antigen in striated muscle. *Clin. Exp. Immunol.* **10,** 453–461.

Aarli, J. A., Mattsson, C. and Heilbronn, E. (1975). Antibodies against nicotinic acetylcholine receptor and skeletal muscle in human and experimental myasthenia gravis. *Scand. J. Immunol.* **4,** 849–852.

Abramsky, O., Aharonov, A., Webb, C., Teitelbaum, D. and Fuchs, S. (1974). Lymphocyte sensitization to acetylcholine receptor in myasthenia gravis. *Israel J. Med. Sci.* **10,** 1167.

Abramsky, O., Aharonov, S., Webb, C. and Fuchs, S. (1975a). Cellular immune response to acetylcholine receptor-rich fraction in patients with myasthenia gravis. *Clin. Exp. Immunol.* **19,** 11–16.

Abramsky, O., Aharonov, A. Teitelbaum, D. and Fuchs, S. (1975b). Myasthenia gravis and acetylcholine receptor: effect of steroids on clinical course and cellular immune response to acetylcholine receptor. *Arch. Neurol.* **32,** 684–687.

Aharonov, A., Gurari, D. and Fuchs, S. (1974). Immunochemical characterization of *Naja naja siamensis* toxin and of a chemically modified toxin. *Eur. J. Biochem.* **45,** 297–303.

Aharonov, A., Kalderon, N., Silman, I. and Fuchs, S. (1975a). Preparation and immunochemical characterization of a water-soluble acetylcholine receptor fraction from the electric organ tissue of the electric eel. *Immunochemistry* **12,** 765–771.

Aharonov, A., Abramsky, O., Tarrab-Hazdai, R. and Fuchs, S. (1975b). Humoral antibodies to acetylcholine receptor in patients with myasthenia gravis. *Lancet* **4,** 340–342.

Aharonov, A., Tarrab-Hazdai, R., Abramsky, O. and Fuchs, S. (1975c). Immunological relationship between acetylcholine receptor and thymus: a possible significance in myasthenia gravis. *Proc. Nat. Acad. Sci. U.S.* **72,** 1456–1459.

Albuquerque, E. X., Lebeda, F. J., Appel, S. H., Almon, R., Kauffman, F. C., Mayer, R. F., Norahashi, T. and Yeh, Y. Z. (1976). Effects of normal and myasthenic serum factors on innervated and chronically denervated mammalian muscles. *Ann. N.Y. Acad. Sci.* **274,** 475–492.

Almon, R. R., Andrew, C. G. and Appel, C. H. (1974). Serum globulin in myasthenia gravis: inhibition of α-bungarotoxin binding to acetylcholine receptors. *Science* **186,** 55–57.

Almon, R. R. and Appel, S. H. (1976a). Serum acetylcholine receptor antibodies in myasthenia gravis. *Ann. N.Y. Acad. Sci.* **274**, 235–243.

Almon, R. R. and Appel, S. H. (1976b). Interaction of myasthenic serum globulin with the acetylcholine receptor. *Biochem. Biophys. Acta* **393**, 66–77.

Bender, A., Ringle, A., Engel, W., Daniels, M. and Vogel, Z. (1975). Myasthenia gravis: a serum factor blocking acetylcholine receptors of the human neuromuscular junction. *Lancet* **1**, 607–609.

Berg, K. (1953). Biological assays in myasthenia gravis. Scand. *J. Clin. Lab. Invest.* (Supp 5) **5**, 1–42.

Bergström, K., Frankson, C., Matell, G., von Reis, G. (1973). The effect of thoracic duct. lymph drainage in myasthenia gravis. *Eur. Neurology* **9**, 157–167.

Biesecker, G. (1973). Molecular properties of the cholinergic receptor purified from *Electrophorus electricus. Biochemstry* **12**, 4403–4409.

Cantor, H. and Boyse, E. A. (1975). Functional subclasses of T lymphocytes bearing differently antigens. I. The generation of functionally distinct T-cell subclasses is a differentiative process independent of antigen. *J. Exp. Med.*

Castleman, B. (1966). The pathology of the thymus gland in myasthenia gravis. *Ann. N.Y. Acad. Sci.* **135**, 496–503.

Chang, C. and Lee, C. Y. (1966), Electrophysiological study of neuromuscular blocking action of cobra neurotoxin. *Br. J. Pharmacol. Chemother.* **28**, 172–181.

Chang, H. W. (1974). Purification and characterization of acetylcholine receptor from *Electrophorus electricus. Proc. Nat. Acad. Sci. U.S.A.* **71**, 2113–2117.

Clementi, F., Conti-Tronconi, B., Berti, F. and Folco, G. (1975). The immunological response against two specific components of the cholinergic synapse: the receptor protein (AChR) and acetylcholinesterase (AChE). Sixth Int. Congress of Pharmacology, Helsinki, p. 305.

Desmedt, J. E. (1973) The neuromuscular disorder in myasthenia gravis. *In* "New Developments in Eng. and Clin. Neurophysiol" (J. E. Desmedt, ed.), pp. 305–342. Karger, Basel.

Drachman, D. B., Fambrough, D. M. and Satyamurti, S. (1973) Reduced acetylcholine receptors in myasthenia gravis. *Transact. Am. Neurol. Ass.* **98**, 93–95.

Drachman, D. B., Kao, I., Pestronk, A. and Toyka, K. V. (1976). Myasthenia gravis as a receptor disorder. *Ann. N.Y. Acad. Sci.* **274**, 226–230.

Duchen, L. W., Heilbronn, E. and Tonge, D. A. (1975). Functional denervation of skeletal muscle in the mouse after the local injection of a postsynaptic blocking fraction of *Naja naja siamensis* venom. *J. Physiol.* **250**, 26–27.

Eldefrawi, M. E. and Eldefrawi, A. T. (1973). Purification and molecular properties of the acetylcholine receptor from *Torpedo* electroplax. *Arch. Biochem. Biophys.* **159**, 362–373.

Elmqvist, D., Hofmann, W. W., Kugelberg, J. and Quassel, D. M. J. (1964) An electrophysiological investigation of neuromuscular transmission in myasthenia gravis. *J. Physiol.* **174**, 417–434.

Elmqvist, D., Mattsson, C., Heilbronn, E., Lundh, H. and Libelius, R. (1976). Neuromuscular transmission in rabbits immunized with acetylcholine receptor protein. *Arch. Neurol.* (in press).

Engel, A. G. and Santa, T. (1971). Histometric analysis of the ultrastructure of the neuromuscular junction in myasthenia gravis and in the myasthenic syndrome. *Ann. N.Y. Acad. Sci.* **183**. 46–63.

Engel, A. G., Tsujikata, M., Lindstrom, J. M. and Lennon, V. A. (1976). The motor-end plate in myasthenia gravis and in experimental autoimmune myasthenia gravis. A quantitative ultrastructural study. *Ann. N.Y. Acad. Sci.* **274**, 60–79.

Fambrough, D. M. (1974) Acetylcholine receptors. Revised estimates of extra junctional receptor density in denervated rat diaphragm. *J. Gen. Physiol.* **64**, 468–472.

Fambrough, D. M., Drachman, D. B. and Satyamurti, S. (1973) Neuromuscular junction in myasthenia gravis, Decreased acetylcholine receptors. *Science* **182**, 293–295.

Field, E. J. *et al.* (1973). Lymphocyte sensitization in myasthenia gravis. *In* Xth International Congress of Neurology, Amsterdam, Excerpta Medica, Vol. 296, 67.

Foldes, F., Heilbronn, E. and Mattsson, C. (unpublished).

Goldstein, G. (1968). The thymus and neuromuscular function. A substance in thymus which causes myositis and myasthenic neuromuscular block in guinea pigs. *Lancet* **11**, 119–122.

Goldstein, G. (1974). Isolation of bovine thymin: A polypeptide hormone of the thymus. *Nature* **247**, 11–14.

Goldstein, G. and Hofmann, W. W. (1971). Experimental myasthenia gravis. *Res. Publ. Assoc. Res. Nerv. Ment. Dis.*

Goldstein, G. and Whittingham, S. (1966). Experimental autoimmune thymitis. An animal model of human myasthenia gravis. *Lancet* **2**, 315–318.

Goldstein, G. and Schlesinger, D. H. (1975). Thymopoietin and myasthenia gravis: neostigmine-responsive neuromuscular block produced in mice by a synthetic peptide fragment of thymopoietin. *Lancet* **2**, 256–258.

Gordon, A., Brandine, G. and Hucho, F. (1974). Investigation of the *Naja naja siamensis* toxin binding site of the cholinergic receptor protein from *Torpedo* electric tissue. *F.E.B.S. Lett.* **47**, 204–208.

Green, D. P. L., Miledi, R. and Vincent, A. (1975a). Neuromuscular transmission after immunization against acetylcholine receptors. *Proc. R. Soc. Lond. B* **189**, 57–58.

Green, D. P. L., Miledi, R., Peres de la Mora, M. and Vincent, A. (1975b). Acetylcholine receptors. *Phil. Trans. R. Soc. Lond. B.* **270**, 551–559.

Hartzell, H. C. and Fambrough, D. M. (1972) Acetylcholine receptors. Distribution and extrajunctional density in rat diaphragm after denervation correlated with acetylcholine sensitivity. *J. Gen. Physiol.* **60**, 248–262.

Heilbronn, E., Karlsson, E. and Widlund, L. (1973). Purification of a membrane constituent, possibly the nicotinic cholinergic receptor from the electric organ of *Torpedo marmorata*. *In* "La Transmission Cholinergique de l'Exitation" (M. Fardeau, M. Israel and R. P. Manaranche, eds), pp. 151–168. INSERM Colloquium, 1972.

Heilbronn, E., Mattsson, C. and Stålberg, E. (1974). Immune response in rabbits to a cholinergic protein—possibly a model for myasthenia gravis. Excerpta Medica. Internat. Series No. 360. Recent Advances in myologi. Proc. of the 3rd Int. Congr. on muscle diseases, Newcastle upon Tyne, 15–21 Sept., 1974.

Heilbronn, E. and Mattsson, C. (1974). The nicotinic cholinergic receptor protein: improved purification method, preliminary amino acid composition and observed autoimmuno response. *J. Neurochem.* **22**, 315–317.

Heilbronn, E., Mattsson, C. and Elfman, L. (1975a). Biochemical and physical properties of the nicotinic ACh receptor from Torpedo marmorata. In "Proceedings of the Ninth F.E.B.S. Meeting, Budapest, 1974" (G. Gardos, ed.) pp. 29–38. North. Holland, Amsterdam.

Heilbronn, E., Mattsson, C., Stålberg, E. and Hilton-Brown, P. (1975b). Neurophysiological signs of myasthenia in rabbits after receptor antibody development. *J. Neurol. Sci.* **24**, 59–64.

Heilbronn, E., Mattsson, C. and Stålberg, E. (1976b). Bacteria necessary to produce experimental autoimmune myasthenia gravis in acetylcholine receptor immunized rabbits? *Nature Lond.* **262**, 509.

Heilbronn, E., Mattsson, C. and Stålberg, E. (1975c). Immune response in rabbits to a cholinergic receptor protein: possibly a model for myasthenia gravis. Excerpta Medica Int. Congress Series No. 360, Recent Advances in Myology, Proc. third Int. Congr. on Muscle Diseases, Newcastle upon Tyne, 15–21 Sept., 1974, Excerpta Medica, Amsterdam, 486–492.

Heilbronn, E., Mattsson, C., Thornell, L-E., Sjöström, M., Stålberg, E., Hilton-Brown, P. and Elmqvist, D. (1976a). Experimental myasthenia gravis in rabbits. Biochemical, immunological, electrophysiological and morphological aspects. *Ann. N.Y. Acad. Sci.* **274**, 337–353.

Ito, F., Kanamori, N. and Kuroda, H. (1974). Structural and functional assymetries of myelinated branches in the frog muscle spindle. *J. Physiol.* **241**, 389–406.

Kalden, J. R., Williamson, W. G., Johnston, R. J. and Irvine, W. J. (1969). Studies on experimental autoimmune thymitis in guinea pigs. *Clin. Exp. Immunol.* **5**, 319–340.

Karlin, A., and Cowburn, D. (1973). The affinity-labelling of partially purified acetylcholine receptor from electric tissue of Electrophorus. *Proc. Nat. Acad. Sci. U.S.* **70**, 3636–3640.

Karlsson, E., Heilbronn, E. and Widlund, L. (1972). Isolation of the nicotinic acetylcholine receptor by biospecific chromatography on insolubilized *Naja naja* neurotoxin. *F.E.B.S. Letters* **28**, 197–201.

Katz, B. and Miledi, R. (1973). The effect of α-bungarotoxin on acetylcholine receptors. *Brit. J. Pharmacol.* **49**, 138–139.

Kawahami, S. and Mori, R. (1972). Experimental myasthenia in mice: the role of the thymus and lymphoid cells. *Clin. Exp. Immunol.* **12**, 447–454.

Keynes, G. (1954). Surgery of the thymus gland. Second (and third) thoughts. *Lancet* **266**, 1197–1202.

Klett, R., Fulpius, B., Cooper, D., Smith, M., Reich, E. and Possani, L. (1973). The acetylcholine receptor. Purification and characterization of a macromolecule isolated from *Electrophorus electricus*. *J. Biol. Chem.* **248**, 6841–6853.

Lambert, E. H. and Elmqvist, D. (1971). Quantal components of end-plate potentials in the myasthenic syndrome. *Ann. N.Y. Acad. Sci.* **183**, 183–199.

Lee, C. Y. (1972). Chemistry and pharmacology of polypeptide toxins in snake venoms. *Ann. Rev. Pharmacol.* **12**, 265–286.

Lennon, V., Lindstrom, J. and Seybold, M. (1975). Experimental autoimmune myasthenia (EAMG): A model of myasthenia gravis in rats and guinea pigs. *J. Exp. Med.* **141**, 1365–1375.

Lennon, V. A., Lindstrom, J. M. and Seybold, M. E. (1976). Experimental autoimmune myasthenia gravis: Cellular and humoral immune response. *Ann. N.Y. Acad. Sci.* **274**, 283–299.

Lindstrom, J. Immunological studies on acetylcholine receptors. *J. Supramolecular Biol.* (in press).

Lindstrom, J., Lennon, V., Seybold, M. and Wittingham, S. (1976). Experimental autoimmune MG and MG. Biochemical and Immunochemical Aspects. *Ann. N.Y. Acad. Sci.* **274**, 254–274.

Mattsson, C. and Heilbronn, E. (1975a). The nicotinic acetylcholine receptor, a glycoprotein. *J. Neurochem.* **25**, 899–901.

Mattsson, C. and Heilbronn, E. (1975b). Immunization of rabbits with purified nicotinic acetylcholine receptor protein. *Croatica Chem. Acta.* **47**, 489–495.

Mattsson, C. and Heilbronn, E., Bock, E. and Ramlau, J. (1976). α-toxin binding proteins in the electric organ of *Torpedo marmorata*. A biochemical and immunochemical study of membrane fractions and solubilized material. Abstract., 10th Int. Congr. Biochem.

Meunier, J., Sealock, R., Olsen, R. and Changeux, J. (1974). Purification and properties of the cholinergic receptor proteins from *Electrophorus electricus* electric tissue. *Eur. J. Biochem.* **45**, 371–394.

Metcalf, D. (1965). Delayed effect of thymectomy in adult life on immunological competence. *Nature Lond.* **208**, 1336.

Michaelson, D. and Raftery, M. (1974). Purified acetylcholine receptor: Its reconstitution to a chemically excitable membrane. *Proc. Nat. Acad. Sci. U.S.* **71**, 4768–4772.

Michaelson, D., Vandlen, R., Bode, J., Moody, T., Schmidt, J. and Raftery, M. A. (1974). Some molecular properties of an isolated acetylcholine receptor iontranslocation protein. *Arch. Biochem. Biophys.* **165**, 796–804.

Miller, J. F. A. P. and Osoba, D. (1967). Current concepts of immunological function of the thymus. *Physiol. Rev.* **47**, 437–520.

Mori, R. and Kawanami, S. (1973). Destruction of cultured thymus cells by autologues lymphocytes from a patient with myasthenia gravis. *Lancet* **1**, 210.

Moore, W. M., Holloday, L. A., Puett, D. and Brady, R. N. (1974). On the conformation of the acetylcholine receptor protein from *Torpedo nobiliana*. *F.E.B.S. Lett.* **45**, 145–149.

Namba, T., Brown, S. B. and Grob, D. (1970). Neonatal myasthenia gravis: report of two cases and review of the literature. *Pediatrics* **45**, 488–504.

Nastuk, W. L., Strauss, A. J. L. and Osserman, K. E. (1959). Search for a neuromuscular blocking agent in the blood of patients with myasthenia gravis. *Amer. J. Med.* **26**, 394–409.

Olsen, R. W., Meunier, J. C. and Changeux, J. P. (1972). Progress in the purification of the cholinergic receptor protein from *Electrophorus electricus* by affinity chromatography. *F.E.B.S. Lett.* **28**, 96–100.

Papatestas, A. E., Alpert, L. I., Osserman, K. E., Osserman, R. S. and Kark, A. E. (1971). Studies in myasthenica gravis: Effects of thymectomy. Results on 185 patients with nonthymomatous and thymomatous myasthenia gravis 1941–1969. *Amer. J. Med.* **50**, 465–474.

Patrick, J. and Lindström, J. (1973). Autoimmune response to acetylcholine receptor. *Science* **180**, 871–872.

Patrick, J., Lindström, J., Culp, B. and McMillan, J. (1973). Studies on purified eel acetylcholine receptor and anti-acetylcholine receptor antibody. *Proc. Nat. Acad. Sci. U.S.A.* **70**, 3334–3338.

Patrick, J., Boulter, J. and O'Brien, J. (1975). An acetylcholine receptor preparation lacking the 42 000 dalton component. *Biochem. Biophys. Res. Comm.* **64**, 219–225.

Perlo, V., Poskaner, D., Schwab, R., Viets, H., Osserman, K. and Genkins, G. (1966). Myasthenia gravis: Evaluation of treatment in 1355 patients. *Neurology* **61**, 431–439.

Raftery, M. A., Bode, J., Vandlen, R., Michaelson, D., Deutsch, J., Moody, T., Ross, M. J. and Stroud, R. M. (1975). *In* "Protein-gland Interactions (Horst Sund and Gideon Blauer, eds), pp. 328–355. Walter de Grytes, Berlin and New York.

Ringel, S. P., Bender, A. N., Festoff, B. W., Engel, W. K., Vogel, Z. and Daniels, M. P. (1975). Ultrastructural demonstration and analytical application of extrajunctional receptors of denervated human and rat skeletal muscle fibres. *Nature* **255**, 730.

Rule, A. H. and Kornfeld, P. (1971). Studies in myasthenia gravis: biologic aspects. *Mt. Sinai J. Med.* **38**, 538–572.

Santa, T., Engel, A. G. and Lambert, E. H. (1972). Histometric study of neuromuscular junction ultrastructure. *Neurology* **22**, 71–82.

Schechter, B., Nir, E., and Levine, S. (1974). *Israel J. Med. Sci.* **10**, 1170.

Schlesinger, D. H. and Goldstein, G. (1975). The amino acid sequence of thymopoietin II. *Cell* **5**, 361–366.

Schlesinger, D. H., Goldstein, G., Scheid, M. P. and Boyse, E. A. (1975). Chemical synthesis of a peptide fragment of thymopoietin II that induces selective T cell differentiation. *Cell* **5**, 367–370.

Schmidt, J. and Raftery, M. (1973). Purification of acetylcholine receptors from *Torpedo californica* electroplax by affinity chromatography. *Biochemistry* **12**, 852–856.

Simpson, J. (1960). Myasthenia gravis: A new hypothesis. *Scot. Med. J.* **5**, 419–436.

Simpson, J. A. (1966). *Ann N.Y. Acad. Sci.* **135**, 506.

Smiley, J. D., Bradley, J., Daly, D. and Zift, M. (1969). Immunoglobin synthesis *in vitro* by human thymus. Comparison of myasthenia gravis and normal thymus. *Clin. Exp. Immunol.* **4**, 387.

Straaus, A. J. L., Seegal, B. C., Hju, K. G. Burkholder, P. H., Nortuk, W. L. and Osserman, K. E. (1960). Immunofluorescence demonstration of a muscle-binding complement fixing serum globulin fraction in myasthenia gravis. *Proc. Soc. Exp. Biol. Med. (N.Y.)* **105**, 184–191.

Stålberg, E. and Ekstedt, J. (1973). Single fibre EMG and microphysiology of the motor unit in normal and diseased human muscle. *In* "New developments in EMG and clinical neurophysiology" (J. E. Desmedt, ed.), Vol. 1, pp. 113–129. Karger, Basel.

Stålberg, E., Ekstedt, J. and Broman, A. (1974). Neuromuscular transmission in myasthenia gravis studied with single fibre electromyography. *J. Neurol. Neurosurg. Psychiatry* **37**, 540–547.

Stålberg, E., Trontelj, J. and Schwartz, M. (1976). Single fibre recording of the jitter phenomenon in patients with myasthenia gravis and in members of their families. *Ann. N.Y. Acad. Sci.* **274**, 189–202.

Sugiyama, H., Bendo, P., Meunier, J. and Changeux, J. (1973). Immunological characterization of the cholinergic receptor protein from *Electrophorus electricus*. *F.E.B.S. Lett.* **35**, 124–128.

Tarrab-Hazdai, R., Aharonov, A., Abramsky, O., Silman, I. and Fuchs, S. (1975a). Animal model for myasthenia gravis: acetylcholine receptor-induced myasthenia in rabbits, guinea pigs and monkeys. *Israel J. Med. Sci.* (in press).

Tarrab-Hazdai, R., Aharonov, A., Abramsky, O., Silman, I. and Fuchs, S. (1975b). Experimental autoimmune myasthenia induced in monkeys by purified acetylcholine receptor. *Nature* **256**, 128–130.

Tarrab-Hazdai, R., Aharonov, A., Abramsky, O., Israel, Y. and Fuchs, S. (1975c). Passive transfer of experimental autoimmune myasthenia by lymph node cells in inbred guinea pigs. *J. Exp. Med.* **142**, 785–789.

Taylor, R. B. (1965). Decay of immunological responsiveness after thymectomy in adult life. *Nature Lond.* **208**, 1334–1335.

Toyka, K. V., Drachman, D. B., Pestronk, A. and Kao, I. (1975). Myasthenia gravis: Passive transfer from man to mouse. *Science* **190**, 397–399.

Thornell, L-E, Sjöström, M., Mattsson, C. H., and Heilbronn, E. (1976). Morphological observations on motor end-plates in experimental myasthenia in rabbits. *J. Neurol. Sci.* (in press).

Uddgärd, A. and Heilbronn, E. (1976). Comparison of chemical and immunological properties of acetylcholine receptors from *Torpedo* electric organ and mammalian muscle. *Abstr. 10th Int. Congr. Biochem.*

Valderama, R., Weill, C. L., McNamee, M. and Karlin, A. (1976). Isolation and properties of acetylcholine receptors from *Electrophorus* and *Torpedo*. *Ann. N.Y. Acad. Sci.* **274**, 108–115.

van der Geld, H., Feltkamp, T. E. W., van Loghem, J. J., Oosterhuis, H. J. G. H. and Biemond, A. (1963). Multiple antibody production in myasthenia gravis. *Lancet* **2**, 373–375.

van der Geld, H. W. R. and Strauss, A. J. L. (1966). Myasthenia gravis. Immunological relationship between striated muscle and thymus. *Lancet* **1**, 57–60.

van der Velde, R. L. and Friedman, N. B. (1966). The thymic "myoid zellen" and myasthenia gravis. *J. Amer. Med. Assoc.* **198**, 287–288.

Weill, C. L., McNamee, M. G. and Karlin, A. (1974). Affinity labeling of purified acetylcholine receptor from *Torpedo californica*. *Biochem. Biophys. Res. Com.* **61**, 997–1003.

Woolf, A. L. (1964). Morphology of the myasthenic neuromuscular junction. *Ann. N.Y. Acad. Sci.* **135**, 35–36.

Zacks, S. I., Bauer, W. C. and Blumberg, J. M. (1962). The fine structure of the myasthenic neuromuscular junction. *J. Neuropath. Exp. Neurol.* **21**, 335.

7

Isolation of Receptor Proteins from the Neuromuscular Junction in Arthropods

G. G. Lunt

Biochemistry Department, University of Bath, England

I   Introduction

The chemical theory of synaptic transmission presupposes the presence of a specific receptor in the post-synaptic membrane which recognizes the neurotransmitter. Our knowledge of neurotransmitter synthesis, storage and release has progressed rapidly, but it is only in the last ten years that research on the identification of the receptors for these transmitters has made significant advances. Attempts to isolate receptors started with the assumption that they are proteins which constitute an integral part of the postsynaptic membrane. Furthermore, it was assumed that the proteins possess a specific binding site for the neurotransmitter and that this site retains its binding properties when the protein is removed from the membrane environment. The most well characterized of the isolated receptors is the nicotinic cholinergic receptor from the electric organ of fishes such as *Electrophorus electricus* or species of *Torpedo*. This receptor is found to be an intrinsic membrane protein and therefore dissolution of the postsynaptic membrane is a prerequisite to its isolation. The isolated protein binds the natural transmitter, acetylcholine, and its pharmacological characteristics closely parallel those of the native receptor *in situ*. There is now little doubt that this isolated protein constitutes the essential part of the physiological acetylcholine receptor (see Chapter 1 and recent reviews by Changeux, 1975; De Robertis, 1975; Karlin, 1974).

The successful isolation of the cholinergic receptor has been greatly aided by the availability of a wide range of well-characterized pharmacologically active agents. The properties of the isolated receptor preparations can, therefore, be constantly checked and compared with the properties of the receptor *in vivo*. Outstanding among these agents are the snake venom α-toxins which show a highly specific curare-like action at the vertebrate neuromuscular junction *in vivo* (Lee and Chang, 1966) and show specific high-affinity binding to the isolated receptor (Changeux *et al.*, 1970). A further important factor is the availability of a rather unusual tissue—the electroplax—which has greatly aided the isolation of this receptor. The electroplax of the electric fishes offers the advantage of a pharmacologically homogeneous receptor population and the receptor-rich innervated plasma membrane can be readily separated from the non-innervated face of the cell. The receptor proteins constitute a very high proportion of the total proteins of the innervated membranes.

The situation at the arthropod neuromuscular junction is considerably more complex. Two neurotransmitters are known to be involved at this site; glutamate, acting as an excitatory transmitter and γ-amino butyrate (GABA) which has an inhibitory role (Chapter 8 and 9; and reviews by Gerschenfeld, 1973; Lunt, 1975). The pharmacological characteristics of these two systems are not well defined and there are, to date, no analogues to the highly specific cholinergic α-toxins. Putative receptors can be identified at present primarily by their ability to bind the transmitter in a specific manner and by investigating their interactions with a small group of drugs thought to act at the level of the postsynaptic glutamate and GABA receptors. A major complication is the presence of active uptake systems for the two transmitters; therefore ways must be sought of distinguishing between binding to the postsynaptic receptor site and to constituents of the uptake systems.

The early attempts at receptor isolation from vertebrate central nervous tissue made by De Robertis and his colleagues (see De Robertis, 1971; 1975 for review) were prompted by the availability of receptor-enriched synaptic membrane preparations. Similarly the ease of preparation of the innervated membranes of electroplax has been an important factor in the isolation of the cholinergic receptor. There are, however, no such preparations available from arthropod muscle. Donnellan *et al.* (1974) attempted to carry out a subcellular fractionation of flight muscle from *Sarcophaga* but were unable to obtain a fraction enriched in nerve terminals. The initial homogenate contained intact terminals in large numbers but these were disrupted during the fractionation procedure. No attempts were made to monitor glutamate binding to any of the membranous subfractions. Recently the problem of obtaining an enriched preparation of insect muscle plasma membrane has been re-examined in our laboratory by James (unpublished experiments). He finds that homogenization of muscles from *Schistocerca gregaria* in 0·25M sucrose at a concentration of 10% (wt. fresh tissue/v) leads to marked clumping of the membranes and subsequent fractionation by differential centrifugation is not possible. When the muscles are homogenized at a concentration of 2% (w/v) however, no clumping occurs and after a series of centrifugation steps a membrane fraction is obtained which electron microscopy shows to be reasonably homogeneous and free from mitochondria. This membrane preparation shows stereospecific binding of L-glutamate and work is progressing to establish whether the binding is to the synaptic receptor site. The possible presence of a glutamate transport system associated with the insect muscle plasma membrane cannot be overlooked and it may be that constituents of such a system could contribute to the observed glutamate binding.

II   Receptor isolation

Two basic methods of achieving the dissolution of synaptic membranes and the subsequent solubilization of receptor proteins have been used. The group of De Robertis first reported the use of chloroform:methanol mixtures to extract hydrophobic protein-lipid complexes or proteolipids from vertebrate cerebral cortex which showed some of the properties expected for a cholinergic receptor (De Robertis *et al.*, 1967). Since then receptor-proteolipids from a wide range of tissues have been isolated and partially characterized (see De Robertis, 1971; 1975 for reviews). An alternative and more conventional approach to the isolation of receptors was based on the solubilization of membrane proteins by aqueous detergent solutions and using this procedure Changeux and his collaborators have isolated the cholinergic receptor from the electroplax of *Electrophorus electricus* (see Changeux, 1975, for review).

The binding properties of the proteolipid receptors are usually measured by co-chromatographing the protein with radio-labelled transmitter or drug on the lipophilic gel Sephadex LH20. Co-elution of the ligand with a particular peak of protein is taken as an indication of specific binding and a number of putative receptors have been investigated in this way, among them, the nicotinic cholinergic receptor from electric tissue (La Torre *et al.*, 1970), an adrenergic receptor from spleen capsule (Fiszer de Plazas and De Robertis, 1972), a cholinergic receptor from housefly heads (Cattell and Donnellan, 1972) and glutamate and GABA receptors from shrimp muscle (Fiszer de Plazas and De Robertis, 1973).

The study of the interactions between proteolipids and ligands in organic solvents presents many problems. There is not a clear understanding of the nature of the interactions between the ligands and the protein and the basis of the separation of individual proteins on Sephadex LH20 is not well defined. Addition of a ligand to the chloroform: methanol solution of proteolipids prior to chromatography appears to result in a true reversible equilibrium being established. However, when the proteolipid-ligand complex is chromatographed on LH20 Sephadex, this binding appears to be irreversible, i.e. no dissociation of bound ligand occurs during the chromatography. It has been suggested (see De Robertis, 1975 for discussion) that within the LH20 Sephadex column the highly apolar environment favours the polar interactions between ligand and binding site and this disturbs the equilibrium greatly; so much so that the binding is effectively irreversible. Critics of the method point out that under such conditions weak, non-specific interactions may be so favoured as to lead to binding of ligands to sites other than the true receptor site. Levinson and Keynes (1972) made a preliminary study of the behaviour of proteolipids in the LH20 Sephadex system and have suggested that the observed associations may be entirely coincidental and not related to specific receptor sites. Recently, however, a more careful study has been made of the behaviour of proteolipids and ligands and of their interactions and it is clear that, while the chromatographic conditions can greatly influence the behaviour of the proteins, the interaction between ligand and protein is specific and is certainly not a fortuitous association (Donnellan *et al.*, 1975). There is, however, little doubt that the investigation of ligand receptor interaction is greatly facilitated by having the receptor in aqueous solution; the problems presented by the detergents necessary to maintain that solution being considerably fewer than those stemming from the organic solvents of the proteolipid solutions.

A THE GLUTAMATE RECEPTOR

1 Glutamate-binding proteolipids

The proteolipid isolation procedures of De Robertis and his colleagues have been applied to the whole muscle of the locust *Schistocerca gregaria* (Lunt, 1973; James *et al.*, 1974) and of the shrimp *Artemesia longinaris* (Fiszer de Plazas and De Robertis, 1973, 1974).

The total intrathoracic muscles from locusts were extracted with chloroform:methanol 2:1 (v/v) and the lipid extract was then chromatographed on a column of Sephadex LH20 after the addition of radiolabelled glutamate. Figure 1 shows the elution pattern obtained and the presence of two peaks of proteolipid eluted with chloroform that show glutamate-binding. In all experiments of this kind, it is important to establish that free radio-labelled ligand does not elute from the column in an identical position to the bound ligand. Control experiments show that free radio-labelled glutamate can only be eluted from the Sephadex LH20 by the application of chloroform:methanol 2:1 (v/v), therefore there is no question of free glutamate contributing to the bound peak. It is not yet clear whether there are differences between the two glutamate-binding peaks; their specific binding activity appears to be the same and in some experiments no separation is achieved. It was found that neither peak shows binding of glutamine or aspartate under the same chromatographic conditions (Lunt, 1973). Binding studies have been carried out over a wide range of glutamate concentrations and a dissociation constant K_d of $1·45 \times 10^{-5}M$ is found for the glutamate-binding site of the combined binding peaks. At saturation, achieved at $10^{-4}M$, the protein binds 8 n mol glutamate/mg; equivalent to 1 mol of glutamate/125 000 g of protein.

Proteolipids which also show high-affinity glutamate-binding (Fiszer de Plazas and De Robertis, 1973, 1974) have been isolated from the muscles of the shrimp, *Artemesia longinaris*. A single peak of proteolipid which eluted from the Sephadex LH20 in chloroform binds glutamate with a K_d of $1·3 \times 10^{-5}M$. Saturation of binding occurs at a glutamate concentration of $10^{-4}M$ at which point the protein binds 3·0 n mol glutamate/mg; this is equivalent to 1 mol of glutamate/333 000 g of protein. It is interesting to note that although the K_d of $1·45 \times 10^{-5}M$ for the glutamate-binding proteolipid from locust is almost identical with the value for the shrimp protein, the apparent molecular weight of the binding protein is considerably lower.

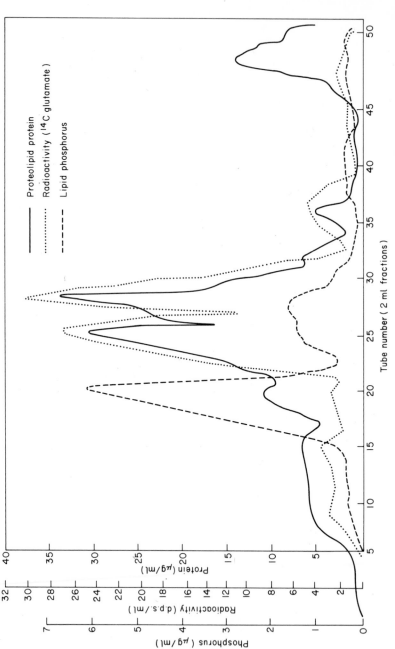

Fig. 1. Chromatography of locust (*Schistocerca gregaria*) muscle proteolipids on Sephadex LH20—glutamate binding. Fresh muscle was homogenized in 19 volumes of chloroform:methanol (2:1) (v/v) and the homogenate filtered. 1 μCi of L-[U-^{14}C] glutamate was added to 2 ml of the concentrated filtrate and after standing for 15 min the sample was chromatographed. Elution was made with 30 ml chloroform followed by 20 ml each of chloroform:methanol, 15:1, 10:1, 6:1 and 4:1 (v/v). Protein, lipid phosphorus and radioactivity were measured in each 2 ml fraction.

A glutamate-binding proteolipid has also been isolated from housefly muscle. Chromatography of a chloroform:methanol 2:1 (v/v) extract of housefly thorax revealed a single peak of protein eluted with chloroform which showed glutamate-binding. The peak was eluted slightly before the binding peaks from the locust muscle, but bound comparable amounts of glutamate (James, *et al.*, 1974). Recently, a preliminary study has been made of a glutamate-binding proteolipid from housefly leg muscle (Lunt and De Robertis, unpublished findings). The fly legs were homogenized in distilled water to give a 10% (w/v) homogenate which was then filtered to remove the chitinous exoskeleton. The filtrate was then lyophillized and proteolipid extracts made from the dry material. Chromatography on LH20 Sephadex in the presence of radio-labelled glutamate gave the pattern shown in Fig. 2 in which a single sharp peak of glutamate emerges with a peak of protein in the chloroform eluate. A large preparation of this proteolipid fraction was then made and the binding of glutamate measured by the chromatographic procedure. Saturation of binding occurs at a concentration of

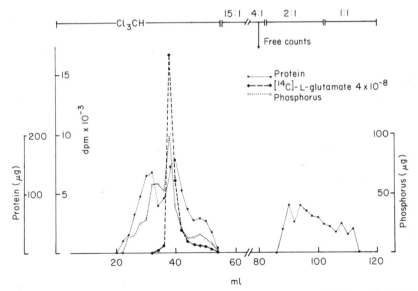

Fig. 2. Chromatography of fly (*Musca domestica*), leg muscle proteolipids on Sephadex LH20-glutamate binding. A proteolipid extract was prepared from lyophillized fly leg muscle and after the addition of radio-labelled glutamate chromatographed as in Fig. 1. Free glutamate emerges from the column only after the addition of chloroform:methanol 2:1 (v/v).

5×10^{-5}–10^{-4}M at which point the protein binds 2·8 nmol glutamate/ mg; this is equivalent to 1 mol of glutamate/357 000 g protein. The K_d of the binding site is 8×10^{-6}M and, as in the case of the proteolipids from locust and shrimp muscle, there appears to be a single type of binding site.

It is clearly not possible at this stage to unequivocally identify these glutamate-binding proteolipids with the postsynaptic glutamate receptor. It is known that there are high-affinity glutamate uptake systems associated with the sheath cells of the insect neuromuscular junction (Faeder and Salpeter, 1970) and there may be glutamate transport systems associated with the mitochondria of the muscles. There is at present little information available on the affinities of these systems for glutamate relative to that of the synaptic receptor and the possibility remains that the glutamate binding proteolipids may be derived from sources other than the postsynaptic membrane. There is however some pharmacological evidence which supports the hypothesis that these proteolipids may in part be representative of the postsynaptic glutamate receptor.

2   *The glutamate-binding proteolipids—pharmacological properties*

As stated earlier the characterization of glutamate receptors is greatly hindered by the lack of compounds having well-characterized pharmacological effects at the glutamate—mediated synapse. There have been reports of compounds having potent glutamate agonist properties at the invertebrate neuromuscular junction (Shinozaki and Shibuya, 1974), but it is still not certain whether these agents exert their effect solely at the level of the postsynaptic glutamate receptor. The compounds, α-methyl ester of DL-glutamic acid and the diethyl ester of L-glutamate have been reported to prevent the excitatory effect of L-glutamate in vertebrate neurones (McLennan et al., 1971), and Wheal and Kerkut (1974) have shown the diethyl ester to be an effective reversible antagonist of glutamate at the neuromuscular junction of the Hermit crab, *Eupagurus bernhardus*. Roberts (1974) has reported that in rat cerebral cortex synaptic membranes the diethyl ester produces a marked inhibition of the binding of glutamate to the receptor site but without effect on the binding of glutamate to uptake sites. It is interesting therefore to note that Fiszer de Plazas and De Robertis (1974) find the diethyl ester to be a strong inhibitor of glutamate-binding to the shrimp muscle proteolipid. At a concentration of 10^{-4}M diethyl ester the binding of

3.0×10^{-7}M glutamate was reduced by 78%; a 62% reduction of glutamate-binding was achieved with 1×10^{-4}M α-methyl DL-glutamate. These findings lend some support to the suggestion that the binding-proteolipid constitutes part of the postsynaptic glutamate receptor. Clearly however there is still a need for well-characterized agonists and antagonists which will compete specifically for the post-synaptic glutamate receptor.

A compound which may prove to be of great use in characterizing the glutamate receptor is 2-amino 4-phosphono butyric acid (APB). Figure 3 shows the compound to be closely structurally related to glutamate. Preliminary studies showed that APB inhibits the binding of glutamate to proteolipids from locust and housefly muscle (James et al., 1974), and it also inhibits the binding of glutamate to the proteolipid from housefly leg (Lunt and De Robertis, unpublished observations). Further studies (Cull-Candy et al., 1976) have confirmed that DL-APB is an effective inhibitor of glutamate-binding to the proteolipid from locust muscle and Table 1 shows the effect of DL-APB on the binding of 10^{-5}M glutamate to this preparation.

TABLE 1

The effect of DL-2 amino 4-phosphono butyric acid on the
binding of glutamate to a proteolipid from locust (*Schistocerca gregaria*) muscle

APB concentration M	Glutamate concentration M	% reduction of glutamate binding
10^{-7}	10^{-5}	42·1
10^{-6}	10^{-5}	53·2
10^{-4}	10^{-5}	80·1
10^{-3}	10^{-5}	96·3

In each experiment, proteolipids were prepared from about 2g of the total intrathoracic muscles of adult locusts. The proteolipid extract was incubated with the APB for 1h before the addition of radio-labelled glutamate after which chromatography on Sephadex LH20 was carried out as in Fig. 1.

An apparent inhibition constant K_i has been calculated from these data and a value of 1.09×10^{-4}M is obtained. It is found that pre-treatment of the proteolipid with DL-APB for periods of up to 1h before adding the glutamate produces greater inhibition than simultaneous exposure to DL-APB and glutamate (James et al., 1974). The effects of

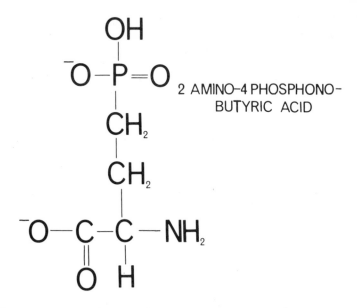

2 AMINO-4 PHOSPHONO-
BUTYRIC ACID

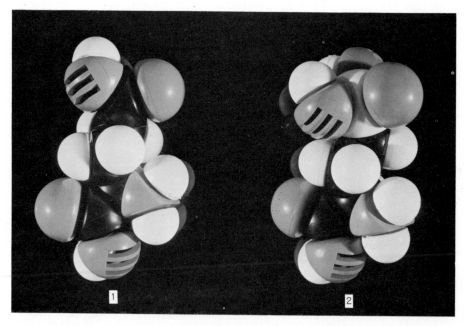

Fig. 3. 2-amino 4-phosphono butyric acid. The CPK space-filling models show (1) glutamate and (2) 2-amino 4-phosphono butyric acid. The close structural similarity of the two molecules can be seen.

APB on the binding of glutamate to the proteolipid are consistent with it behaving as a competitive inhibitor of glutamate-binding and the question arises as to the physiological effects of APB at the neuromuscular junction. Iontophoretic application of APB to a preparation of locust *extensor tibia* suggests that it is a glutamate antagonist at this site (Cull-Candy *et al.*, 1976).

3 *Detergent-solubilised glutamate-binding proteins*

Extraction of a total membrane fraction of locust muscle with aqueous solutions of deoxycholate as described by Changeux *et al.* (1970) results in the solubilization of about 20% of the muscle membrane proteins. Binding of radio-labelled glutamate and other ligands to the detergent-solubilized fraction was measured by equilibrium dialysis using the procedure employed by Changeux and co-workers for the assay of the detergent-solubilized cholinergic receptor. The detergent extract showed high affinity glutamate-binding, a K_d of $5 \times 10^{-7}M$ being obtained (Lunt, 1973). No binding of glutamine or aspartate to this preparation is seen until ligand concentrations of $10^{-4}M$ are reached, and even at $10^{-2}M$ considerably less ligand is bound than in the case of glutamate.

B THE GABA RECEPTOR

1 *GABA-binding proteolipids*

Proteolipids isolated from shrimp muscle (Fiszer de Plazas and De Robertis, 1973; De Robertis and Fiszer de Plazas, 1974) and from house-fly leg muscle (Lunt and De Robertis, unpublished observations) show specific GABA binding as measured by the Sephadex LH20 column procedure. In the case of the shrimp muscle proteins, it was found that the GABA binding protein was not completely separated from the glutamate binding protein. Precipitation of the proteolipids with diethyl ether at $-20°C$ results in a clear separation of the two binding activities. Chromatography of the diethyl ether precipitate reveals the presence of the GABA binding proteolipid while the glutamate binding protein is found exclusively in the diethyl ether supernatant. The GABA binding proteolipid shows saturation of binding of radio-labelled GABA at about 11·5 n mol GABA/mg protein with a single type of site having a K_d of $8 \times 10^{-6}M$. In the case of the proteolipid from housefly leg the GABA binding was seen with two peaks of protein; as may be seen in Fig. 4, the specific binding activity of the second peak of protein is

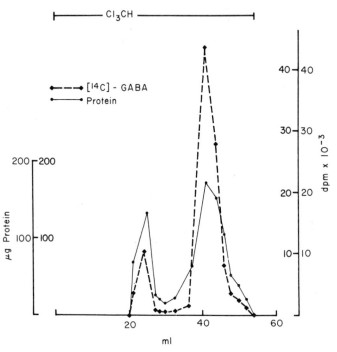

Fig. 4. Chromatography of fly (*Musca domestica*), leg muscle proteolipids on Sephadex LH20-GABA binding. The proteolipids were extracted as in Fig. 2 and chromatographed after the addition of radio-labelled GABA. Two binding peaks are seen, the second of which elutes in an identical position to the glutamate binding peak seen in Fig. 2. As in the case of glutamate, free GABA was only eluted after the addition of chloroform:methanol 2:1 (v/v).

many times higher than that of the first. However, the second peak of protein appears to be identical with that showing glutamate-binding as shown in Fig. 2. Binding studies carried out over a wide range of GABA concentrations show that saturation of binding occurs at about 3·2 n mol/mg protein, considerably lower than the value for the shrimp muscle proteolipid. The binding protein appears to have a single type of binding site with a K_d of about $6 \times 10^{-6}M$, a value close to that of the shrimp muscle protein. At the present time it is not possible to separate the glutamate and GABA binding activities of the housefly leg proteolipid fraction. In the case of the shrimp muscle De Robertis and Fiszer de Plazas (1974) have shown that L-glutamate does not bind to the GABA binding proteolipid. It may be therefore that the single peak of proteolipid isolated from housefly leg is composed of two separate binding proteins having similar solubility and chromatographic properties but showing different binding characteristics.

Some studies of the pharmacological properties of the shrimp muscle proteolipids have been made (Fiszer de Plazas and De Robertis, 1973; De Robertis and Fiszer de Plazas, 1974). Bicuculline was found to greatly inhibit the binding of GABA, 10^{-5}M bicuculline reducing the binding of 8×10^{-6}M GABA by 80%. Picrotoxin under the same conditions produced only a 23% reduction in GABA binding. Muscimol which is reported to have a GABA-like effect on vertebrate neurons (Johnston et al., 1968) reduces the binding to the same extent as picrotoxin.

It is however quite clear that these agents do not exert their effects solely by interaction with the postsynaptic GABA receptor (see discussion below). Therefore, although the observed inhibition confirms the specific nature of the GABA binding, it does not provide unequivocal proof that the binding protein is the postsynaptic GABA receptor.

2 Muscle membrane GABA binding sites

In an attempt to define more precisely the nature of the postsynaptic GABA receptor, Olsen and his colleagues have measured GABA binding to membranous sub-fractions of invertebrate muscle and of vertebrate brain (Olsen, 1975; Olsen et al., 1975a, b). Particulate membrane fractions of crayfish (Procambarus clarkii) tail muscle were examined for GABA binding by an equilibrium dialysis procedure. A low level of high-affinity GABA binding is seen which involves a single class of binding site with a K_d of about 10^{-6}M. Binding was also seen with preparations from the muscles of lobster, crabs and shrimps but not with cockroach (Periplaneta americana) muscle. The concentration of high-affinity sites is of the order of 200 pmol/g tissue, a figure which is very similar to the value of 247 pmol/g tissue obtained by De Robertis and Fiszer de Plazas (1974) for the concentration of GABA-binding proteolipid sites in shrimp muscle. These values are considerably higher than the 1–10 pmol/g tissue of cholinergic receptor sites found in vertebrate skeletal muscle and may be a reflection of the multiple innervation of the arthropod muscle.

Olsen and his colleagues have carried out many experiments to try to distinguish between postsynaptic receptor sites and GABA transport sites. Uptake of GABA into membrane bound vesicles has been eliminated from the experiments as a possible contributory factor to their

TABLE 2

Characteristics of receptor preparations from arthropod muscle

Organism	Receptor preparation	Specific binding activity nmol glutamate/mg protein	K_d(M)	Amount	Pharmacological properties	Ref.
Glutamate						
Schistocerca gregaria total intrathoracic muscles	proteolipid peak from Sephadex LH20	8·0	$1·45 \times 10^{-5}$	10 µg/mg total muscle protein	shows no binding of glutamine or aspartate glutamate binding inhibited by APB	1, 2
Musca domestica total leg muscles	proteolipid peak from Sephadex LH20	2·8	8×10^{-6}	—	glutamate binding inhibited by APB	3
Artemesia longinaris tail muscles	proteolipid peak from Sephadex LH20	3·0	$1·3 \times 10^{-5}$	0·3µg/mg total muscle protein	shows no binding of glutamine, aspartate or GABA. Glutamate binding inhibited by methyl glutamate and diethyl glutamate	4, 5
Schistocerca gregaria total intrathoracic muscles	deoxycholate-solubilized fraction of total muscle membranes	16·6	5×10^{-7}	0·2 µg/mg total muscle protein	shows no binding of glutamine or aspartate	1

GABA

Artemesia longinaris tail muscles	proteolipid peak from Sephadex LH20	11·5	8×10^{-6}	0·25 µg/mg total muscle protein	shows no binding of glutamate. GABA binding inhibited by bicuculline, picrotoxin and musicimol	4, 6
Musca domestica total leg muscles	proteolipid peak from Sephadex LH20	3·2	6×10^{-6}	—	GABA binding inhibited by bi-cuculline and picrotoxin	3
Procambarus clarkii abdominal and claw muscles	particulate membrane fraction	200 pmol/g fresh tissue	10^{-6}	—	GABA binding inhibited by D-glutamate, imidazole acetic acid, sulphydryl reagents; weakly inhibited by bicuculline, chlorpromazine, imipramine. GABA binding insensitive to Na concentration	7

Refs: 1, Lunt (1973). 2, James *et al.* (1974). 3, Lunt and De Robertis, unpublished findings. 4, Fiszer de Plazas and De Robertis (1973). 5, Fiszer de Plazas and De Robertis (1974). 6, De Robertis and Fiszer de Plazas (1974). 7, Olsen *et al.* (1975b).

observed GABA binding. They find that the high-affinity binding is unaffected by repeated freezing and thawing or vigorous homogenization, neither do metabolic inhibitors show any effect. All of these procedures greatly reduce GABA binding to a mouse brain membrane fraction and Olsen concludes that in this case the observed "binding" is, in fact, uptake into a closed vesicular fraction. Interestingly, they find that in the crayfish muscle preparation GABA binding is not very sensitive to sodium concentration and is still seen at zero sodium levels. This insensitivity of the postsynaptic receptor site to sodium is the basis for distinguishing between the glutamate receptor and glutamate transport sites in rat cerebral cortex synaptic membranes recently described by Roberts (1974, 1975). Olsen and co-workers find little effect of bicuculline and picrotoxin on their GABA binding preparation, a finding which contrasts with the effects seen on the GABA binding proteolipid (De Robertis and Fiszer, 1974). However, the site at which these GABA blocking agents exert their physiological effects is not known. In crustacean muscle an inhibition of GABA-induced conductance changes is produced by both bicuculline and picrotoxin but their effects are non-competitive with GABA (Takeuchi and Onodera, 1972; Takeuchi and Takeuchi, 1969). Bicuculline is reported to compete with GABA for the synaptic receptor site on vertebrate central neurons (Curtis et al., 1970) and inhibits GABA binding to putative receptor sites in rat brain (Zukin et al., 1974). It is, however, quite likely that these agents have different effects on the arthropod system. Certainly, in the case of the cholinergic receptor, the pharmacological properties of the preparation from fly-head are quite different from those of any of the vertebrate preparations (Cattell and Donnellan, 1972; Donnellan et al., 1975). The extrapolation of data obtained in vertebrate systems to the arthropods must be made with caution and it may well be that the biochemical characterization of the arthropod receptors must await advances in our knowledge of the pharmacology and physiology of these systems. The current data on the receptor characteristics are summarized in Table II.

III   Acknowledgements

I am grateful to my colleague Richard James who has carried out much of the original work described here. Thanks are also due to J. F. Donnellan and to R. W. Olsen for making copies of their papers available prior to publication. The work carried out with E. De Robertis was done under the auspices of the Royal Society Latin American Exchange Programme in conjunction with CONICET, Argentina.

IV References

Cattel, K. J. and Donnellan, J. F. (1972). The isolation of an acetylcholine- and decamethonium-binding protein from housefly heads. *Biochem. J.* **128**, 187–189.

Changeux, J.-P. (1975). The cholinergic receptor protein from fish electric organ. *In* "Handbook of Psychopharmacology" (S. Snyder and L. Iversen, ed) (in press).

Changeux, J.-P., Kasai, M., Huchet, M. and Meunier, J.-C. (1970). Extraction à partir du tissu électrique du gymnote d'une protéine présentant plusieurs propiétés caractéristique du récepteur physiologique de l'acetylcholine. *C.R. Acad. Sci. Paris* **270**, Série D, 2864–2867.

Changeux, J.-P., Kasai, M. and Lee, C. Y. (1970). Use of a snake venom toxin to characterize the cholinergic receptor protein. *Proc. Nat. Acad. Sci. U.S.A.* **67**, 1241–1247.

Cull-Candy, S. G., Donnellan, J. F., James, R. W. and Lunt, G. G. (1976). 2-amino-4-phosphonobutyric acid as a glutamate antagonist on locust muscle. *Nature Lond.* **262**. 408–409.

Curtis, D. R., Duggan, A. W., Felix, D. and Johnston, G. A. R. (1970). Bicuculline and central GABA receptors. *Nature Lond.* **228**, 676–677.

De Robertis, E. (1971). Molecular biology of synaptic receptors. *Science* **171**, 963–971.

De Robertis, E. (1975). Synaptic receptors. "Isolation and Molecular Biology." Marcel Dekker, New York.

De Robertis, E., Fiszer, S. and Soto, E. (1967). Cholinergic binding capacity of proteolipids from isolated nerve ending membranes. *Science* **158**, 928–929.

De Robertis, E. and Fiszer de Plazas, S. (1974). Isolation of hydrophobic proteins binding neurotransmitter amino acids: aminobutyric acid receptor of the shrimp muscle. *J. Neurochem.* **23**, 1121–1125.

Donnellan, J. F., Jenner, D. W. and Ramsey, A. (1974). Subcellular fractionation of freshfly flight muscle in attempts to isolate synaptosomes and to establish the localization of glutamate enzymes. *Insect Biochem.* **4**, 243–265.

Donnellan, J. F., Jewess, P. J. and Cattell, K. J. (1975). Subcellular localisation and properties of a cholinergic receptor isolated from housefly heads. *J. Neurochem.* (in press).

Faeder, I. R. and Salpeter, M. M. (1970). Glutamate uptake by a stimulated insect nerve muscle preparation. *J. Cell. Biol.* **46**, 300–307.

Fiszer de Plazas, S. and De Robertis, E. (1972). Isolation of a proteolipid from spleen capsule binding ($\pm$)-[^{3}H] norepinephrine. *Biochim. Biophys. Acta* **266**, 246–254.

Fiszer de Plazas, S. and De Robertis, E. (1973). Hydrophobic proteins isolated from crustacean muscle having glutamate and γ-amino butyrate receptor properties. *F.E.B.S. Letts.* **33**, 45–48.

Fiszer de Plazas, S. and De Robertis, E. (1974). Isolation of hydrophobic proteins binding neurotransmitter aminoacids. Glutamate receptor of the shrimp muscle. *J. Neurochem.* **23**, 1115–1120.

Gerschenfeld, H. M. (1973). Chemical transmission in invertebrate central nervous systems and neuromuscular junctions. *Physiol. Revs.* **53**, 1–119.

James, R. W., Lunt, G. G. and Donnellan, J. F. (1974). Glutamate receptors from insect muscle. Proc. 9th F.E.B.S. Meeting, p. 258. Hung. Biochem. Soc. Budapest.

Johnston, G. A. R., Curtis, D. R., Groat, W. C. and Duggan, A. W. (1968). Central actions of ibotenic acid and muscimol. *Biochem. Pharmac.* **17**, 2488–2489.

Karlin, A. (1974). The acetylcholine receptor: progress report. *Life Sciences* **14**, 1385–1415.

La Torre, J. L., Lunt, G. S. and De Robertis, E. (1970). Isolation of a cholinergic proteolipid receptor from electric tissue. *Proc. Nat. Acad. Sci. U.S.A.* **65**, 716.

Lee, C. Y. and Chang, C. C. (1966). Modes of action of purified toxins from elapid venoms on neuromuscular transmission. *Mem. Inst. Butantan Simp. Internac.* **33**, 555–572.

Levinson, S. R. and Keynes, R. D. (1972). Isolation of acetylcholine receptors by chloroform-methanol extraction: Artifacts arising in use of Sephadex LH20 columns. *Biochim. Biophys. Acta* **288**, 241–247.

Lunt, G. G. (1973). Hydrophobic proteins from locust (*schistocerca gregaria*) muscle with glutamate receptor properties. *Comp. Gen. Pharmacol.* **4**, 75–79.

Lunt, G. G. (1975). Synaptic transmission in Insects. *In* "Insect Biochemistry and Function" (D. J. Candy and B. A. Kilby, eds.), pp. 285–306. Chapman and Hall, London.

McLennan, H., Marshall, K. C. and Huffman, R. D. (1971). The antagonism of glutamate action at central neurones. *Experientia* **27**, 1116.

Olsen, R. W. (1975). Approaches to study of GABA receptors. *In* "GABA in nervous system function" (E. Roberts, T. N. Chase and D. Tower, eds). Raven Press, New York.

Olsen, R. W., Bayless, J. D. and Ban, M. (1975a). Potency of inhibitors for γ-amino butyric acid uptake by mouse brain subcellular particles at $0°$. *Molec. Pharm.* **11**, 558–565.

Olsen, R. W., Lee, J. M. and Ban, M. (1975b). Binding of γ-amino butyric acid to crayfish muscle and its relationship to receptor sites. *Molec. Pharm.* **11**, 566–577.

Roberts, P. J. (1974). Glutamate receptors in the rat central nervous system. *Nature Lond.* **252**, 399–401.

Roberts, P. J. (1975). Glutamate binding to synaptic membranes—detection of postsynaptic receptor sites. *J. Physiol.* **247**, 44–45P.

Shinozaki, H. and Shibuya, J. (1974). A new potent excitant, Quisqualic acid: Effects on crayfish neuromuscular junction. *Neuropharmacology* **13**, 665–672.

Takeuchi, A. and Onodera, K. (1972). Effect of bicuculline on the GABA receptor of the crayfish neuromuscular junction. *Nature New Biol.* **236**, 55–56.

Takeuchi, A. and Takeuchi, N. (1965). Localised action of gamma-aminobutyric acid on the crayfish muscle. *J. Physiol.* **177**, 225–238.

Wheal, H. V. and Kerkut, G. A. (1974). The effect of diethyl ester L-glutamate on evoked excitatory junction potentials at the crustacean neuromuscular junction. *Brain Res.* **82**, 338–340.

Zukin, S. R., Young, A. B. and Snyder, S. H. (1974). Gamma-aminobutyric acid binding to receptor sites in the rat central nervous system *Proc. Nat. Acad. Sci. U.S.A.* **71**, 4802–4807.

8

Excitatory and Inhibitory Transmitter Actions at the Crayfish Neuromuscular Junction

A. Takeuchi

Dept. of Physiology, School of Medicine, Jutendo University, Hongo, Tokyo, Japan

I Introduction

Since the specific inhibitory and excitatory actions of amino acids were first proposed about 20 years ago (Hayashi, 1954, 1956; Bazemore *et al.*, 1956, 1957), much attention has been paid to the action of amino acids on the peripheral as well as on the central nervous systems. For several reasons, much effort has been paid to this problem. First, it has been postulated that some of these amino acids may act as a neural transmitter. γ-Amino butyric acid (GABA) has been identified as an inhibitory transmitter at the crustacean neuromuscular junction. Furthermore, it is very likely but still not conclusively shown, that GABA acts as the inhibitory transmitter in the central nervous systems. Glutamic acid is a leading candidate for the excitatory transmitter in the crustacean and insect neuromuscular junctions and in vertebrate central synapses. Second, the action of transmitter is a special instance of a chemoreceptor process. The molecular structure of amino acids is relatively simple and studies on the action of amino acids may provide information about the basic mechanisms of drug-receptor interactions. It is further hoped that such studies may give fundamental information about the mechanism of permeability changes in the membrane.

For the basic studies on the actions of amino acids, the crayfish nerve muscle preparation is an adequate material, because its anatomy is relatively simple and the synaptic mechanisms have been well established (Fatt and Katz, 1953a, b; Boistel and Fatt, 1958; Hoyle and Wiersma, 1958a, b; Dudel and Kuffler, 1961a, b, c). Although the role of the invertebrate neuromuscular junction differs from that of other synapses, the basic mechanisms may probably be the same and detailed knowledge about the responses at the neuromuscular junction may be useful in understanding the synaptic mechanism in neurones.

This chapter will mainly be concerned with biophysical aspects of the transmitter action. Biochemical and pharmacological studies on amino acids have been presented in several recent reviews (Johnson, 1972; Otsuka, 1972; Gerschenfeld, 1973; Curtis and Johnston, 1974; Krnjević, 1974).

II   Innervation of the crayfish muscle

The neuromuscular system of crustacea is characterized by the fact that the whole muscle is provided by very few axons. Histological and electrophysiological observations show that muscle fibres receive one to four excitatory and one to two inhibitory axons. The pattern of innervation differs according to the species and individual muscles (Wiersma, 1961; Bullock and Horridge, 1965). The excitatory and inhibitory axons usually run side by side and they divide profusely and make endings at many places on the muscle fibre (Van Harreveld, 1939; Fatt and Katz, 1953b). Electron miscroscopic observation showed that, in the fast muscles, synapses are present in sarcolemmal invaginations a few microns away from the muscle fibre surface, while in the tonic muscles they are near the surface of the muscle fibre (Atwood and Morin, 1970).

When the muscle receives more than one motor axon, it often generates a fast or a slow contraction depending on which axon is stimulated. Different responses of the muscle fibre to each type of nerve stimulation are due to several factors: contractile characteristics, membrane properties of the muscle fibre and the release pattern of transmitter. Different transmitters are not likely to be involved for slow and fast contractions. Varieties of responses of the muscle fibre have been discussed by Atwood (1967).

III   Excitatory synapses

A   EXCITATORY POSTSYNAPTIC POTENTIAL

The excitatory postsynaptic potential (e.p.s.p.) was recorded with intracellular microelectrodes by Fatt and Katz (1953a, b) and Hoyle and

Wiersma (1958a) in various species of crustacea. The excitatory transmission has been most thoroughly studied on the opener muscle of the crayfish walking leg which is supplied with only two axons: one excitatory and one inhibitory. Single muscle fibres are innervated at 40–60 synapses (Van Harreveld, 1939; Dudel and Kuffler, 1961a). The amplitude of the e.p.s.p. is usually small, of the order of a few mV and rarely exceeds 10 mV. On repetitive stimulation, the e.p.s.p.s show remarkable facilitation which is a characteristic of the tonic muscle.

Dudel and Kuffler (1961a, b) recorded the e.p.s.p. with a microelectrode placed extracellularly at the neuromuscular junction. With this technique the synaptic current through the individual synaptic membrane could be measured (del Castillo and Katz, 1956a). It was found that the release of the transmitter from the nerve terminal is quantal and is facilitated by preceding nerve impulses. These observations are similar to those observed at the vertebrate neuromuscular junction (del Castillo and Katz, 1956b). The quantum content at the individual neuromuscular junction, was relatively small compared with that at the vertebrate neuromuscular junction. However, the quantum content differs greatly between different muscles and even between different regions of the same muscle (Atwood, 1967; Bittner, 1968).

It has been established that acetylcholine (ACh) is not involved in crustacean neuromuscular transmission (Katz, 1936; Grundfest et al., 1959). The concentration of free glutamate and aspartate is very high in the nervous system of invertebrates, compared to that in peripheral nerves of vertebrates (Lewis, 1952; Kravitz et al., 1963; Marks et al., 1970), and these free amino acids are likely to play an important role in the nervous function.

In 1959 Robbins reported that bath-applied glutamate enhanced neurally evoked contractions of crayfish muscle at relatively low concentrations (about $2 \times 10^{-5} M$) and, at higher concentrations (about $10^{-4} M$), a muscle contraction was induced. Van Harreveld and Mendelson (1959) observed that depolarization accompanied the muscle contraction. It was suggested that glutamate was acting at the neuromuscular junction.

B LOCALIZATION OF GLUTAMATE RECEPTOR

A precise localization of the drug action has been made by iontophoretic application of glutamate from a micropipette. This method, developed by Nastuk (1953) and del Castillo and Katz (1955), allows the application of drugs to very localized portions of the muscle cell. Usually glutamate is ejected as an anion, by passing inward current pulses through the glutamate-filled micropipette. When brief pulses of L-glutamate were applied to the surface of the muscle fibre, it produced a transient depolarization (glutamate potential), but only when applied to a circumscribed area on the muscle fibre. If the location of a neuro-

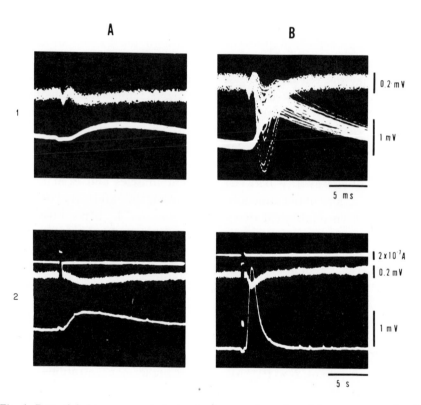

Fig. 1. Potential changes recorded at a single synaptic region of the opener muscle of the crayfish. 1. Upper traces, extracellularly recorded potential changes obtained by stimulating the inhibitory nerve (A) and the excitatory nerve (B) at a stimulation rate of 40 Hz; lower traces, simultaneous intracellular i.p.s.p. and e.p.s.p. 2. Upper traces, monitored injection currents for iontophoretic application of GABA (A) and L-glutamate (B); middle traces, extracellularly recorded potential changes of GABA (A) and L-glutamate (B) from a double-barrel micropipette; lower traces, simultaneous intracellular records of GABA potential (A) and glutamate potential (B). All records were obtained at the same synapse. (Takeuchi and Takeuchi, 1965.)

muscular junction was determined by the recording of e.p.s.p.s with an extracellular microelectrode, it was found that the glutamate sensitive spot critically coincided with the excitatory neuromuscular junction (Fig. 1) (Takeuchi and Takeuchi, 1964). Localization of glutamate sensitivity to the excitatory neuromuscular junction has also been observed in insect muscle fibres (Beránek and Miller, 1968; Usherwood and Machili, 1968; see also Chapter 9).

If the sensitive area is mapped by applying the same dose of glutamate to various spots along a muscle fibre, it is found to occupy discrete areas with the size of about 30–40 μm in diameter. The area covered by the applied glutamate depends on the diffusion distance from the tip of the glutamate-filled micropipette to the synaptic region. Correction for diffusion distance showed that the diameter of the sensitive area would be of the order of 10 μm. The minimal dose of L-glutamate which produced a depolarization of about 1 mV was about 5×10^{-17} mole (Takeuchi and Takeuchi, 1972). This is of the same order of magnitude as the amount of ACh which is necessary to produce an ACh potential ($1 \cdot 5 \times 10^{-17}$ mole) in the rat diaphragm muscle (Krnjević and Miledi, 1958). The separation between sensitive spots in crustacean fibres was of the order of 100 μm, but sometimes the sensitive areas were found in more close proximity. When large doses of glutamate were applied, irregular humps were sometimes observed on the falling phase of the glutamate potential. This may be due to the diffusion of glutamate to more distant synaptic regions (Takeuchi and Takeuchi, 1964; Beránek and Miller, 1968; Usherwood and Machili, 1968). Correspondingly, electron microscopic observations showed that the distribution of synapses on the crayfish muscle fibre is not uniform over the muscle surface; synapses are found singly, in pairs or in clusters (Atwood and Morin, 1970).

If the glutamate pipette enters a fibre by penetrating the membrane, a glutamate potential is no longer produced (Takeuchi and Takeuchi, 1964), indicating that the glutamate receptor is located on the outer surface of the membrane, as in the case of the ACh receptor (del Castillo and Katz, 1955).

In frog and snake muscles ACh sensitivity is critically localized to the neuromuscular junction (McMahan et al., 1972; Kuffler and Yoshikami, 1975), but a small and distinct sensitivity extends diffusely to the extra-synaptic region of the frog, rat and snake muscles (Miledi, 1960; Katz and Miledi, 1964; Feltz and Mallart, 1971; Kuffler and Yoshikami,

1975). In crayfish muscle, the extrasynaptic glutamate sensitivity has not been accurately tested. However, recent observations suggest the existence of extrasynaptic receptors which are different in character from those at the synaptic region. Shinozaki and Shibuya (1974b) found that when kainic acid (an anthelmintic principle obtained from red algae with a molecular structure similar to glutamate) is added to the bath solution (about 10^{-4}M), it enhances by up to five times the depolarization evoked by bath-applied L-glutamate, while the amplitude of e.p.s.p.s remains unchanged. Kainic acid itself does not produce a potential change.

In contrast to bath application, if L-glutamate is applied iontophoretically from a micropipette placed close to the synaptic region (probably within about 15 μm), the amplitude of the glutamate potential is not changed by kainic acid. As the distance between the tip of the micropipette and the synapse was increased, the glutamate potential was augmented by kainic acid up to three times that of control (Takeuchi and Onodera, 1975). When L-glutamate is applied from points distant from the synapse, more of the ejected L-glutamate will diffuse to extrasynaptic regions. If kainic acid sensitizes an extrasynaptic receptor which is not responsive to L-glutamate in the normal solution, the above observations may be explained. This possibility is further supported by the observation that if the sensitive area to L-glutamate is mapped on the muscle fibre, it spreads to the extrasynaptic region in the presence of kainic acid, although the response at the extrasynaptic region is much smaller than that at the synapse. Although the mode of action of kainic acid is not known, one possibility is that the binding of kainic acid to the receptor modifies its conformation, so that it binds L-glutamate. Another possibility is that the extrasynaptic receptor binds L-glutamate but that kainic acid causes a more favourable conformation for triggering the ionic conductance. In spinal neurones of cat and frog, kainic acid is an excitant up to 200 times more potent than L-glutamate (Johnston et al., 1973; Biscoe et al., 1975).

Recently, it has been reported that in the tonic muscle of the abdominal flexor of the crayfish the iontophoretic application of ACh, but not of L-glutamate, produces depolarization at the neuromuscular junction, and it has been suggested that in this muscle the transmitter may be ACh (Futamachi, 1972). However, no such effect of ACh could be detected, when the same muscle was reinvestigated (Takeuchi and Onodera, unpublished observation).

C THE REVERSAL POTENTIAL AND IONIC MECHANISMS

The permeability changes produced by the transmitter substance are reflected in the reversal potential or in the transmitter equilibrium potential (Ginsborg, 1967). It is usually difficult to measure the reversal potential for the excitatory transmitter action, largely for technical reasons. In crustacean and insect muscles some other factors make the measurement of the reversal potential even more difficult. Some of the difficulties are (a) the membrane resistance shows a marked non-linearity when the membrane potential is changed (Ozeki et al., 1966) and (b) the muscle fibre receives a widely distributed innervation and it is not possible to shift the membrane potential by applying current at a single point on the muscle fibre to measure the reversal potential (Burke and Ginsborg, 1956). These difficulties are overcome by inserting a metal electrode longitudinally into the muscle fibre and by employing the voltage clamp technique (Takeuchi and Onodera, 1973; Onodera and Takeuchi, 1975).

A stainless steel wire of about 70 μm in diameter was inserted longitudinally into the muscle fibre through the exoskeleton. About 80–90% of the muscle fibre length was kept equipotential with this technique. When the membrane potential was clamped at the resting potential, excitatory stimulation caused an inward current (excitatory postsynaptic current, e.p.s.c.). The properties of e.p.s.c. were essentially similar to those observed at the frog end-plate (Fig. 2). The amplitude

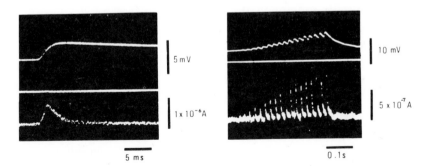

Fig. 2. E.p.s.p. and e.p.s.c. of the opener muscle of the crayfish claw. The membrane potential was clamped with a metal wire inserted longitudinally into the muscle fibre. Upper traces are e.p.s.p.s without voltage clamp. Middle traces show the clamped membrane potential and lower traces are e.p.s.c. (Onodera and Takeuchi, 1975).

of e.p.s.c. changed almost linearly with the membrane potential. The reversal potential measured by extrapolation was 10–20 mV positive with respect to the bath solution.

Iontophoretic application of L-glutamate to the voltage clamped muscle produced an inward current at resting potential (glutamate current). The reversal potential for L-glutamate was identical with that for the natural transmitter measured from the same muscle fibre (Takeuchi and Onodera, 1973; Dudel, 1974; Onodera and Takeuchi, 1975). Similar results were also observed by recording the extracellular potential changes produced by nerve stimulation and by the application of L-glutamate to a single junction (Taraskevich, 1971; Dudel, 1974).

It has been shown that the ions involved in the e.p.s.p. of crustacean muscles are mainly sodium (Dudel and Kuffler, 1960; Ozeki and Grundfest, 1967). More systematic measurements were made on the voltage clamped muscle fibre. Changes in the sodium concentration of the bath solution shifted the reversal potential in the direction expected from the Nernst equation, while changes in the potassium and chloride concentrations had little or no effect on the reversal potential. Therefore, in the crayfish muscle, the ions involved appear to be mainly sodium, and the contribution of potassium and chloride is small (Takeuchi and Onodera, 1973; Onodera and Takeuchi, 1975). Changing the external concentrations of sodium, potassium and chloride showed the same effects on the reversal potential of the glutamate current (Onodera and Takeuchi, 1976).

When all the sodium in the bath solution was replaced with impermeant cations, such as choline or Tris, the application of L-glutamate still produced a small but significant inward current at resting potential. Since the reversal potential for this current was at least several tens of mV more positive than the resting potential, this current is very likely to be carried by calcium. This is supported by the observation that the glutamate current in sodium-free solution was about doubled by increasing the external concentration of calcium from 3·6 to 36 mM. Thus the activated synaptic membrane may be permeable mainly to sodium and slightly to calcium (Takeuchi and Onodera, 1973; Dudel, 1974; Onodera and Takeuchi, 1975, 1976).

If sodium and calcium permeate through the activated membrane, a question arises: Do these ions pass through the membrane via the same channel or separate channels? When the calcium concentration was increased keeping the sodium concentration constant, the glutamate current was suppressed, instead of augmented as in the sodium-free solution (Dudel, 1974; Onodera and Takeuchi, 1976). Thus it may be tentatively assumed that calcium ions permeate through the sodium channel and interfere with the passage of sodium ions. Similar interactions have been observed at the frog end-plate membrane activated by ACh (Takeuchi, N., 1963). Although the role of calcium which enters through the synaptic membrane is not clear, calcium permeability may be a phenomenon presented at excitatory synaptic membranes in general. At the squid giant synapse it was observed that a synaptic potential can be obtained under a condition where calcium is practically the only cation in the external medium (Katz and Miledi, 1969), and that the presynaptic stimulation or application of L-glutamate produced the light emission from the postsynaptic axon after it had been injected with aequolin (Kusano et al., 1975). In this synapse, however, the reversal potential for the natural transmitter differs from that for L-glutamate; the reversal potential being $+28$ mV for the e.p.s.p. and -22 mV for L-glutamate (Miledi, 1969), which suggests that the natural transmitter is different from glutamate.

D POTENTIATION AND DESENSITIZATION OF GLUTAMATE RESPONSES

The L-glutamate-induced contraction and depolarization are transient and the responses decline while the muscle is perfused with glutamate (Van Harreveld and Mendelson, 1959; Robbins, 1959). This phenomenon or "desensitization" is observed in many types of drug action and has most thoroughly been studied with the action of ACh on the vertebrate neuromuscular junction (Thesleff, 1955; Katz and Thesleff, 1957; Manthey, 1966; Nastuk and Parson, 1970; Rang and Ritter, 1970; Miledi and Potter, 1971; see also Chapter 5). Katz and Thesleff (1957) assumed a refractory state of the receptor and Miledi and Potter (1971) observed that the binding of radioactive α-bungarotoxin was reduced by desensitization of the receptor.

While brief test pulses of L-glutamate were applied by a micropipette to a sensitive spot at constant intervals, conditioning doses of varying amount of L-glutamate were applied to the same spot from another glutamate pipette. After the start of steady efflux, the test responses showed at first an increase in amplitude, which was followed by a decrease. The increase in amplitude (potentiation) was more pronounced when the conditioning dose was relatively small. As the conditioning dose of L-glutamate was increased, the potentiation gave way to the desensitization. The onset and decline of desensitization was also measured by the double-pulse technique. It was found that desensitization developed in about 0·5–1·0 s after the conditioning pulse and that it lasted for a few seconds (Takeuchi and Takeuchi, 1964; Dudel, 1975a, b). These observations are similar to those of the ACh response at the vertebrate neuromuscular junction (Katz and Thesleff, 1957). Although the mechanism of desensitization is not clear, it may possibly be related to changes in receptor properties (see Chapter 5).

The receptor desensitization was used to identify the neuroreceptor. If the conditioning dose of L-glutamate desensitizes the glutamate response as well as the e.p.s.p., the receptor which responds to L-glutamate is likely to be identical with the neuroreceptor. It was found that the steady application of L-glutamate to the neuromuscular junction reversibly suppressed the amplitude of the e.p.s.p.s recorded extracellularly from the synapse (Takeuchi and Takeuchi, 1964).

Potentiation of the glutamate response may be related to the nonlinear summation of the glutamate action. This was tested by applying approximately equal doses of glutamate to the same spot from each barrel of a double-barrel glutamate pipette, separately or simultaneously. For smaller doses the combined response showed more than a simple additive effect and the dose-response relation showed a sigmoidal relation (Takeuchi and Takeuchi, 1964; Dudel, 1975b). Potentiation and a sigmoidal dose-response relation have previously been observed for the ACh receptor (Katz and Thesleff, 1957).

A sigmoidal dose-response relation was also suggested from the time course of the glutamate response produced by iontophoretic application. If the time course of the concentration of glutamate at the synapse is calculated from the diffusion equation, the time course of the glutamate potential (Takeuchi and Onodera, 1975) or the glutamate current (Dudel, 1975b; Onodera and Takeuchi, 1976) was always faster than that of the calculated concentration. A better fit was obtained, when square (Takeuchi and Onodera, 1975; Onodera and Takeuchi, 1976) or higher orders of the concentration (Dudel, 1975b) were plotted. The same observation has also been made for the glutamate response at the goldfish Mauthner cell (Diamond and Roper, 1973). These observations suggest that the glutamate response is a higher order reaction or that there is a cooperativity in producing the glutamate response.

When an isolated single muscle fibre of the crayfish is rapidly perfused with various concentrations of L-glutamate and the dose-response relation is measured, a sigmoidal curve is also obtained, the half maximal concentration being about 1×10^{-4}M and the Hill coefficient about 2 (Takeuchi and Iimura, unpublished observation).

E SPECIFICITY OF THE GLUTAMATE RECEPTOR

Bath-application of a variety of acidic amino acids demonstrated that L-glutamate is the most potent among related substances in crustacean and insect muscles (Robbins, 1959; Van Harreveld and Mendelson, 1959; Usherwood and Machili, 1968). Iontophoretic application to sensitive spots on the crayfish muscle from a double barrel micropipette confirmed these observations (Takeuchi and Takeuchi, 1964, 1970 and unpublished observations). One barrel of a double barrel micropipette was filled with L-glutamate and the other with the test drug. L-glutamate and test drugs were applied separately to the sensitive spot and then applied simultaneously to test the interaction between drugs. Drugs tested included D-glutamic acid, L-aspartic acid, L-α-amino-adipic acid, L-glutamine, α-ketoglutaric acid, L-cysteoic acid, DL-homo-cysteic acid and DL-β-hydroxyglutamic acid. It was found that these drugs are much less potent than L-glutamate. For example D-glutamate and L-aspartate are at least 250–400 times less effective than L-gluta-mate (Fig. 3). Simultaneous application of test drugs with L-glutamate

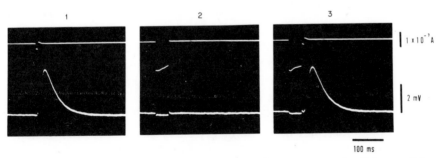

Fig. 3. Effects of L-glutamate and L-aspartate on the opener muscle of the crayfish. 1. Depolarization produced by L-glutamate ejected from one barrel of a double-barrel micropipette. 2. L-aspartate was ejected at the same spot from another barrel. 3. L-glutamate and L-aspartate at the same dose as in 1 and 2 when applied simultaneously to the same spot, little change in the glutamate potential was observed. Upper traces are monitored ejection currents of drugs. (Takeuchi and Takeuchi, unpublished.)

caused no appreciable change in the glutamate potential. This suggests that these drugs are not bound to the glutamate receptor and that the steric conformation of the drug is an important factor for binding. The high specificity of the glutamate receptor might be attributed to the fact that the L-glutamate molecule is bound to the receptor at three points: one amino and two carboxylic groups.

In contrast to the results with iontophoretic application, when L-aspartate was applied to the bath, L-aspartate could produce depolarization of crayfish and lobster muscle fibres, although L-aspartate was two to four times less potent than L-glutamate. The responses to L-aspartate were slower than those to L-glutamate and the depolarization produced by L-aspartate had little effect on the e.p.s.p. (Slater, cited in Kravitz et al., 1970). These results are still not explained, but have some similarities to the glutamate response in the presence of kainic acid. Depolarization produced by bath-applied L-aspartate might be due to its action on extrasynaptic receptors.

A similar argument might also be applied to the action of quisqualic acid which is 500–1000 times more effective than L-glutamate on the crayfish muscle, when applied to the bath. The time course of depolarization is substantially slower than that of L-glutamate (Shinozaki and Shibuya, 1974a). To elucidate a possible contribution of extrasynaptic receptors, a comparison of sensitivities using the iontophoretic micromethod may be necessary. (Actually they used iontophoretic application, but sensitivity was not measured.) Quisqualic acid is a very potent excitant in rat and frog spinal neurones (Biscoe et al., 1975).

If there is a specific antagonist to the glutamate receptor, this drug would be a very useful tool in characterizing the neuroreceptor. Several drugs have been proposed, e.g. the esters of L-glutamic acid, in particular γ-methylester, which reversibly block the e.p.s.p. and the glutamate potential, but do not affect the GABA response at the crayfish neuromuscular junction (Lowagie and Gerschenfeld, 1974). In spinal and cuneate neurones, however, the methylesters have stimulating actions (Haldeman and McLennan, 1972). In central neurones and the neuromuscular junctions different receptors or different processes in the later stages of the transmitter action may be involved.

F REMOVAL OF TRANSMITTER

After release from the nerve terminal, the transmitter is removed from the synaptic cleft via one of three processes: diffusion, enzymatic destruction or uptake. When L-glutamate was iontophoretically applied from a micropipette which was placed close to the synapse, the rise and fall of the glutamate potential was similar to that of the e.p.s.p. (Takeuchi and Takeuchi, 1972). It suggests that the activation and inactivation processes of L-glutamate may be similar to those of the neural transmitter. At the crustacean neuromuscular junction the mechanism of inactivation has not been determined. Diffusion may be an important process, but it is not known whether this process alone is rapid enough to remove the transmitter from the synaptic region, because the synapses are in many cases in the cleft a few microns away from the surface (Atwood and Morin, 1970). In locust muscle some inhibitors of glutamic decarboxylase (particularly phenylhydrazine) potentiates the neurally evoked contraction (Usherwood and Machili, 1968). In crayfish, however, the application of phenylhydrazine had no appreciable effect on the e.p.s.p. (Takeuchi and Takeuchi, unpublished observation). Since a specific uptake system for glutamate is found in the lobster nerve-muscle preparation, which is sodium dependent (Iversen and Kravitz, 1968), this process could serve to terminate the action of the transmitter. However, no positive evidence has, so far, been reported.

IV   Inhibitory synapses

A distinctive feature of the crustacean neuromuscular junction is the presence of inhibitory axons. There are two types of inhibitory actions: a presynaptic and a postsynaptic. The relative importance of these inhibitions seems to differ in individual muscles. For example in the opener muscle of the walking legs of the crayfish, presynaptic inhibition is a dominant mechanism (Dudel and Kuffler, 1961c), while in the abdominal slow flexor muscle the presynaptic mechanism seems to be absent (Kennedy and Evoy, 1966).

The evidence that GABA is inhibitory transmitter at the inhibitory neuromuscular junction of the crustacea is quite convincing (for recent reviews see Kravitz, 1967; Otsuka, 1972; Gerschenfeld, 1973). Some of this evidence is: the presence of high concentration of GABA in the inhibitory axon, but not in the excitatory axon of the lobster (Kravitz et al., 1963); glutamic decarboxylase is found in high concentration in inhibitory axons and is inhibited by GABA (Kravitz et al., 1965); GABA is released upon inhibitory stimulation but not upon excitatory stimulation (Otsuka et al., 1966); the action of GABA is the same as that of the neural transmitter (e.g. Kuffler, 1960; Dudel, 1965a; Takeuchi and Takeuchi, 1965). It has also been shown that there is a specific uptake system for GABA which is sodium dependent (Iversen and Kravitz, 1968).

V Postsynaptic inhibition

A INHIBITORY POSTSYNAPTIC POTENTIAL

The inhibitory postsynaptic potential (i.p.s.p.) was recorded intra-cellularly by Fatt and Katz (1953a) in crustacean muscle and an equivalent circuit of the synaptic membrane similar to that of the frog end-plate was proposed. The characteristics of the i.p.s.p. are essentially the same as those of the e.p.s.p., the main difference being their reversal potentials. In some crustacean inhibitory synapses the reversal potential is almost the same as the resting potential, and inhibitory stimulation produces no potential change (Fatt and Katz, 1953a). In other synapses, it is either at a slightly depolarized level (Dudel and Kuffler, 1961c; Takeuchi and Takeuchi, 1965) or at a hyperpolarized level (Moto-kizawa et al., 1969). Since the activated postsynaptic membrane is almost exclusively permeable to chloride and if chloride is passively distributed between the external and internal solutions, the reversal potential is expected to be identical with the resting potential. A difference between the reversal potential and the resting potential suggests that some mechanism operates to keep the concentration from an equilibrium value.

The release mechanism for the inhibitory transmitter from the nerve terminal is much less understood than that of the excitatory transmitter. This is mainly due to the small size of the i.p.s.p. However, the spontaneous appearance of i.p.s.p.s suggests that the release mechanism may also be quantal (Grundfest and Reuben, 1961; Takeuchi and Takeuchi, 1966b).

B LOCALIZATION OF GABA RECEPTOR

When GABA was applied iontophoretically to the muscle fibre, it was found that the action of GABA is critically localized to the inhibitory neuromuscular junction (Fig. 1). Injection of GABA into the interior of the muscle fibre was ineffective in producing a specific GABA response. The GABA receptor seems to be located on the outer surface of the muscle membrane, as in the case of ACh and glutamate receptors. The GABA sensitive spot was usually found very close to the glutamate sensitive spot (Takeuchi and Takeuchi, 1965). Since the molecular structure of GABA and glutamate have some similarities, it was conceivable that GABA was bound to the same receptor as glutamate and thus inhibited the glutamate action (Watkins, 1965). This scheme was attractive, but the experimental results suggest that different receptors are involved for the glutamate and GABA actions. This was tested in two ways. (a) After desensitization of the glutamate response and e.p.s.p.s by a prolonged application of glutamate, application of GABA or inhibitory stimulation still produced normal responses (Takeuchi and Takeuchi, 1965). (b) When the membrane potential was clamped and glutamate and GABA were applied to the neuromuscular junction, the glutamate current was not influenced by the application of GABA. The results indicate that the application of GABA opens channels which are in parallel to those activated by L-glutamate (Takeuchi and Takeuchi, 1966a).

When e.p.s.p.s and i.p.s.p.s recorded from the same synapse were compared, it was found that the rise and fall of the e.p.s.p. were much faster than those of the i.p.s.p. (Fig. 1) (Fatt and Katz, 1953a; Takeuchi and Takeuchi, 1965). The different time courses may be attributed to a difference either in the time course of the transmitter release or in the duration of the transmitter action. Experimental results suggest that the latter possibility is more likely, because when GABA and L-glutamate were applied to the same spot from a double-barrel micropipette, the rise and fall of the GABA potential were several times slower than those of the glutamate potential (Fig. 1). Similar differences in the time course between glutamate and GABA potentials were found in the Mauthner cell of the goldfish (Diamond and Roper, 1973). Binding of GABA to the receptor or the process at later stages may be slower than those in the glutamate response.

C CONDUCTANCE INCREASE AND IONS INVOLVED

The permeability changes produced by GABA and the inhibitory transmitter at the crustacean muscle were first demonstrated by Boistel and Fatt (1958) by measuring the reversal potential. The i.p.s.p. was measured from a muscle fibre, while membrane potential was varied by current injection. The removal of chloride (replacing it by organic anions) shifted the reversal potential in the depolarizing direction, while changes in the potassium concentration from 0 to 20 mM produced little or no change in the reversal potential. The action of GABA was affected in a similar way by changing the potassium and chloride concentrations. The comparison of the reversal potentials for GABA and for the inhibitory transmitter was one of the crucial tests in identifying GABA as the neural transmitter. When both the GABA potential and the i.p.s.p. were recorded from the same muscle fibre and the membrane potential was varied, they reversed their signs at the same membrane potential (Fig. 4) (Kuffler, 1960; Takeuchi and Takeuchi, 1965).

A quantitative measurement of the conductance increase was obtained by adding a known concentration of GABA to the external solution. (This type of experiment was possible, because the GABA receptor of the crayfish muscle shows no appreciable desensitization; Takeuchi and Takeuchi, 1965, 1975a.) The threshold concentration was about 10^{-5}M and the membrane conductance reached its maximal value at about 5×10^{-4}M, half maximal concentration being about 5×10^{-5}M. The absolute value of the membrane conductance varied from fibre to fibre, but it increased up to ten times of the resting membrane conductance in lobster and crayfish muscle (Grundfest et al., 1959; Takeuchi and Takeuchi, 1967). A similar value for the conductance increase was also observed, when the inhibitory nerve was stimulated at high frequency (150 Hz) (Dudel and Kuffler, 1961c). Although the GABA receptor of the crayfish muscle showed no appreciable desensitization, this may not be a general property of the GABA receptor. In crab muscle the GABA receptor in the post-synaptic membrane showed a remarkable degree of desensitization, while no desensitization was observed for presynaptic inhibition (Epstein and Grundfest, 1970; Parnas et al., 1975).

When the external chloride concentration was decreased the conductance change produced by GABA decreased almost linearly with the chloride concentration. When all chloride was removed, GABA and

Fig. 4. Reversal potential of i.p.s.p. and GABA potential. N, i.p.s.p.s produced by stimulating the inhibitory nerve at a stimulation rate of 30 Hz; GABA, potentials produced by iontophoretic ejection of GABA to the sensitive area of the same muscle fibre. Uppermost trace is the monitored ejection current. Bottom trace was recorded at the resting potential (-70 mV), and the muscle fibre was depolarized in 2 mV steps. The reversal potential was about -66 mV. (Takeuchi and Takeuchi, 1965.)

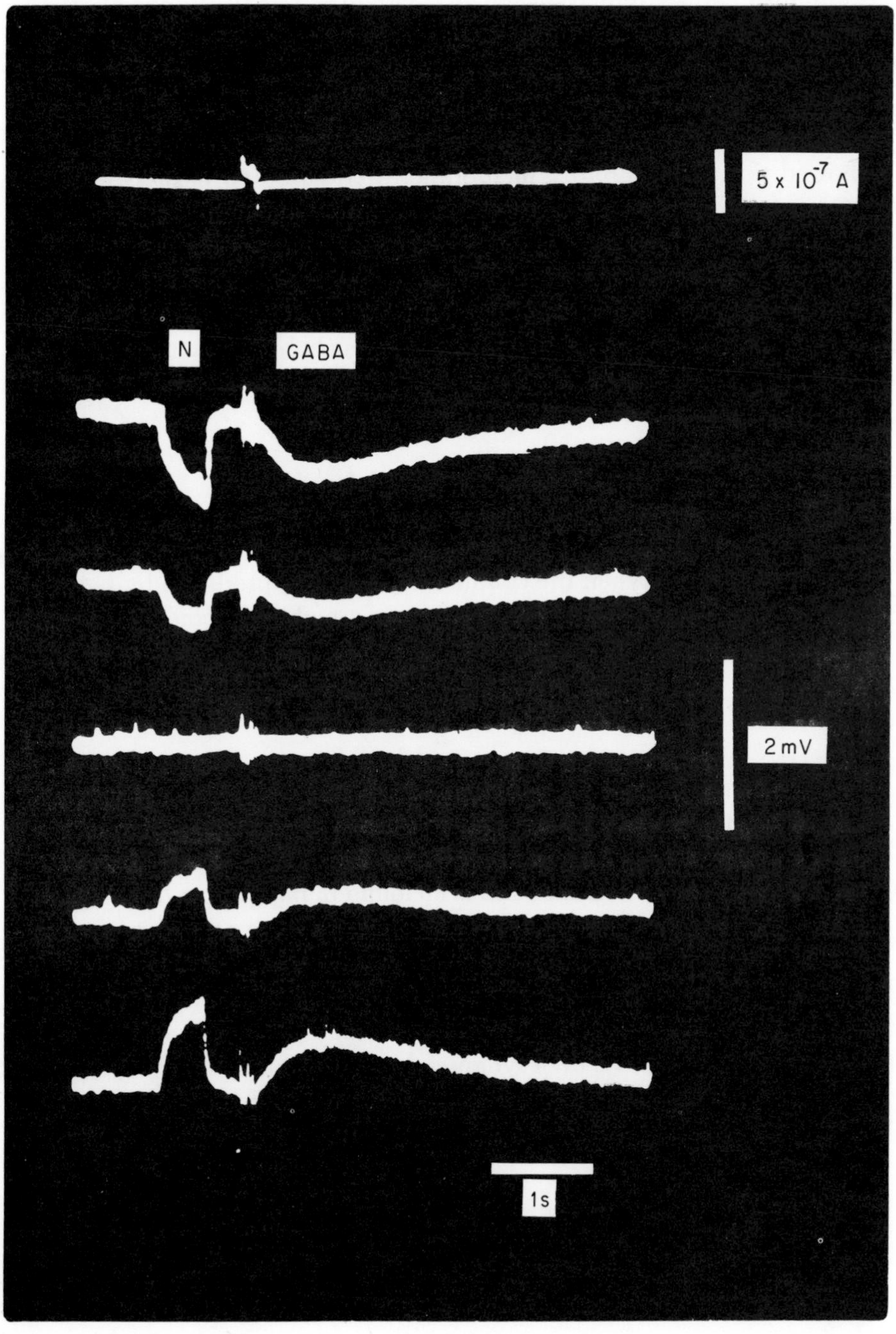

5 x 10⁻⁷ A
N
GABA
2 mV
1s

inhibitory nerve stimulation failed to produce a conductance change. Changes in sodium and potassium concentrations caused no effect on the GABA conductance, indicating that the inhibitory postsynaptic membrane is impermeable to cations. Replacing chloride with various anions of relatively small size, such as bromide, iodide or nitrate, showed that the inhibitory membrane in crayfish and lobster muscles had a variable degree of permeability to these anions (Takeuchi and Takeuchi, 1967; Motokizawa *et al.*, 1969). The inhibitory postsynaptic membrane, thus, appears to have the properties of an anion selective membrane. It is suggested that positive charges at the synaptic membrane control the permeation of ions (Takeuchi and Takeuchi, 1971a, b).

D BINDING TO THE GABA RECEPTOR

The first step of the transmitter action may be binding to the receptor. The transmitter-receptor complex then induces a permeability change of the membrane. When the conductance increase produced by GABA was measured in crayfish, the relation was highly sigmoidal and the response (y) is fitted by the equation

$$y = \frac{K_a}{K_a + A^n} \quad n > 1$$

where A is the concentration, K_a and n are constants. Hill coefficient (n) was 2–3 (Takeuchi and Takeuchi, 1967, 1969; Feltz, 1971). This relationship suggests some sort of cooperative phenomenon. A sigmoidal dose-response relation has been found in many types of synapses, although in many of these synapses the responses are depolarization or contraction and there is non-linear factors between the conductance change and the response (Werman, 1969; Rang, 1971). More recently has the dose-conductance relationship of the transmitter action been measured in various types of synapses and it was found to be sigmoidal (ACh receptor in frog end-plate, Rang, 1971; GABA receptor in stretch receptor, Swagel *et al.*, 1973; glutamate receptor in crayfish, Dudel, 1975b).

One approach to characterize the properties of the receptor is to test the structure-activity relation of various inhibitory substances, as has been done on the stretch receptor neurones (Kuffler and Edwards, 1958; Edwards and Kuffler, 1959). The conductance change of crayfish muscle produced by various drugs was measured and a dose-conductance relation was obtained (Takeuchi and Takeuchi, 1975a). The concentration necessary to produce a conductance change and the maximal conductance produced varied depending on the molecular structure of the drugs (Fig. 5). Lengthening or shortening the chain length of GABA decreased both the maximal conductance and the apparent affinity of the drug (estimated from the concentration which produced a half maximal conductance change). Addition of a neutral group to GABA decreased mainly its apparent affinity. Attachment of side groups possibly restricts the binding of the drug to the receptor.

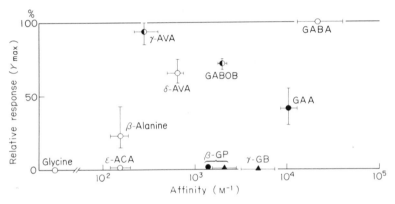

Fig. 5. Structure-activity relationship of GABA and related substances on the crayfish muscle. Ordinate: the maximal conductance increase produced by the substances relative to that produced by GABA in the same muscle fibre. Abscissa: reciprocal of the concentration of drugs which produced a half maximal conductance increase. Bars indicate the range. $\bigcirc$: ω-Amino acids $H_2N—(CH_2)_n—COOH$; Glycine ($n = 1$); β-Alanine (2); GABA (3); δ-Aminovaleric acid (δ-AVA) (4); ε-Aminocaproic acid (ε-ACA) (5). $\bullet$: ω-Guanidino acids $H_2N—C—NH—(CH)_n—COOH$;

$$\overset{\|}{NH}$$

Guanidinoacetic acid (GAA) ($n = 1$); β-guanidinopropionic acid (β-GP) (2); γ-Guanidinobutyric acid (γ-GB) (3). Affinity of β-GP was measured from dose-conductance curves obtained in bromide solution. $\blacktriangle$: Affinities calculated from the kinetics of the competitive antagonism with GABA. $\mathbb{C}$: γ-Amino-β-Hydroxybutyric acid (GABOB) $H_2N—CH_2—CH—CH_2—COOH$: γ-Aminovaleric acid (γ-AVA)

$$\overset{|}{OH}$$

$H_2N—CH—(CH_2)_2—COOH$. It should be noted that ω-Guanidino acids behave

$$\overset{|}{CH_3}$$

as a separate group among related substances. (Takeuchi and Takeuchi, 1975a.)

When the amino group was replaced with a guanidino group, the conductance increase produced by these ω-guanidino acids was very low, but the affinity was relatively high. For example, β-guanidino-propionic acid or γ-guanidinobutyric acid produced no conductance increase on the postsynaptic membrane. When β-guanidinopropionic acid or γ-guanidinobutyric acid was applied together with GABA, these drugs competitively inhibited the GABA action (Dudel, 1965a; Feltz, 1971; Takeuchi and Takeuchi, 1975a). ω-Guanidino acids appear to be bound to the receptor, but they do not produce a conductance change. It is suggested that negative and positive charges separated by a three carbon chain is optimal for binding to the GABA receptor (Edwards and Kuffler, 1959; Takeuchi and Takeuchi, 1975a). However, the conductance change is not explained by the chain length alone. Other factors, such as the charge distribution within the terminal group, may intervene between the binding of drug molecule and the conductance change.

Analysis of the competitive inhibition of β-guanidinopropionic acid and γ-guanidinobutyric acid suggests that the binding of both GABA and antagonists follows simple Michaelis-Menten kinetics and that each receptor binds a single drug molecule (Takeuchi and Takeuchi, 1975a). This is supported by the observation that apparently no cooperativity is involved in the binding of radioactive GABA to a particulate fraction of crayfish muscle which presumably contained GABA receptor (Olsen, 1975).

Picrotoxin, which is a strong convulsant in both vertebrates and invertebrates, depresses the GABA action. But this drug is not a competitive inhibitor and its action is likely to be related to the ionophore, rather than to the GABA receptor (Takeuchi and Takeuchi, 1969; Takeuchi, 1975). In this connection it was found that picrotoxin does not inhibit the binding of radio active GABA to the isolated GABA receptor (Olsen, 1975).

E   RELATION BETWEEN BINDING AND CONDUCTANCE CHANGE

Two lines of evidence indicate a discrepancy between the mode of action of GABA. Competitive inhibition can be explained by assuming that one molecule of GABA is bound to the receptor, while the concentration-conductance relationship shows a sigmoidal relationship. A sigmoidal dose-conductance relation has been explained either by an allosteric model which postulates the cooperativity among neighbouring receptors (Changeux *et al.*, 1967; Karlin, 1967), or by the binding to receptor of more than one transmitter molecule to gate one ionic channel (Takeuchi and Takeuchi, 1967, 1969; Brookes and Werman, 1973; Olsen, 1975). (In the allosteric model the competitive inhibitor usually changes both the maximal value and slope of the log dose-response curve and the parallel shift of the log dose-response curve is obtained only under limited conditions.)

The actions of ω-guanidino acids, in particular β-guanidinopropionic acid, may possibly provide a clue to this problem. Competitive inhibition indicates that β-guanidinopropionic acid can be bound to the GABA receptor, but fails to produce a conductance change. When, however, this drug is applied to the muscle in a solution in which chloride was replaced with iodide, the membrane conductance was increased up to about 30% of that produced by GABA. The dose-conductance curve of β-guanidinopropionic acid in the iodide solution was sigmoidal, the Hill coefficient being about 2, while the Hill coefficient for GABA in the iodide solution was decreased to about 1·5 (Takeuchi and Takeuchi, 1975b). It is conceivable that in the iodide solution the receptors which are bound by β-guanidinopropionic acid could be different from those bound in chloride solution. However, in competition experiments the apparent affinity in the iodide solution was approximately the same as that measured in chloride solution. Therefore, it is likely that the same receptors are involved in both chloride and iodide solutions. The above results suggest that the receptors activated by β-guanidinopropionic acid make the membrane permeable to bromide and to iodide but not to chloride ions.

The selectivity of a charged membrane for various ion species has been discussed (Eisenman, 1965). It has been shown that in a charged membrane the permeability is controlled by the field strength of the charge sites. The membrane prefers chloride to iodide, when the field strength is high, while it prefers iodide rather than chloride, when the field strength is low. A high field strength is brought about when the size of the charge or the spacing between the charges is small. This explanation may be applicable to the permeation at the inhibitory synaptic membrane (Takeuchi and Takeuchi, 1971a, b).

A working hypothesis which tentatively could explain the above results is briefly described. The following facts are assumed. (a) Four receptor molecules are situated at the outer surface of the membrane surrounding an ionic channel (tetramer is assumed to conform to a Hill coefficient (n) of 2). (b) Binding of ligands with receptors produces positive charges at the ion channel. (c) The permeation of ions is controlled in a manner described for a fixed charged membrane (Eisenman, 1965). The field strength or the density of the charges determines the preference for permeant ions. (d) The field strength or charge density at a channel is determined by the number of receptors occupied and by the nature of the ligand bound to the receptor. For example, the charge density is higher, when GABA is bound to the receptors, than when β-guanidinopropionic acid occupies the receptors. (e) Binding of ligands with the receptor follows the Michaelis-Menten kinetics.

When four receptor molecules are occupied by GABA, the channel can pass chloride and iodide. On the other hand when four receptor molecules are bound by β-guanidinopropionic acid, the field strength at the channel is not strong enough to pass chloride but allows the passage of iodide. Similarly the channels are permeable to iodide but not to chloride, when a smaller number of receptors are occupied by GABA. Correspondingly, in the iodide solution, Hill coefficient is about 1·5 for GABA and it was close to 2 for β-guanidinopropionic acid (Takeuchi and Takeuchi, 1975b), while in the chloride solution, Hill coefficient is close to 2 for GABA, and β-guanidinopropionic acid produces little or no conductance change. Another support for this model is the effect of pH. If a decrease in pH enhances the charge density of the channel (Takeuchi and Takeuchi, 1967), it is expected that the occupation of a smaller number of receptors could induce a conductance change. It was found that at lower pH the Hill coefficient showed a tendency to decrease (Takeuchi and Takeuchi, unpublished obervations).

This model is only tentative and further experimentation is certainly necessary. In particular, it is assumed that the degree of change in the receptor molecule differs depending on the species of drug bound to the receptor. This assumption may not be improper, because it has been observed that the gating time varies depending on the species of drug bound to the ACh receptor (Katz and Miledi, 1972, 1973).

VI Presynaptic inhibition

When Fatt and Katz (1953a) investigated the inhibitory action at the crustacean neuromuscular junction, it was found that the increase in the postsynaptic membrane conductance associated with the inhibitory stimulation accounted for only 5% of the reduction of the excitatory potential. There were two possibilities to explain this discrepancy: (a) the inhibitory transmitter competitively blocked the action of the excitatory transmitter as does d-tubocurarine the ACh receptor and (b) the inhibitory stimulation decreased the amount of excitatory transmitter released by stimulation. Initially, the first possibility seemed more likely. However, Dudel and Kuffler (1961c) showed that the latter possibility is the one operating at the crayfish neuromuscular junction. Statistical analyses of the amplitude distribution of extracellular e.p.s.p. recorded from a single junction showed that the quantum content of the e.p.s.p. decreased by inhibitory stimulation, while the quantum size remained almost constant (Dudel and Kuffler, 1961c).

Presynaptic inhibition was observed, when the inhibitory nerve spike preceded the excitatory nerve spike by about 1–2 milliseconds. The delay of the appearance of presynaptic inhibition suggested the presence of a synaptic delay characteristic of a chemical synapse (Dudel and Kuffler, 1961c). The chemical nature of presynaptic inhibition is further supported by the action of GABA (Dudel and Kuffler, 1961c; Dudel, 1965a; Takeuchi and Takeuchi, 1966a). When GABA was iontophoretically applied to a single junction, while recording the extracellular e.p.s.p. from the same junction, the appearance of extracellular e.p.s.p. decreased, and at larger doses of GABA the transmission was eventually blocked. Statistical analysis showed that the quantum content was decreased and that the quantum size was not changed by the action of GABA (Dudel, 1965a; Takeuchi and Takeuchi 1966a).

When an extracellular microelectrode is placed at the synapse, it records synaptic current as well as the presynaptic nerve spike. Inhibitory stimulation decreases the amplitude of the excitatory spike (Dudel, 1963, 1965b). The application of GABA decreased the excitatory presynaptic nerve spike, while GABA did not influence the inhibitory nerve spike. Presumably GABA receptors are present on excitatory nerve terminals but not on the inhibitory terminal (Dudel, 1965c; Takeuchi and Takeuchi, 1966a).

Presynaptic inhibition is caused by an increase in the chloride conductance and not by a potential change of the nerve terminal. The application of GABA in normal solution produced no change in the frequency of spontaneous miniature e.p.s.p.s (Dudel, 1965c; Takeuchi and Takeuchi, 1966b). However, when GABA was applied in low chloride solution, the frequency of spontaneous miniature e.p.s.p. was remarkably increased, suggesting that the nerve terminal was depolarized by GABA (Takeuchi and Takeuchi, 1966b). Recently this observation has been extended to the action of the natural transmitter. Shortly after a reduction of the chloride concentration, inhibitory stimulation induced e.p.s.p., instead of producing inhibition (Hironaka, 1975).

There are some differences in the pharmacological properties of presynaptic inhibition and postsynaptic inhibition. A remarkable example is the action of ω-guanidino compounds. Effects of β-guanidino-propionic acid or γ-guanidinobutyric acid on the presynaptic terminal are almost as pronounced as those of GABA, while they produce no conductance increase in the postsynaptic membrane (Kuffler, 1960; Dudel, 1965a; Takeuchi and Takeuchi, 1975a). The relative potency of these agents at the presynaptic terminal is similar to that observed in stretch receptor neurones of the crayfish (Edwards and Kuffler, 1959; McGeer et al., 1961) and in central neurones (Purpura et al., 1959; Curtis et al., 1961; Krnjević and Phillis, 1963). It is not known whether the selective specificity for GABA at the inhibitory postsynaptic membrane of the crayfish muscle in comparison with a lesser specificity at neural elements (stretch receptor, central neurones and the excitatory nerve terminal), indicates that different receptors are involved or that the receptor is the same but that different ionophores are present (Takeuchi and Takeuchi, 1975b).

VII   Conclusion

Increasing evidence supports the view that L-glutamate and GABA are neural transmitters in the crustacean neuromuscular synapses as well as in other synapses, such as those in some of the vertebrate central neurones. Thus, the amino acid system may possibly be as important as the cholinergic and adrenergic transmitter systems. However, the understanding of the action of amino acids has lagged behind the knowledge of the cholinergic and the adrenergic systems. For example, although the release of glutamate by excitatory stimulation has been suggested in experiments by Kravitz et al. (1970), the amount of glutamate and the action of calcium on the release process needs to be elucidated. Removal of L-glutamate and GABA from the synaptic region is still to be investigated. Specific antagonists to L-glutamate and

GABA are useful tools in studying the mode of action of amino acids. However, antagonists equivalent to d-tubocurarine or to α-bungarotoxin for the ACh receptor have not been found. For such studies the crustacean neuromuscular synapse may serve as a valuable model.

VIII Acknowledgements

The author is indebted to Drs N. Takeuchi and K. Onodera for their collaboration in the experiments performed in this laboratory, and to Misses I. Ishii and M. Iimura for their technical assistance.

IX References

Atwood, H. L. (1967). Crustacean neuromuscular mechanism. *Amer. Zoologist* **7**, 527–551.

Atwood, H. L. and Morin, W. A. (1970). Neuromuscular and axoaxonal synapses of the crayfish opener muscle. *J. Ultrastruct. Res.* **32**, 351–369.

Bazemore, A., Elliott, K. A. C. and Florey, E. (1956). Factor I and γ-aminobutyric acid. *Nature Lond.* **178**, 1052–1053.

Bazemore, A., Elliott, K. A. C. and Florey, E. (1957). Isolation of factor I. *J. Neurochem.* **1**, 334–339.

Beránek, R. and Miller, P. L. (1968). The action of iontophoretically applied glutamate on insect muscle fibres. *J. Exp. Biol.* **49**, 83–93.

Biscoe, T. J., Evans, R. H., Headley, P. M., Martin, M. and Watkins, J. C. (1975). Domoic and quisqualic acids as potent amino acid excitants of frog and rat spinal neurones. *Nature Lond.* **255**, 166–167.

Bittner, G. D. (1968). Differentiation of nerve terminals in the crayfish opener muscle and its functional significance. *J. Gen. Physiol.* **51**, 731–758.

Boistel, J. and Fatt, P. (1958). Membrane permeability change during inhibitory transmitter action in crustacean muscle. *J. Physiol.* **144**, 176–191.

Brookes, N. and Werman, R. (1973). The cooperativity of γ-aminobutyric acid action on the membrane of locust muscle fibres. *Mol. Pharmacol.* **9**, 571–579.

Bullock, T. H. and Horridge, G. A. (1965). "Structure and function in the nervous system of invertebrates." Freeman, San Francisco.

Burke, W. and Ginsborg, B. L. (1956). The action of the neuromuscular transmitter on the slow fibre membrane. *J. Physiol.* **132**, 599–610.

Changeux, J-P., Thiéry, J., Tung, Y. and Kittel, C. (1967). On the cooperativity of biological membranes. *Proc. Nat. Acad. Sci. U.S.* **57**, 335–341.

Curtis, D. R. and Johnston, A. R. (1974). Amino acid transmitters in the mammalian central nervous system. *Ergeb. Physiol.* **69**, 97–188.

Curtis, D. R., Phillis, J. W. and Watkins, J. C. (1961). Actions of amino acids on the isolated hemisected spinal cord of the toad. *Br. J. Pharmacol.* **16**, 262–283.

del Castillo, J. and Katz, B. (1955). On the localization of acetylcholine receptors. *J. Physiol.* **128**, 157–181.

del Castillo, J. and Katz, B. (1956a). Localization of active spots within the neuro-muscular junction of the frog. *J. Physiol.* **132**, 630–649.

del Castillo, J. and Katz, B. (1956b). Biophysical aspects of neuromuscular trans-mission. *Progr. Biophys.* **6**, 121–170.

Diamond, J. and Roper, S. (1973). Analysis of Mauthner cell responses to ionto-phoretically delivered pulses of GABA, glycine and L-glutamate. *J. Physiol.* **232**, 113–128.

Dudel, J. (1963). Presynaptic inhibition of the excitatory nerve terminal in the neuromuscular junction of the crayfish. *Pflügers Arch. Ges. Physiol.* **277**, 537–557.

Dudel, J. (1965a). Presynaptic and postsynaptic effects of inhibitory drugs on the crayfish neuromuscular junction. *Pflügers Arch. Ges. Physiol.* **283**, 104–118.

Dudel, J. (1965b). The mechanism of presynaptic inhibition at the crayfish neuro-muscular junction. *Pflügers Arch. Ges. Physiol.* **284**, 66–80.

Dudel, J. (1965c). The action of inhibitory drugs on nerve terminals in crayfish muscle. *Pflügers Arch. Ges. Physiol.* **284**, 81–94.

Dudel, J. (1974). Nonlinear voltage dependence of excitatory synaptic current in crayfish muscle. *Pflügers Arch. Ges. Physiol.* **352**, 227–241.

Dudel, J. (1975a). Potentiation and desensitization after glutamate induced post-synaptic currents at the crayfish neuromuscular junction. *Pflügers Arch. Ges. Physiol.* **356**, 317–327.

Dudel, J. (1975b). Kinetics of postsynaptic action of glutamate pulses applied ion-tophoretically through high resistance micro-pipettes. *Pflügers Arch. Ges. Physiol.* **356**, 329–346.

Dudel, J. and Kuffler, S. W. (1960). Excitation at the crayfish neuromuscular junction with decreased membrane conductance. *Nature Lond.* **187**, 246–247.

Dudel, J. and Kuffler, S. W. (1961a). The quantal nature of transmission and spon-taneous miniature potentials at the crayfish neuromuscular junction. *J. Physiol.* **155**, 514–529.

Dudel, J. and Kuffler, S. W. (1961b). Mechanism of facilitation at the crayfish neuro-muscular junction. *J. Physiol.* **155**, 530–542.

Dudel, J. and Kuffler, S. W. (1961c). Presynaptic inhibition at the crayfish neuro-muscular junction. *J. Physiol.* **155**, 543–562.

Edwards, C. and Kuffler, S. W. (1959). The blocking effect of γ-aminobutyric acid and the action of related compounds on single nerve cells. *J. Neurochem.* **4**, 19–30.

Eisenman, G. (1965). Some elementary factors involved in specific ion permeation. *In* "Proc. Int. Union Physiol. Sci.," Vol. 4, pp. 489–506, Excerpta Medica Founda-tion, Amsterdam.

Epstein, R. and Grundfest, H. (1970). Desensitization of gamma-aminobutyric acid (GABA) receptors in muscle fibers of the crab *Cancer borealis. J. Gen. Physiol.* **56**, 33–45.

Fatt, P. and Katz, B. (1953a). The effect of inhibitory nerve impulses on a crustacean muscle fibre. *J. Physiol.* **121**, 374–388.

Fatt, P. and Katz, B. (1953b). Distributed "end-plate potentials" of crustacean muscle fibres. *J. Exp. Biol.* **30**, 433–439.

Feltz, A. (1971). Competitive interaction of β-guanidino propionic acid and γ-amino-butyric acid on the muscle fibre of the crayfish. *J. Physiol.* **216**, 391–401.

Feltz, A. and Mallart, A. (1971). An analysis of acetylcholine response of junctional and extrajunctional receptors of frog muscle fibres. *J. Physiol.* **218**, 85–100.

Futamachi, K. J. (1972). Acetylcholine: Possible neuromuscular transmitter in crustacea. *Science* **175**, 1373–1375.

Gerschenfeld, H. M. (1973). Chemical transmission in invertebrate central nervous systems and neuromuscular junctions. *Physiol. Rev.* **53**, 1–119.

Ginsborg, B. L. (1967). Ion movements in junctional transmission. *Pharmacol. Rev.* **19**, 289–316.

Grundfest, H. and Reuben, J. P. (1961). Neuromuscular synaptic activity in lobster. *In* "Nervous Inhibitions" (E. Florey, ed.), pp. 92–104. Pergamon Press, Oxford.

Grundfest, H., Reuben, J. P. and Rickles, W. H. Jr. (1959). The electrophysiology and pharmacology of lobster neuromuscular synapses. *J. Gen. Physiol.* **42**, 1301–1323.

Haldeman, S. and McLennan, H. (1972). The antagonistic action of glutamic acid diethylester towards amino acid-induced and synaptic excitations of central neurones. *Brain Res.* **45**, 393–400.

Hayashi, T. (1954). Effects of sodium glutamate on the nervous system. *Keio J. Med.* **3**, 183–192.

Hayashi, T. (1956). "Chemical physiology of excitation in muscle and nerve." Nakayama Shoten, Tokyo.

Hironaka, T. (1975). Excitatory potentials induced by stimulation of the inhibitory axon at the crustacean neuromuscular junction. *Jap. J. Physiol.* **25**, 79–91.

Hoyle, G. and Wiersma, C. A. G. (1958a). Excitation at neuromuscular junctions in crustacea. *J. Physiol.* **143**, 403–425.

Hoyle, G. and Wiersma, C. A. G. (1958b). Inhibition at neuromuscular junctions in crustacea. *J. Physiol.* **143**, 426–440.

Iversen, L. L. and Kravitz, E. A. (1968). The metabolism of γ-aminobutyric acid (GABA) in the lobster nervous system: uptake of GABA in nerve-muscle preparations. *J. Neurochem.* **15**, 609–620.

Johnson, J. L. (1972). Glutamic acid as a synaptic transmitter in the nervous system. A review. *Brain Res.* **37**, 1–19.

Johnston, G. A. R., Curtis, D. R., Davies, J. and McCulloch, R. M. (1973). Spinal interneurone excitation by conformationally restricted analogues of L-glutamate acid. *Nature Lond.* **248**, 804–805.

Karlin, A. (1967). On the application of "a plausible model" of allosteric proteins to the receptor for acetylcholine. *J. Theor. Biol.* **16**, 306–320.

Katz, B. (1936). Neuromuscular transmission in crabs. *J. Physiol.* **87**, 199–221.

Katz, B. and Miledi, R. (1964). Further observations on the distribution of acetylcholine-reactive sites in skeletal muscle. *J. Physiol.* **170**, 370–388.

Katz, B. and Miledi, R. (1969). Tetrodotoxin-resistant electric activity in presynaptic terminals. *J. Physiol.* **203**, 459–487.

Katz, B. and Miledi, R. (1972). The statistical nature of the acetylcholine potential and its molecular components. *J. Physiol.* **224**, 665–699.

Katz, B. and Miledi, R. (1973). The characteristics of "end-plate noise" produced by different depolarizing drugs. *J. Physiol.* **230**, 707–717

Katz, B. and Thesleff, S. (1957). A study of the "desensitization" produced by acetylcholine at the motor end-plate. *J. Physiol.* **138**, 63–80.

Kennedy, D. and Evoy, W. H. (1966). The distribution of pre- and postsynaptic inhibition at crustacean junctions. *J. Gen. Physiol.* **49**, 457–468.

Kravitz, E. A. (1967). Acetylcholine, gamma aminobutyric acid and glutamic acid: physiological and chemical studies related to their roles as neurotransmitter agents. *In* "The Neurosciences", A Study Program (G. C. Quarton, T. Melnechuk and F. O. Schmitt, eds), pp. 433–444. Rockefeller Univ., New York.

Kravitz, E. A., Kuffler, S. W. and Potter, D. D. (1963). Gamma-aminobutyric acid and other blocking compounds in crustacea. III. Their relative concentrations in separated motor and inhibitory axons. *J. Neurophysiol.* **26**, 739–751.

Kravitz, E. A., Kuffler, S. W., Potter, D. D. and Van Gelder, N. M. (1963). Gamma-aminobutyric acid and other blocking compounds in crustacea. II. Peripheral nervous system *J. Neurophysiol.* **26**, 729–738.

Kravitz, E. A., Molinoff, P. B. and Hall, Z. W. (1965). A comparison of the enzyme and substrates of gamma-aminobutyric acid metabolism in lobster excitatory and inhibitory axons. *Proc. Nat. Acad. Sci. U.S.* **54**, 778–782.

Krnjević, K. (1974). Chemical nature of synaptic transmission in vertebrates. *Physiol. Rev.* **54**, 418–540.

Krnjević, K. and Miledi, R. (1958). Acetylcholine in mammalian neuromuscular transmission. *Nature Lond.* **182**, 805–806.

Krnjević, K. and Phillis, J. W. (1963). Iontophoretic studies of neurones in the mammalian cerebral cortex. *J. Physiol.* **165**, 274–304.

Kuffler, S. W. (1960). Excitation and inhibition in single nerve cells. *Harvey Lectures* **54**, 176–218.

Kuffler, S. W. and Edwards, C. (1958). Mechanisms of gamma-aminobutyric acid (GABA) action and its relation to synaptic inhibition. *J. Neurophysiol.* **21**, 589–610.

Kuffler, S. W. and Yoshikami, D. (1975). The distribution of acetylcholine sensitivity at the postsynaptic membrane of vertebrate skeletal twitch muscles: iontophoretic mapping in the micron range. *J. Physiol.* **244**, 703–730.

Kusano, K., Miledi, R. and Stinnakre, J. (1975). Postsynaptic entry of calcium induced by transmitter action. *Proc. Roy. Soc. Lond.* B **189**, 49–56.

Lewis, P. R. (1952). The free amino-acids of invertebrate nerve. *Biochem. J.* **52**, 330–338.

Lowagie, C. and Gerschenfeld, H. M. (1974). Glutamate antagonist at a crayfish neuromuscular junction. *Nature Lond.* **248**, 533–535.

Manthey, A. A. (1966). The effect of calcium on the desensitization of membrane receptors at the neuromuscular junction. *J. Gen. Physiol.* **49**, 963–976.

Marks, N., Datta, R. K. and Lajtha, A. (1970). Distribution of amino acids and of exo- and endopeptidases along vertebrate and invertebrate nerves. *J. Neurochem.* **17**, 53–63.

McGeer, E. G., McGeer, P. L. and McLennan, H. (1961). The inhibitory action of 3-hydroxytyramine, gamma-aminobutyric acid (GABA) and some other compounds towards the crayfish stretch receptor neuron. *J. Neurochem.* **8**, 36–49.

McMahan, U. J., Spitzer, N. C. and Peper, K. (1972). Visual identification of nerve terminals in living isolated skeletal muscle. *Proc. Roy. Soc.* B **181**, 421–430.

Miledi, R. (1960). Junctional and extra-junctional acetylcholine receptors in skeletal muscle fibres. *J. Physiol.* **151**, 1–23.

Miledi, R. (1969). Transmitter action in the giant synapse of the squid. *Nature Lond.* **223**, 1284–1286.

Miledi, R. and Potter, L. T. (1971). Acetylcholine receptors in muscle fibres. *Nature Lond.* **233**, 599–603.

Motokizawa, F., Reuben, J. F. and Grundfest, H. (1969). Ionic permeability of the inhibitory postsynaptic membrane of lobster muscle fibers. *J. Gen. Physiol.* **54**, 437–461.

Nastuk, W. L. (1953). Membrane potential changes at a single muscle endplate produced by acetylcholine, with an electrically controlled microjet. *Fedn Proc.* **12**, 102.

Nastuk, W. L. and Parsons, R. L. (1970). Factors in the inactivation of postjunctional membrane receptors of frog skeletal muscle. *J. Gen. Physiol.* **56**, 218–249.

Olsen, R. W. (1976). Approaches to study of GABA receptors. *In* "GABA in Nervous System Function" (E. Roberts, T. N. Chase and D. Tower, eds), pp. 287–304. Raven Press, N.Y.

Onodera, K. and Takeuchi, A. (1975). Ionic mechanism of the excitatory synaptic membrane of the crayfish neuromuscular junction. *J. Physiol.* **252**, 295–318.

Onodera, K. and Takeuchi, A. (1976). Permeability changes produced by L-glutamate at the excitatory postsynaptic membrane of the crayfish muscle. *J. Physiol.* **255**, 669–685.

Otsuka, M. (1972). γ-aminobutyric acid in the nervous system. *In* "The Structure and Function of Nervous System" (G. H. Bourne, ed.), Vol. 4, pp. 249–289. Academic Press, New York and London.

Otsuka, M., Iversen, L. L., Hall, Z. W. and Kravitz, E. A. (1966). Release of gamma-aminobutyric acid from inhibitory nerves of lobster. *Proc. Nat. Acad. Sci. U.S.A.* **56**, 1110–1115.

Ozeki, M., Freeman, A. R. and Grundfest, H. (1966). The membrane components of crustacean neuromuscular systems. II. Analysis of interactions among the electrogenic components. *J. Gen. Physiol.* **49**, 1335–1349.

Ozeki, M. and Grundfest, H. (1967). Crayfish muscle fiber: Ionic requirements for depolarizing synaptic electrogenesis. *Science* **155**, 478–481.

Parnas, I., Rahamimoff, R. and Sarner, Y. (1975). Tonic release of transmitter at the neuromuscular junction of the crab. *J. Physiol.* **250**, 275–286.

Purpura, D. P., Girado, M., Smith, T. G., Callan, D. A. and Grundfest, H. (1959). Structure activity determinants of pharmacological effects of amino acids and related compounds on central synapses. *J. Neurochem.* **3**, 238–268.

Rang, H. P. (1971). Drug receptors and their function. *Nature Lond.* **231**, 91–96.

Rang, H. P. and Ritter, J. M. (1970). On the mechanism of desensitization at cholinergic receptors. *Molec. Pharmacol.* **6**, 357–382.

Robbins, J. (1959). The excitation and inhibition of crustacean muscle by amino acids. *J. Physiol.* **148**, 39–50.

Shinozaki, H. and Shibuya, I. (1974a). A new potent excitant, quisqualic acid: effects on crayfish neuromuscular junction. *Neuropharmacol.* **13**, 665–672.

Shinozaki, H. and Shibuya, I. (1974b). Potentiation of glutamate-induced depolarization by kainic acid in the crayfish opener muscle. *Neuropharmacol.* **13**, 1057–1065.

Swagel, M. W., Ikeda, K. and Roberts, E. (1973). Effects of GABA, imidazol acetic acid, and related substances on conductance of crayfish abdominal stretch receptor. *Nature New Biol.* **246**, 91–92.

Takeuchi, A. (1976). Studies of inhibitory effects of GABA in invertebrate nervous systems. *In* "GABA in Nervous System Function" (E. Roberts, T. N. Chase and D. Tower, eds), pp. 255–267. Raven Press, N.Y.

Takeuchi, A. and Onodera, K. (1973). Reversal potentials of the excitatory transmitter and L-glutamate at the crayfish neuromuscular junction. *Nature New Biol.* **242**, 124–126.

Takeuchi, A. and Onodera, K. (1975). Effects of kainic acid on the glutamate receptors of the crayfish muscle. *Neuropharmacology* **14**, 619–625.

Takeuchi, A. and Takeuchi, N. (1964). The effect on crayfish muscle of iontophoretically applied glutamate. *J. Physiol.* **170**, 296–317.

Takeuchi, A. and Takeuchi, N. (1965). Localized action of gamma-aminobutyric acid on the crayfish muscle. *J. Physiol.* **177**, 225–238.

Takeuchi, A. and Takeuchi, N. (1966a). A study of the inhibitory action of γ-aminobutyric acid on neuromuscular transmission in the crayfish. *J. Physiol.* **183**, 418–432.

Takeuchi, A. and Takeuchi, N. (1966b). On the permeability of the presynaptic terminal of the crayfish neuromuscular junction during synaptic inhibition and the action of γ-aminobutyric acid. *J. Physiol.* **183**, 433–449.

Takeuchi, A. and Takeuchi, N. (1967). Anion permeability of the inhibitory postsynaptic membrane of the crayfish neuromuscular junction. *J. Physiol.* **191**, 575–590.

Takeuchi, A. and Takeuchi, N. (1969). A study of the action of picrotoxin on the inhibitory neuromuscular junction of the crayfish. *J. Physiol.* **205**, 377–391.

Takeuchi, A. and Takeuchi, N. (1970). Effects of some amino acids on the excitatory and inhibitory postsynaptic membranes of the crayfish muscle. *J. Physiol. Soc. Japan*, **32**, 405–406.

Takeuchi, A. and Takeuchi, N. (1971a). Anion interaction at the inhibitory postsynaptic membrane of the crayfish neuromuscular junction. *J. Physiol.* **212**, 337–351.

Takeuchi, A. and Takeuchi, N. (1971b). Variations in the permeability properties of the inhibitory postsynaptic membrane of the crayfish neuromuscular junction when activated by different concentrations of GABA. *J. Physiol.* **217**, 341–358.

Takeuchi, A. and Takeuchi, N. (1972). Actions of transmitter substances on the neuromuscular junctions of vertebrates and invertebrates. *Advan. Biophys.* **3**, 45–95.

Takeuchi, A. and Takeuchi, N. (1975a). The structure-activity relationship for GABA and related compounds in the crayfish muscle. *Neuropharmacol.* **14**, 627–634.

Takeuchi, A. and Takeuchi, N. (1975b). Permeability changes of the crayfish muscle produced by β-guanidinopropionic acid and related substances. *Neuropharmacol.* **14** 635–641.

Takeuchi, N. (1963). Effects of calcium on the conductance change of the end-plate membrane during the action of transmitter. *J. Physiol.* **167**, 141–155.

Taraskevich, P. S. (1971). Reversal potentials of ʟ-glutamate and the excitatory transmitter at the neuromuscular junction of the crayfish. *Biochim. biophys. Acta* **241**. 700–704.

Thesleff, S. (1955). The mode of neuromuscular block caused by acetylcholine, nicotine, decamethonium and succinylcholine. *Acta Physiol. Scand.* **34**, 218–231.

Usherwood, P. N. R. and Machili, P. (1968). Pharmacological properties of excitatory neuromuscular synapses in the locust. *J. Exp. Biol.* **49**, 341–361.

Van Harreveld, A. (1939). The nerve supply of doubly and triply innervated crayfish muscles related to their function. *J. Comp. Neurol.* **70**, 267–284.

Van Harreveld, A. and Mendelson, M. (1959). Glutamate-induced contractions in crustacean muscle. *J. Cell. Comp. Physiol.* **54**, 85–94.

Watkins, J. C. (1965). Pharmacological receptors and general permeability phenomena of cell membrane. *J. Theor. Biol.* **9**, 37–50.

Werman, R. (1969). An electrophysiological approach to drug-receptor mechanisms. *Comp. Biochem. Physiol.* **30**, 997–1017.

Wiersma, C. A. G. (1961). The neuromuscular system. *In* "The Physiology of Crustacea" (T. H. Waterman, ed.), Vol. II, pp. 191–240. Academic Press, New York and London.

9

L-Glutamate Receptors in Locust Muscle Fibres

S. G. Cull-Candy

Dept. of Biophysics, University College, London, England

I Introduction

Glutamate receptors in invertebrate muscle have been studied mainly in crustaceans and insects where glutamate, or a closely related substance, may be the excitatory neuromuscular transmitter (see Kravitz *et al.*, 1970; Gerschenfeld, 1973). Because of the close phylogenetic relationship between these two classes, it is not surprising that many properties of their neuromuscular physiology and pharmacology are strikingly similar. It was the now classical studies on crayfish muscle which initially demonstrated a highly localized depolarizing action of L-glutamate when applied iontophoretically to the excitatory junctional regions (Takeuchi and Takeuchi, 1963, 1964). This prompted a number of workers to investigate the possibility that glutamate receptors occur at the neuromuscular junction of the locust and other insects.

It was found that L-glutamate applied iontophoretically to locust muscle has a depolarizing action at discrete sites in the muscle membrane (Beránek and Miller, 1968; Usherwood and Machili, 1968). When glutamate is applied iontophoretically from conventional micropipettes (i.e. less than 50 MΩ) to the muscle fibre surface, the sensitivity of the muscle membrane decreases sharply to zero, or near zero, in regions away from the nerve-muscle junction (Usherwood, 1969b). In order to detect extrajunctional sensitivity it is necessary to reduce the background leakage of glutamate from the micropipette to a minimum. This is best achieved by using high resistance micropipettes (150–300 MΩ) with appropriate braking current (see Peper and

263

McMahan, 1972; Dreyer and Peper, 1974). Under these conditions the extrajunctional membrane demonstrates a low uniform sensitivity to glutamate which results from the presence of extrajunctionally situated glutamate receptors (Cull-Candy and Usherwood, 1973).

Recent investigations concerning the properties of junctional glutamate receptors in locust muscle are reviewed in this chapter, with particular reference to the pharmacological properties of the receptors, their distribution in the muscle membrane and the ionic permeability changes they mediate. Where possible these properties are compared with those of junctional glutamate receptors in crustacean muscle. In addition some recent experiments aimed at determining certain general characteristics of the extrajunctional glutamate receptors in locust muscle are described. Emphasis is placed, partly, on the evidence that two types of extrajunctional glutamate receptors are present in the muscle membrane. With this in mind, aspects of the pharmacology of the receptors are described. Also the distribution of extrajunctional glutamate receptors and the effects of denervation-type changes on these receptors are considered. Finally some properties of the ionic conductances mediated by extrajunctional receptors are described.

The methods employed in these experiments are reported elsewhere in detail (Cull-Candy, 1976). Muscle fibres were impaled with independent recording and polarizing microelectrodes filled with 2M potassium citrate. Extrajunctional receptors were activated with L-glutamate (Na salt, 0.1M, pH 8) or D, L-ibotanate (Na salt, 0.1 M, pH 7) applied from single or double barrel micropipettes of high resistance (150–300 MΩ). Double barrel micropipettes which contained glutamate in one barrel and ibotenate in the other enabled investigation of receptors at the same extrajunctional site.

II   Junctional glutamate receptors

A   JUNCTIONAL RECEPTOR PROPERTIES

When L-glutamate is applied iontophoretically from conventional micropipettes, to the surface of muscle fibres, it produces transient depolarizations at sites which correspond closely to sites of excitatory innervation on the multiply innervated locust muscle fibres (Usher-wood and Machili, 1968; Beránek and Miller, 1968). A postsynaptic origin of the response to glutamate is evident from the smoothly graded depolarization, lacking in quantal components. Also, the depolarizing response to glutamate applied iontophoretically to the muscle membrane is still obtained after short term denervation of the muscle (Beránek and Miller, 1968; Usherwood and Machili, 1968) or after treatment of the nerve muscle system with Black Widow Spider Venom, which depletes locust excitatory nerve terminals of synaptic vesicles and exhausts the spontaneous miniature potentials in locust (Cull-Candy et al., 1973) and crayfish muscle (Kawai et al., 1972). Interestingly glutamate and aspartate may exert a presynaptic action, increasing the frequency of spontaneous miniature potentials at locust nerve muscle junctions (Dowson and Usherwood, 1973), although it is unlikely that this contributes in any way to the response to glutamate applied iontophoretically.

A small conditioning iontophoretic pulse of ACh can potentiate the response to a subsequent ACh pulse at the vertebrate end plate (see Katz and Thesleff, 1957). A similar phenomenon occurs when glutamate is applied locally to crayfish or locust muscle (Takeuchi and Takeuchi, 1964; Dudel, 1975a; Walther and Usherwood, 1972; Usherwood and Cull-Candy, 1975). Potentiation gives way to desensitization when larger doses of glutamate are applied (Takeuchi and Takeuchi, 1964; Dudel, 1975a; Usherwood and Machili, 1968; Beránek and Miller, 1968). The recovery half time from partial desensitization is less than 1 s for the action of glutamate on the junctional receptors in crayfish and locust muscle (Dudel, 1975a; Cull-Candy, unpublished).

Bath application of L-glutamate 10^{-5} M, to locust muscle, completely abolishes neurally evoked contractures by desensitizing junctional receptors (Usherwood and Machili, 1968). In addition, bath application of glutamate 10^{-5}–10^{-4} M completely abolishes spontaneous excitatory miniature potentials with an accompanying increase in membrane "noise". Glutamate "noise" occurs at nerve-muscle junctions of crayfish (Crawford and McBurney, 1975) and locust (Anderson et al., 1976) and is presumed to have an origin similar to ACh "noise" in vertebrate muscle (see Katz and Miledi, 1972).

B PHARMACOLOGICAL PROPERTIES OF JUNCTIONAL GLUTAMATE RECEPTORS

1 *Action of agonists and antagonists at junctional receptors*
The junctional glutamate receptor in locust muscle is stereospecific for

the L-isomer of glutamate, being much less sensitive to D-glutamate or L-aspartate (Usherwood and Machili, 1968; Beránek and Miller, 1968; McDonald and O'Brien, 1972; Clements and May, 1974). In this respect junctional glutamate receptors in crayfish and locust muscle are similar (Takeuchi and Takeuchi, 1964) and resemble glutamate receptors in mammalian cortical neurones (Krnjević and Phyllis, 1963), and to a lesser extent those in spinal interneurones (Curtis and Watkins, 1960, 1963). However, from comparisons of the structure activity relationships of other amino acids, the glutamate receptors in locust muscle appear to show marked differences from glutamate receptors in vertebrate central neurones (see Curtis and Watkins, 1960, 1963; Clements and May, 1974; Usherwood and Cull-Candy, 1975). In addition, studies with conformationally restricted glutamate analogues show that kainic acid is only a weak excitant of junctional glutamate receptors in locust, lobster and crayfish muscle (Clements and May, 1974; Constanti and Nistri, 1975; Shinozaki et al., 1975), while the sodium salt of ibotenic acid has no action (Lea and Usherwood, 1973a; Cull-Candy and Usherwood, 1973). These glutamate analogues are more potent than L-glutamate at exciting mammalian spinal interneurones (Johnson et al., 1974). On the other hand quisqualic acid is an effective agonist both at peripheral and central cites exciting glutamate receptors in crayfish muscle (Shinozaki and Shibuya, 1974), locust muscle (Anderson et al., 1976) and spinal neurones (Biscoe et al., 1975). Caution is necessary however, when comparing the hypothetical conformation of the junctional glutamate receptors in locust muscle with those in mammalian neurones, in view of the suggestion that more than one type of amino acid receptor may occur in neurones (Johnston et al., 1974) as presently described for locust muscle.

It seems likely that a receptor containing three charged centres, somewhat similar to that suggested for mammalian central neurones (Curtis and Watkins, 1960), occurs at junctional sites in locust muscle (Clements and May, 1974; Usherwood and Cull-Candy, 1975). To interact with this receptor the amino acids should have two acidic groups and one basic group all of which are charged and unsubstituted. Two or three carbon atoms should separate the distal acidic group from the basic amino group, the other acidic group being situated α with respect to the amino group (i.e. linked directly to the carbon which carries the amino group). This in fact defines the configuration of the

L-glutamate molecule which is ionized at physiological pH : COO^-— $CHNH_3^+$—CH_2—CH_2—COO^- (see Usherwood and Machili, 1968; Clements and May, 1974).

Several investigations of possible antagonists of the excitatory action of glutamate at junctional receptors in locust and crayfish muscle have been reported (see Clements and May, 1974; Lowagie and Gerschenfeld, 1974). Of the drugs perfused over crayfish muscle the most potent antagonists, both of the junctional response to iontophoretically applied glutamate and of the excitatory junctional response to the transmitter, are the esters of L-glutamic acid (Lowagie and Gerschenfeld, 1974). L-glutamic acid γ-methylester is the most effective, while the diethylester is the least effective. On spinal and cuneate neurones the reverse is true (Haldeman and McLennan, 1974; Curtis et al., 1972).

The diethylester does not antagonize the response to glutamate applied topically to locust or lobster muscle (Clements and May, 1974; Nistri and Constanti, 1975). However, in locust muscle a number of β-carbolines (i.e. Harmala alkaloids) antagonize the excitatory action of iontophoretically applied glutamate, although they also have a presynaptic effect (Dowson et al., 1975).

DL-2-amino, 4-phosphonobutyric acid (DL-APB) inhibits the binding of glutamate to a receptor proteolipid isolated from locust muscle (James et al., 1974; Cull-Candy et al., 1976). If glutamate and DL-APB are iontophoretically applied to junctional glutamate receptors from adjacent barrels of a micropipette, when the application of DL-APB precedes glutamate application by a sufficiently brief time interval, the response to glutamate is abolished (Cull-Candy et al., 1976). Coulomb dose-response curves for glutamate, applied iontophoretically, demonstrate a parallel shift to the right in the presence of a constant pulse of DL-APB.

In all these cases, when the antagonist is applied to the bathing medium, the concentration required is high compared with antagonists for ACh at the vertebrate end plate.

2 *Coulomb dose-response relationships*

There is a non-linear relationship between the size of the glutamate pulse (i.e. dose and the resultant junctional response in locust muscle (Walther and Usherwood, 1972; Cull-Candy et al., 1976) and crayfish muscle (Takeuchi and Takeuchi, 1964; Dudel, 1975a). Small pulses of glutamate produce responses which show a more than linear addition while a less than linear addition is seen for responses to large pulses. This is reminiscent of the relationship for carbachol on ACh receptors when investigated under the same conditions (Katz and Thesleff, 1957).

When the amplitude of the junctional depolarization is plotted, as a function of the coulomb strength of glutamate applied, on double logarithmic co-ordinates, the relationship has a limiting slope of about two for locust and crayfish muscle (Walther and Usherwood, 1972; Takeuchi and Takeuchi, 1964). A similar relationship is obtained for glutamate receptors in crayfish muscle when current rather than potential is the response measured (Dudel, 1975a). However, a limiting slope of 4–6 has been reported for the relationship when glutamate is applied from high resistance micropipettes (Dudel, 1975b).

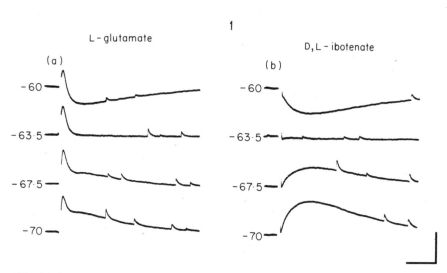

Fig. 1.1. Comparison of the extrajunctional responses to L-glutamate and D, L-ibotenate applied iontophoretically from a twin barrel micropipette to the same region of extrajunctional membrane. (a). At the resting potential, -60 mV, glutamate produces a DH-response. The H-component is annulled at -63.5 mV when a D-component alone is seen. The H-component reverses at membrane potentials greater than -63.5 mV. A two-component depolarization occurs at -67.5 mV and -70 mV. (b) At the resting potential, -60 mV, ibotenate produces an H-response. This is annulled at -63.5 mV in comparison with the glutamate induced D-response at -63.5 mV. The H-response reserves at membrane potentials greater than -63.5 mV; *Depolarizing* H-responses occur at -67.5 mV and -70 mV. Note that since the fibres are multiply innervated, spontaneous miniature excitatory potentials occur in junctional and extrajunctional regions in these and subsequent records. (From Cull-Candy, 1976.)

2

D,L – ibotenate   L – glutamate

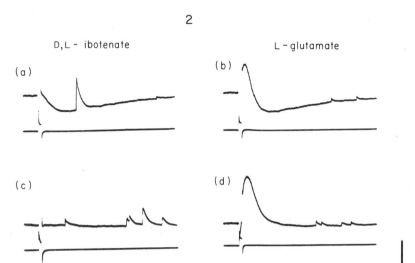

Fig. 1.2. The selective desensitization by D, L-ibotenate of the H-component of the extrajunctional DH-response to L-glutamate. Drugs are applied from a twin barrel micropipette. The drug ejection currents passing through the pipettes are monitored on the lower traces. (a) An H-response to ibotenate. (b) A DH-response to glutamate 1 m later. (c) After ejecting ibotenate at 5 s intervals for 30 s the H-response to ibotenate is abolished as the result of receptor desensitization. (d) 5 s later glutamate generates a D-response, which in the absence of an H-response has increased in amplitude from (b). Calibration: 1 mV and 200 ms, 10 nC. (From Cull-Candy, 1976.)

C SENSITIVITY AND DISTRIBUTION OF JUNCTIONAL RECEPTORS

When investigated with pulses of glutamate (1·0 nC) applied from conventional micropipettes, the sensitivity of the muscle membrane decreases sharply from about 10–12 mV/nC to zero or near zero within 10–35 μm of the junctional site (Beránek and Miller, 1968; Usherwood, 1969b). A high resolution of receptor distribution is obtained when small quantities of the excitant are applied from micropipettes of high resistance (see Peper and McMahan, 1972). Small pulses of glutamate (0.1 nC), applied to junctional receptors from micropipettes of high resistance, produce responses of 100 mV/nC which are lost within 10–15 μm of the junctional sites. Figure 3.1 shows the junctional receptor distribution on the exposed surface of a muscle fibre. Two peaks of high sensitivity to glutamate (i.e. junctional sites) are seen. It is occasionally possible to obtain large depolarizing responses over slightly greater areas of muscle membrane (right hand peak). This may result when the micropipette is moved parallel to the nerve terminal, which may be as long as 30 μm in locust muscle (Rees and Usherwood, 1972).

D IONIC PERMEABILITY CHANGES MEDIATED BY JUNCTIONAL RECEPTORS

The reversal potential for the action of an agonist on chemosensitive receptors is a reflection of the ionic permeability changes occurring. At the locust neuromuscular junction, the extrapolated reversal potential for the action of the excitatory transmitter is reported to be 0 mV (del Castillo et al., 1953). An extrapolated reversal potential of between −10mV and −25mV has also been reported for the response to both glutamate and the excitatory transmitter (Beránek and Miller, 1968). Reversal potentials close to zero are obtained both for glutamate and the transmitter by measuring the junctional current under voltage clamp conditions (Anwyl and Usherwood, 1974). Similar values are obtained by the direct reversal of the junctional potential change produced by glutamate and the excitatory transmitter (Cull-Candy and Usherwood, see Anwyl and Usherwood, 1974; Cull-Candy, 1976). Na^+ and K^+ have been implicated in permeability changes at junctional receptors (Anwyl and Usherwood, 1975).

In crayfish muscle the reversal potential of the junctional currents produced by glutamate and the excitatory transmitter coincide (Taraskevich, 1971; Takeuchi and Onodera, 1973), although a wide range of values are reported (Taraskevitch, 1975; Dudel, 1974). Na^+, K^+ and Ca^{2+} have all been implicated in the conductance changes at junctional receptors (Ozeki and Grundfest, 1967; Takeuchi and Onodera, 1973; Dudel, 1974; Taraskevitch, 1975).

III   Extrajunctional glutamate receptors

To investigate extrajunctional sensitivity it is necessary to use relatively large pulses of glutamate (at least 1–5 nC). This can be useful when the spatial distribution of the receptors is of interest. Relative positions of receptors may be obtained, since these quantities of glutamate travel some distance from the micropipette tip activating first adjacent receptors and then more distant ones.

Extrajunctional receptors have an even distribution in the extrajunctional muscle membrane. They show a low sensitivity; the amplitude of the extrajunctional response is about 200–1000 times less than that produced by superficial junctional receptors in response to an equivalent quantity of glutamate. Differences in sensitivity of a similar magnitude have previously been described for junctional and extrajunctional ACh receptors in frog muscle (Katz and Miledi, 1964a).

Possibly associated with their different sensitivities, junctional and extrajunctional receptors showed different degrees of desensitization when activated by iontophoretically applied glutamate. Extrajunctional receptors require recovery periods of at least 60 s between iontophoretic pulses at any one site. When similar quantities are applied to junctional sites (in the absence of muscle twitch), they do not produce prolonged desensitization. As previously pointed out, the less sensitive a region the faster and more pronounced the desensitization which results from the repetitive application of a chemoexcitant (Miledi, 1960; Katz and Miledi, 1964a), probably as a result of the larger doses of drug required to measurably activate these receptors. Under such circumstances an intrinsic difference between junctional and extrajunctional receptor types cannot necessarily be inferred (Katz and Miledi, 1964a; Cull-Candy and Usherwood, 1973).

Also associated with the low sensitivity of extrajunctional receptors is the observation that larger doses of glutamate activate larger areas of receptors (Cull-Candy and Usherwood, 1973; see also, del Castillo and Katz, 1955; Miledi, 1960), since both the rise time and amplitude of the response increases with larger pulses of glutamate. This is qualitatively similar to the interaction of ACh with extrajunctional receptors in frog muscle (Feltz and Mallart, 1971), where the nearest receptor sites would be readily saturated and the more distant receptors, activated after a longer diffusion time, make an increasingly important contribution to the potential change (methods: Katz and Miledi, 1964a).

IV Pharmacological analysis of extrajunctional glutamate receptors

Two drugs proved to be useful for investigating the pharmacological properties of extrajunctional glutamate receptors in locust muscle: L-glutamate, which has a strong excitatory action on junctional receptors and D,L-ibotenate, and isoxazole with structural similarities to glutamate (Eugster, 1967), but with no apparent action on the junctional glutamate receptors in locust muscle (Lea and Usherwood, 1973a; Cull-Candy and Usherwood, 1973). Ibotenate has previously been reported to activate glutamate receptors in spinal interneurones (Johnston et al., 1968), Renshaw cells (Johnston et al., 1974), central neurones (Shinozaki and Konishi, 1970) and snail neurones (Walker et al., 1971). When applied in the bathing medium ibotenate increases the Cl^- conductance of locust muscle fibres (Lea and Usherwood, 1973b). Extrajunctional glutamate receptors are insensitive to D-glutamate.

A   GLUTAMATE AND IBOTENATE ON EXTRAJUNCTIONAL RECEPTORS

Figure 1.1 shows the responses which result when L-glutamate and D, L-ibotenate are applied to the extrajunctional membrane of a muscle fibre. The drugs are applied to the same extrajunctional region from double barrel micropipettes which contain glutamate in one barrel and ibotenate in the other. Glutamate produces a small biphasic response, depolarization followed by hyperpolarization (DH-response), at the resting potential. Ibotenate produces a hyperpolarization (H-response) which closely resembles the H-component of the DH-response to glutamate. The difference between the response to ibotenate and the response to glutamate is readily seen when the muscle fibre membrane, to which the drugs are applied, is held at different levels of membrane potential. This is accomplished by current injection from a second intracellular microelectrode. Glutamate produces a two component response, the H-component of which is annulled at the same membrane potential level as the single component ibotenate response (i.e. $-63\cdot5$ mV in this fibre). At the equilibrium potential of the H-response glutamate application produces only a D-response. Further hyperpolarization of the membrane causes the second component of the response to glutamate and the single component response to ibotenate to reappear as a depolarizations. These results suggest the coexistence of two types of extrajunctional glutamate receptors: a glutamate sensitive receptor which mediates a depolarization (D-receptor), and a glutamate and ibotenate sensitive receptor which mediates a hyperpolarization (H-receptor) (Cull-Candy, 1976).

If glutamate and ibotenate act on the same receptors to produce the H-response, it is expected that the H-response to glutamate would be desensitized by ibotenate and vice versa. An example of an experiment testing this possibility is shown in Fig. 1.2. Glutamate and ibotenate are applied to the same region of extrajunctional membrane from adjacent barrels of a twin barrel micropipette. Following ibotenate application, the glutamate induced response consists only of a D-response, the H-response having been desensitized. Conversely, iontophoretic application of glutamate completely desensitizes the ibotenate-induced H-response (not shown in figure).

Experiments of this type confirmed that the extrajunctional glutamate receptors, which mediate a hyperpolarization, are also ibotenate sensitive and that the extrajunctional glutamate receptors which mediate a depolarization are insensitive to ibotenate in quantities which completely abolish the H-response. In addition, the possibility is excluded that ibotenate interacts with D-receptors either as an agonist or an antagonist (Cull-Candy, 1976). The idea that the molecular conformation of ibotenate precludes it from activating junctional glutamate receptors is also rendered more favourable (see Lea and Usherwood, 1973a; Usherwood and Cull-Candy, 1975).

B COULOMB DOSE-RESPONSE RELATIONSHIPS

The peak amplitude of both the extrajunctional D- and H-responses is related to the coulomb dose of glutamate or ibotenate released from the micropipette. However, the D- and H-responses appear to result from the activation of receptors which differ in their coulomb dose-response characteristics. Dose-response curves were determined independently for the two types of response (i.e. in each case in the absence of the other response). The dose was measured as the coulomb strength of the current passing through the micropipette, while the response was the amplitude of the potential change observed. The amplitude of the extrajunctional response was plotted as a function of the coulombic strength of glutamate or ibotenate applied, on double logarithmic co-ordinates. The limiting slopes of double logarithmic plots for the H-response to both glutamate and ibotenate are rather similar with values of about 0·75. The limiting slope of double logarithmic plots for the D-response to glutamate is about 1·5 (Cull-Candy, 1976).

The similarity of the value for the H-response produced both by glutamate and by ibotenate is in agreement with the concept that these drugs act on the same receptors to produce the H-response and that the mechanism underlying the interaction of these two drugs with the receptor may be similar or identical. The value obtained for the D-response approaches the values previously determined for junctional glutamate receptors in crayfish muscle (Takeuchi and Takeuchi, 1964; Dudel, 1975a) and in locust muscle (Walther and Usherwood, 1972) which was about 2 (but see Dudel, 1975b).

V   Sensitivity and distribution of extrajunctional receptors

A   DISTRIBUTION OF H-RECEPTORS

1   *Investigated with glutamate*

When H-receptors were mapped close to the junction, preparations were used in which the extrajunctional D-response was not apparent

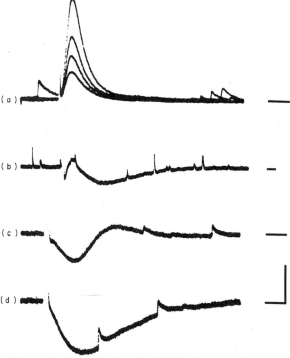

Fig. 2.1. The interaction of extrajunctional H-responses with junctional responses when glutamate is applied from a micropipette. The extrajunctional response in this fibre consists of an H-response. (a) With the micropipette close to the junctional site superimposed depolarizations result from the application of different amounts of glutamate. (b) 10–20 μm from the junctional site, glutamate produces a small junctional depolarization followed by an extrajunctional H-response. (c) 40 μm from the junctional site, the H-response precedes the junctional depolarization. (d) 50 μm from the junctional site, glutamate activates H-receptors but not junctional receptors. Calibration: 1 mV and 200 ms.

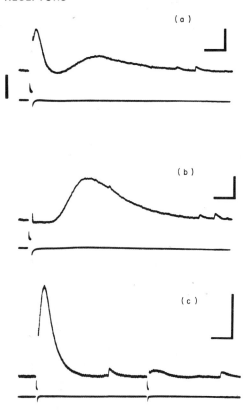

Fig. 2.2 The interaction of extrajunctional D-responses with junctional responses.
(a) With the micropipette 40 μm from a junctional site glutamate application produces an extrajunctional D-response of short latency followed by a slow-rising junctional depolarization. (b) A second application of glutamate 10 s later produces a
larger, slow-rising junctional response. The D-response is abolished as a result of
receptor desensitization. (c) Application of glutamate distant from junctional sites
produces a D-response of short latency. A second glutamate application 1 s later
produces a very reduced response as the D-receptors have become desensitized. Calibration: 1 mV and 200 ms, 5 nC.

(i.e. as a result of shunting by the H-response). In this situation the
responses to glutamate are most readily interpreted. H-receptors occur
in close proximity to the junctional receptors, therefore responses of a
variable shape, which result from the activation of junctional receptors
and extrajunctional H-receptors, can be generated by the appropriate
location of the micropipette within a radius of 10–45 μm from the
junctional site (Cull-Candy and Usherwood, 1973).

Figure 2.2 illustrates an experiment of this type. When the micro-pipette is within 10–20 μm from the junction, a junctional depolariza-tion precedes an extrajunctional hyperpolarization. With the micro-pipette 40 μm from the junction, the situation is reversed so that the extrajunctional hyperpolarization precedes the junctional depolariza-tion. Further movement of the micropipette in a direction away from the junction results in a purely extrajunctional response. It is clearly not possible to discriminate contiguity of junctional and H-receptors from a gradual merging of receptor types.

2   Investigated with ibotenate

Use was made of the fact that ibotenate acts as a specific excitant for H-receptors. This enables further investigation of the H-receptor dis-tribution in the junctional region where glutamate cannot be used as it produces large depolarizing responses which mask the H-response. Glutamate and ibotenate were applied to the same site from adjacent barrels of a double barrel micropipette, or, in a few experiments, from two single micropipettes. The glutamate barrel was necessary to locate the junction.

Judged by their sensitivity, H-receptors have an even distribution in the extrajunctional membrane. The sensitivity to ibotenate at the junc-tional region was similar to that seen in the rest of the membrane. It should be noted that while a significant increase in H-receptor sensitivity would be detected with this method, a significant decrease would be difficult to detect. This limitation arises as the width of the junction (judged by its sensitivity to glutamate) is narrow compared with the distance travelled by large pulses of drug.

B   DISTRIBUTION OF D-RECEPTORS

As no excitant was found which was specific for D-receptors, it was necessary to eliminate the H-response from muscle fibres before the D-receptors were mapped. This was achieved by removing the ions which generate the H-response (see below). D-receptors produce responses very similar in time course to the junctional response. The two types of responses can, however, be readily identified by their different sensitivi-ties to applied glutamate and by the difference in desensitization pro-duced by repetitive application of glutamate pulses.

Experiments of the type illustrated in Fig. 2.2 show that D-receptors occur in close proximity to the junction. With the micropipette positioned 40 μm from the junctional site a fast rising D-response precedes a slower rising junctional response. A second pulse of glutamate confirms that the first peak was an extrajunctional D-response as it is readily desensitized. Contiguity of D-receptors and junctional receptors cannot be distinguished from a merging of receptor types. It is perhaps most tempting to envisage the D-receptors simply increasing in sensitivity to form the junctional site.

C SENSITIVITY OF MUSCLE-TENDON JUNCTION

The sensitivity of the extrajunctional glutamate receptors increases in the region of the muscle-tendon junction. This shows an interesting analogy to the situation in vertebrate muscle where the sensitivity of extrajunctional ACh receptors also increases at the muscle-tendon junction (Miledi, 1962; Katz and Miledi, 1964a).

Application of glutamate, in quantities insufficient to produce appreciable extrajunctional responses will produce depolarizing responses when applied to the muscle-tendon junction. The increased sensitivity is present across the entire width of muscle at the tendon, extending outwards from the tendon by 20–40 μm to form a sensitive band (Fig. 3.3).

Extrajunctional receptors, in the region of the muscle-tendon junction, were tested with glutamate and ibotenate applied from adjacent barrels of a double barrel micropipette. The H-response to ibotenate shows no significant change from that encountered at other extrajunctional sites. On the other hand, the D-response to glutamate shows a gradual increase in sensitivity which begins 30 μm from the tendon (in most fibres investigated) and reaches a peak sensitivity at the actual muscle-tendon junction. Average sensitivities at the muscle tendon junction are about 20 times higher than the sensitivity of the D-receptors at other sites in the extrajunctional membrane.

It is interesting that connective tissue usually covers the muscle surface in the tendon region for a distance which closely follows the area of increased sensitivity. Removal of this tissue does not appreciably alter the area or the degree of sensitivity and it appears, therefore, not to interfere with the iontophoretic application of glutamate to the muscle surface.

D- and H-receptors which occur in the tendon region demonstrate

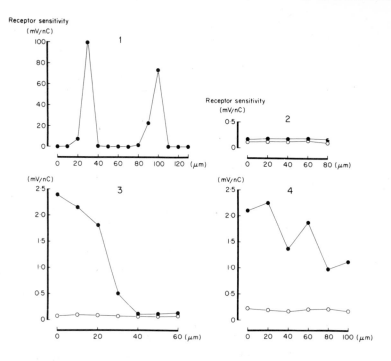

Fig. 3.1. Distribution of junctional receptor sensitivity. L-glutamate (0·1 nC) is applied from a high resistance micropipette moved in 10 µm steps across a muscle fibre of 130 µm width. Note that two separate junctional sites are detected. Closed circles: junctional sensitivity in mV/nC to iontophoretic pulses of glutamate. Abscissa: distance across the muscle fibre surface.

Fig. 3.2. Extrajunctional receptor sensitivity. Sensitivity of D- and H-receptors is measured in response to L-glutamate (1·0 nC) and D, L-ibotenate (1·0 nC) respectively. Drugs are applied from adjacent barrels of a high resistance micropipette which is moved in 20 µm steps across a muscle fibre of 80 µm width. Closed circles, D-receptor sensitivity to glutamate, open circles, H-receptor sensitivity to ibotenate, in mV/nC.

Fig. 3.3. Extrajunctional receptor sensitivity at the muscle-tendon junction. Sensitivity of D- and H-receptors is measured in response to L-glutamate and D, L-ibotenate respectively. Drugs are applied from adjacent barrels of a high resistance micropipette which is moved in 10 µm steps away from the muscle-tendon junction. Closed circles: D-receptor sensitivity to glutamate. Open circles: H-receptor sensitivity to ibotenate, in mV/nC. Abscissa: distance measured from the muscle-tendon junction.

Fig. 3.4. Extrajunctional receptor sensitivity in a muscle fibre 14 days after denervation. Sensitivity of D- and H-receptors is measured in response to iontophoretic pulses of L-glutamate and D, L-ibotenate respectively. The twin barrel micropipette is moved in 20 µm steps across a muscle fibre of 100 µm width. Closed circles: D-receptor sensitivity to glutamate. Open circles: H-receptor sensitivity to ibotenate, in mV/nC. (Partly after Cull-Candy, 1975.)

different rates of desensitization in contrast with D- and H-receptors at other extrajunctional sites which have approximately equal rates of desensitization. Repeated application of glutamate to receptors in the tendon region, in quantities large enough to activate D- and H-receptors, produces a biphasic response which is converted to a D-response as the H-receptors become desensitized.

The increased sensitivity of the D-receptors in the tendon region, without a similar change in H-receptor sensitivity, appears to indicate that the two types of extrajunctional glutamate receptors are independent.

D   DISTRIBUTION OF EXTRAJUNCTIONAL GLUTAMATE RECEPTORS AFTER DENERVATION AND MUSCLE DAMAGE

Following denervation the extrajunctional membrane of vertebrate muscle develops a high degree of sensitivity to ACh (Axelsson and Thesleff, 1959; Miledi, 1960). It is of interest that following denervation of locust muscle extrajunctional sensitivity develops in an apparently analogous manner (Usherwood, 1969b). Mapping the sensitivity of the denervated muscle fibre membrane by moving the glutamate micropipette in about 20 μm steps shows that the extrajunctional membrane attains a sensitivity similar to that of the junctional membrane in control muscles (Usherwood, 1969b). Surprisingly, no such change in sensitivity is seen in crayfish muscle, even following prolonged denervation (Frank, 1974).

Recent studies have shown that local muscle damage can alter the extrajunctional glutamate sensitivity of locust muscle (Cull-Candy, 1975), demonstrating a further similarity to extrajunctional ACh sensitivity in vertebrate muscle (Katz, and Miledi, 1964b). However, as two types of extrajunctional receptors occur in locust muscle, it was feasible that either or both of these could participate in the sensitivity changes seen after denervation or after local damage.

To determine the relative contribution of the receptor types to the increased sensitivity, receptor sensitivities in normal muscles were compared with those in muscles which were locally damaged or denervated. The sensitivities of the D-receptors and the H-receptors were determined by the local application of glutamate and ibotenate respectively from adjacent barrels of a micropipette. In situations where

D-receptor sensitivity was not elevated, their sensitivity was determined after first desensitizing the H-receptors with locally applied ibotenate.

As shown in Fig. 3.4, following denervation the D-receptor sensitivity increases by several fold over that seen in normal control muscles, while the H-receptor sensitivity shows no change, falling within the ranges encountered in innervated control muscles (Fig. 3.2). The increased sensitivity which develops in locally damaged fibres is highest at the site of damage and falls to near normal levels within about 100 μm The D-receptor sensitivity in the region of injury is about one hundred fold greater than that in the rest of the muscle membrane, while H-receptor sensitivity again is within the ranges encountered in control muscle and in the undamaged membrane of the same fibres. From these observations it seems that the extrajunctional D- and H-receptors are able to change independently (Cull-Candy, 1975).

VI Ionic permeability changes mediated by extrajunctional glutamate receptors

A PERMEABILITY CHANGES MEDIATED BY H-RECEPTORS

The H-responses produced when glutamate or ibotenate is applied iontophoretically to the same muscle fibre have very similar or identical equilibrium potential values, as described earlier (Fig. 1.1). This is to be expected if the species of ionophores operating in each instance are the same. The equilibrium potential of the H-response coincides with the Cl^- equilibrium potential for these fibres (see Usherwood and Grundfest, 1965) indicating that H-receptors mediate a conductance increase to Cl^- ions (Cull-Candy and Usherwood, 1973; Cull-Candy, 1976).

1 *Chloride involvement in the H-response*

This was tested by substituting the Cl^- in the bathing medium with the relatively impermeant methyl sulphate anion. The Cl^- concentra-

Fig. 4. The effect of Cl^- free medium on the extrajunctional response to glutamate. When bathed in normal medium the muscle fibre produces a DH-response (the D-component being only just visible). 5 m after removal of Cl^- the H-response reverses; a two component (DH) depolarization is seen. At 70 m, the H-response is abolished leaving a small D-component. In this fibre a small hyperpolarizing H-response, reappears after 80 m. Addition of Cl^- ions to the bathing medium results in the rapid reappearance of an H-response of a slightly larger amplitude than before Cl^- removal. Calibration: 1 mV and 200 ms. (From Cull-Candy, 1976.)

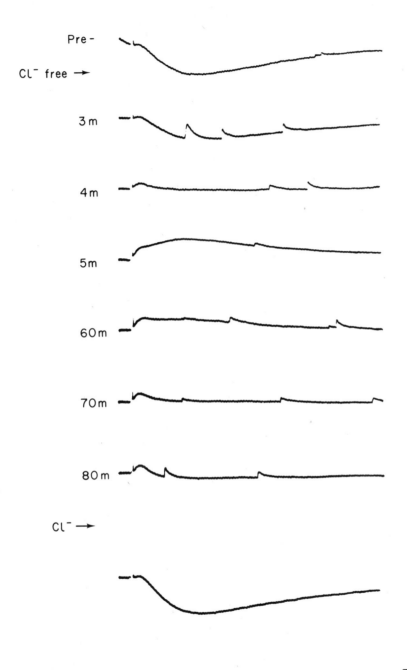

Pre-

Cl⁻ free →

3 m

4 m

5m

60 m

70 m

80 m

Cl⁻ →

tion on the outside of the fibre membrane would be expected to decrease rapidly while the Cl^- concentration inside the fibres would decrease rather more gradually. A rapid reversal of the driving force across the membrane, on the Cl^- ions, (see Boistel and Fatt, 1958; Fatt, 1961) causes a reversal of the H-response (Usherwood and Cull-Candy, 1974) as seen in the experiment shown in Fig. 4. The H-response diminishes with time as Cl^- leaks from the fibres. The small hyper-polarization which is seen after 80 min in this experiment may result from a slight methyl sulphate involvement in the H-response. The re-application of Cl^-, to the bathing medium, increases the Cl^- concentration outside the fibres and the H-response reappears as a hyper-polarization (Cull-Candy, 1976).

2 Picrotoxin action on the H-response

Cl^- permeability in response to GABA action at inhibitory synapses of lobster, crayfish and locust is depressed by picrotoxin (Grundfest et al., 1959; Usherwood and Grundfest, 1965; Takeuchi and Takeuchi, 1969) which acts as a non-competitive GABA antagonist (Takeuchi and Takeuchi, 1969). Complete and reversible block of the H-response to glutamate is obtained with 5×10^{-4} M picrotoxin (Cull-Candy, 1976; see also, Lea and Usherwood, 1973b), a concentration slightly higher than that expected to completely block the junctional response to GABA in crayfish muscle (Takeuchi and Takeuchi, 1969). Picro-toxin has no appreciable effect on the D-response produced by gluta-mate.

B PERMEABILITY CHANGES MEDIATED BY THE D-RECEPTORS

Values for extrapolated equilibrium potentials for the D-responses fell between the K^+-equilibrium potential (Hoyle, 1953; Hagiwara and Watanabe, 1953; Usherwood, 1969a) and the Na^+-equilibrium potential (Wood, 1963). As neither Ca^{2+} nor Cl^- appeared to be significantly involved, this indicates that an increased permeability to Na^+/K^+ is mediated by the D-receptors (Cull-Candy and Usherwood, 1973).

1 Sodium involvement in the D-response

Substituting choline for Na^+ in the bathing medium results in a rever-sible loss of the D-response (Fig. 5). However, some K^+ involvement remains possible, as an increased K^+ permeability would produce no net

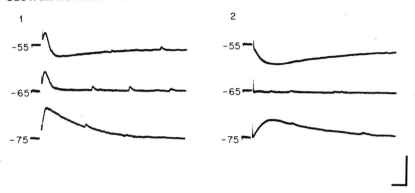

Fig. 5. The effect of Na^+ free medium on the extrajunctional response to ionto-phoretically applied L-glutamate.

Fig. 5.1. In normal bathing medium a typical DH-response to glutamate is seen at the resting potential of -55 mV. At -65 mV the H-response is annulled and at -75 mV a two component depolarization occurs producing a prolongation of the time course of the depolarizing response.

Fig. 5.2. In Na^+ free medium a D-component of the extrajunctional responses is no longer evident over the same range of membrane potentials; thus at -65 mV, the equilibrium potential of the H-component, no response is observed, and at -55 mV and -75 mV slow rise time responses (compared with the rise time of the D-component) are seen. Calibration: 1 mV and 200 ms.

ionic movement at the resting potential which coincides with the K^+ equilibrium potential (Usherwood, 1969a).

2   *Potassium involvement in the D-response*

Because of the problems encountered in altering the K^+ concentration in the bathing medium, the K^+ concentration was increased locally in the vicinity of the receptors under investigation by applying K^+ iontophoretically to the site of glutamate iontophoresis. K^+ and gluta-mate were applied either from two single micropipettes or from adjacent barrels of a twin barrel micropipette. The D-response was markedly increased by K^+ application, while the H-response was unchanged. Identical experiments in which Na^+ was applied locally produced no apparent change in the D-response. This is expected as the Na^+ concentration in the bathing medium is already high compared with the K^+ concentration.

On the basis of the pharmacological differences between the extrajunctional hyperpolarizing and depolarizing responses, two types of extrajunctional glutamate receptors occur in locust muscle, i.e. D-receptors and H-receptors (Cull-Candy, 1976). The difference in the distribution seen in the extrajunctional membrane, from that seen in the region of the muscle-tendon junction, seems to substantiate the existence of the two types of extrajunctional receptors. Further to this the extrajunctional D- and H-receptors are capable of changing independently following transection of the motor nerve supply to the muscle and following local damage of the muscle membrane (Cull-Candy, 1975).

In several of their properties extrajunctional D-receptors are similar to junctional glutamate receptors which occur in the same fibres and in this respect extrajunctional D-receptors are somewhat analogous to the extrajunctional ACh receptors which occur in vertebrate skeletal muscle. Thus extrajunctional D-receptors are pharmacologically similar to junctional glutamate receptors in their insensitivity to the conformationally restricted glutamate analogue ibotenate (Lea and Usherwood, 1973a; Cull-Candy, 1976). In addition, extrajunctional D-responses to glutamate result from a permeability increase to at least some of the ion species involved in the junctional response to glutamate. Perhaps the most striking similarity between extrajunctional D-receptors and extrajunctional ACh receptors is the increased sensitivity which both demonstrate at the muscle tendon junction and following denervation or mechanical damage of the muscle membrane (see Axelsson and Thesleff, 1959; Miledi, 1962; Katz and Miledi, 1964a, b; Usherwood, 1969b; Cull-Candy 1975).

In the ionic conductances which they generate, extrajunctional H-receptors are clearly similar to the GABA-receptors at inhibitory junctions which occur in certain muscle fibres of the locust (Usherwood and Grundfest, 1965). However, H-receptors occur in fibres both with and without inhibitory innervation (as judged by the presence of spontaneous inhibitory miniature potentials) (Usherwood and Cull-Candy, 1974). In addition, extrajunctional H-receptors have not been seen to be detectably sensitive to locally applied GABA (Cull-Candy, 1976). These observations seem to refute the possibility that H-receptors are simply extrajunctional GABA-receptors which also have glutamate sensitivity. Thus, although a relationship between H-receptors and either inhibitory innervation or GABA receptors is possible, at present the significance of the extrajunctional H-receptors remains enigmatic.

VII References

Anderson, C. R., Cull-Candy, S. G. and Miledi, R. (1976). Glutamate and quisqualate noise in voltage-clamped locust muscle fibres. *Nature Lond.* **261**, 151–153.

Anwyl, R. and Usherwood, P. N. R. (1974). Voltage clamp studies of glutamate synapse. *Nature Lond.* **252**, 591–593.

Anwyl, R. and Usherwood, P. N. R. (1975). The ionic permeability changes caused by the excitatory transmitter at the insect neuromuscular junction. *J. Physiol.* **249**, 24P.

Axelsson, J. and Thesleff, S. (1959). A study of supersensitivity in denervated mammalian skeletal muscle. *J. Physiol.* **147**, 178–193.

Beránek, R. and Miller, P. L. (1968). The action of iontophoretically applied glutamate on insect muscle fibres. *J. Exp. Biol.* **49**, 83–93.

Biscoe, T. J., Evans, R. H., Headley, P. M., Martin, M. and Watkins, J. C. (1975). Domoic and quisqualic acids as potent amino acid excitants of frog and rat spinal neurons. *Nature Lond.* **255**, 166–167.

Boistel, J. and Fatt, P. (1958). Membrane permeability change during inhibitory transmitter action in crustacean muscle. *J. Physiol.* **144**, 176–191.

Castillo, J. del, Hoyle, G. and Machne, X. (1953). Neuromuscular transmission in a locust. *J. Physiol.* **121**, 539–547.

Castillo, J. del and Katz, B. (1955). On the localization of acetylcholine receptors. *J. Physiol.* **128**, 157–181.

Clements, A. N. and May, T. E. (1974). Pharmacological studies on a locust neuromuscular preparation. *J. Exp. Biol.* **61**, 421–442.

Constanti, A. and Nistri, A. (1975). Actions of glutamate and kainic acid on the lobster muscle fibre and the frog spinal cord. *Br. J. Pharmacol.* **53**, 437.

Crawford, A. C. and McBurney, R. N. (1975). Glutamate "noise" at the excitatory neuromuscular junctions of a crayfish. *J. Physiol.* **251**, 73–74.

Cull-Candy, S. G. (1975). Effect of denervation and local damage on extrajunctional L-glutamate receptors in locust muscle. *Nature Lond.* **258**, 530–531.

Cull-Candy, S. G. (1976). Two types of extrajunctional L-glutamate receptors in locust muscle fibres. *J. Physiol.* **255**, 449–464.

Cull-Candy, S. G. and Usherwood, P. N. R. (1973). Two populations of L-glutamate receptors on locust muscle fibres. *Nature New Biol.* **246**, 62–64.

Cull-Candy, S. G. Neal, H. and Usherwood, P. N. R. (1973). Action of Black Widow Spider venom on an aminergic synapse. *Nature Lond.* **241**, 353–354.

Cull-Candy, S. G., Donnellan, J. F., James, R. W. and Lunt, G. G. (1976). 2-Amino-4-phosphonobutyric acid as a glutamate antagonist on locust muscle. *Nature Lond.* **262**, 708–709.

Curtis, D. R. and Watkins, J. C. (1960). The excitation and depression of spinal neurones by structurally related amino acids. *J. Neurochem.* **6**, 117–141.

Curtis, D. R. and Watkins, J. C. (1963). Acidic amino acids with strong excitatory actions on mammalian neurones. *J. Physiol.* **166**, 1–14.

Curtis, D. R., Duggan, A. W., Felix, D., Johnston, G. A. R., Tebēcis, A. K. and Watkins, J. C. (1972). Excitation of mammalian central neurones by acidic amino acids. *Brain Res.* **41**, 283–301.

Dowson, R. J. and Usherwood, P. N. R. (1973). The effect of low concentrations of L-glutamate and L-aspartate on transmitter release at the locust excitatory nerve-muscle synapse. *J. Physiol.* **229**, 13–14P.

Dowson, R. J., Clements, A. N. and May, T. E. (1975). The action of some harmala alkaloids on transmission at a glutamate mediated synapse. *Neuropharmacol.* **14**, 235–240.

Dreyer, F. and Peper, K. (1974). Iontophoretic application of acetylcholine: advantages of high resistance micropipettes in connection with an electronic current pump. *Pflügers Arch.* **348**, 263–272.

Dudel, J. (1974). Nonlinear voltage dependence of excitatory synaptic currents in crayfish muscle. *Pflügers Arch.* **352**, 227–241.

Dudel, J. (1975a). Potentiation and desensitization after glutamate induced post-synaptic currents at the crayfish neuromuscular junction. *Pflügers Arch.* **356**, 317–327.

Dudel, J. (1975b). Kinetics of postsynaptic action of glutamate pulses applied iontophoretically through high resistance micropipettes. *Pflügers Arch.* **356**, 329–246.

Eugster, C. H. (1967). Isolation, structure and synthesis of central-active compounds from Amanita muscaria (L. ex Fr.) hooker. *In* "Ethnopharmacologic search for psychoactive drugs" (D. H. Efron, B. Holmstedt and N. S. Kline, eds), pp. 416–418. U.S. public health service publication no. 1645.

Fatt, P. (1961). The change in membrane permeability during the inhibitory process. *In* "Nervous Inhibition" (E. Florey, ed.), pp. 87–91. Pergamon Press, Oxford.

Feltz, A. and Mallart, A. (1971). An analysis of the acetylcholine responses of junctional and extrajunctional receptors of frog muscle fibres. *J. Physiol.* **218**, 85–100.

Frank, E. (1974). The sensitivity to glutamate of denervated muscles of the crayfish. *J. Physiol.* **242**, 371–382.

Gerschenfeld, H. M. (1973). Chemical transmission in invertebrate central nervous systems and neuromuscular junctions. *Physiol. Rev.* **53**, 1–119.

Grundfest, H., Reuben, J. P. and Rickles, W. H. (1959). The electrophysiology and pharmacology of lobster neuromuscular synapses. *J. Gen. Physiol.* **42**, 1301–1323.

Hagiwara, S. and Watanabe, A. (1953). Action potential of insect muscle examined with intra-cellular electrode. *Jap. J. Physiol.* **4**, 65–78.

Haldeman, S. and McLennan, H. (1972). The antagonistic action of glutamic acid diethylester towards amino acid-induced and synaptic excitations of central neurones. *Brain Res.* **45**, 393–400.

Hoyle, G. (1953). Potassium ions and insect nerve muscle. *J. Exp. Biol.* **30**, 121–135.

James, R. W., Lunt, G. G. and Donnellan, J. F. (1974). Glutamate receptors from insect muscle. Proc. 9th F.E.B.S. meeting. Hung. biochem. Soc. Budapest, p. 258.

Johnston, G. A. R., Curtis, D. R., de Groat, W. C. and Duggan, A. W. (1968). Central actions of ibotenic acid and muscimol. *Biochem. Pharmacol.* **17**, 2488–2489.

Johnston, G. A. R., Curtis, D. R., Davies, J. and McCulloch, R. M. (1974). Spinal interneurone excitation by conformationally restricted analogues of L-glutamic acid. *Nature Lond.* **248**, 804–805.

Katz, B. and Miledi, R. (1964a). Further observations on the distribution of acetyl-choline-reactive sites in skeletal muscle. *J. Physiol.* **170**, 379–388.

Katz, B. and Miledi, R. (1964b). The development of acetylcholine sensitivity in nerve-free segments of skeletal muscle. *J. Physiol.* **170**, 389–396.

Katz, B. and Miledi, R. (1972). The statistical nature of the acetylcholine potential and its molecular components. *J. Physiol.* **224**, 665–699.

Katz, B. and Thesleff, S. (1957). A study of the "desensitization" produced by acetylcholine at the motor end-plate. *J. Physiol.* **138**, 63–80.

Kawai, N., Mauro, A. and Grundfest, H. (1972). Effect of black widow spider venom on the lobster neuromuscular junctions. *J. Gen. Physiol.* **60**, 650–664.

Kravitz, E. A., Slater, C. R., Takahashi, K., Bownds, M. D. and Grossfeld, R. M. (1970). Excitatory transmission in invertebrates—glutamate as a potential neuromuscular transmitter compound. *In* "Excitatory Synaptic Mechanisms" (P. Andersen and J. K. S. Jansen, eds), pp. 84–93. Universitetsforlaget, Oslo.

Krnjević, K. and Phyllis, J. W. (1963). Iontophoretic studies of neurones in the mammalian cerebral cortex. *J. Physiol.* **165**, 274–304.

Lea, T. J. and Usherwood, P. N. R. (1973a). The site of action of ibotenic acid and the identification of two populations of glutamate receptors on insect muscle fibres. *Comp. Gen. Pharmacol.* **4**, 333–350.

Lea, T. J. and Usherwood, P. N. R. (1973b). Effect of ibotenic acid on chloride permeability of insect muscle fibres. *Comp. Gen. Pharmacol.* **4**, 351–363.

Lowagie, C. and Gerschenfeld, H. M. (1974). Glutamate antagonists at a crayfish neuromuscular junction. *Nature Lond.* **248**, 533–535.

McDonald, T. J. and O'Brien, R. D. (1972). Relative potencies of L-glutamate analogs on excitatory neuromuscular synapses of the grasshopper, *Romalea microptera*. *J. Neurobiol.* **3**, 277–290.

Miledi, R. (1960). Junctional and extra-junctional acetylcholine receptors in skeletal muscle fibres. *J. Physiol.* **151**, 24–30.

Miledi, R. (1962). Induction of receptors. *In* "Ciba Found. Symp. Enzymes and Drug action," pp. 220–235. Churchill, London.

Nistri, A. and Constanti, A. (1975). Effects of glutamate and glutamic acid diethyl ester on the lobster muscle fibre and the frog spinal cord. *Eur. J. Pharmacol.* **31**, 377–379.

Ozeki, M. and Grundfest, H. (1967). Crayfish muscle fibre: Ionic requirements for depolarizing synaptic electrogenesis. *Science.* **155**, 478–481.

Peper, K. and McMahan, U. J. (1972). Distribution of acetylcholine receptors in the vicinity of nerve terminals on skeletal muscle of the frog. *Proc. R. Soc. Lond.* B **181**, 431–440.

Rees, D. and Usherwood, P. N. R. (1972). Fine structure of normal and degenerating motor axons and nerve-muscle synapses in the locust, *Schistocerca gregaria. Comp. Biochem. Physiol.* **43**, 83–101.

Shinozaki, H. and Konishi, S. (1970). Actions of several anthelmintics and insecticides on rat cortical neurones. *Brain Res.* **24**, 368–371.

Shinozaki, H. and Shibuya, I. (1974). A new potent excitant, quisqualic acid: effects on crayfish neuromuscular junction. *Neuropharmacology* **13**, 665–672.

Shinozaki, H., Shibuya, I. and Natsugoe, K. (1975). Excitatory action of L-glutamate analogues on crayfish synapse. *In* "Sixth International Congress in Pharmacology, Abstracts." p. 25. Finnish Pharmacological Society.

Takeuchi, A. and Onodera, K. (1973). Reversal potentials of the excitatory transmitter and L-glutamate at the crayfish neuromuscular junction. *Nature New Biol.* **242**, 124–126.

Takeuchi, A. and Takeuchi, N. (1963). Glutamate-induced depolarization in crustacean muscle. *Nature Lond.* **198**, 490–491.

Takeuchi, A. and Takeuchi, N. (1964). The effect on crayfish muscle of iontophoretically applied glutamate. *J. Physiol.* **170**, 296–317.

Takeuchi, A. and Takeuchi, N. (1969). A study of the action of picrotoxin on the inhibitory neuromuscular junction of the crayfish. *J. Physiol.* **205**, 377–391.

Taraskevich, P. S. (1971). Reversal potentials of L-glutamate and the excitatory transmitter at the neuromuscular junction of the crayfish. *Biochim. Biophys. Acta* **241**, 700–701.

Taraskevich, P. S. (1975). Dual effect of L-glutamate on excitatory postjunctional membranes of crayfish muscle. *J. Gen. Physiol.* **65**, 677–691.

Usherwood, P. N. R. (1969a). Electrochemistry of insect muscle. In "Advances in Insect Physiology" (J. W. L. Beament, J. F. Treherne and W. B. Wigglesworth, eds), Vol. 6. Academic Press, London and New York.

Usherwood, P. N. R. (1969b). Glutamate sensitivity of denervated insect muscle fibres. *Nature Lond.* **223**, 411–413.

Usherwood, P. N. R. and Cull-Candy, S. G. (1974). Distribution of glutamate sensitivity on insect muscle fibres. *Neuropharmacol.* **13**, 455–461.

Usherwood, P. N. R. and Cull-Candy, S. G. (1975). Pharmacology of somatic nerve-muscle synapses. *In* "Insect Muscle" (P. N. P. Usherwood, ed.), pp. 207–280. Academic Press, London and New York.

Usherwood, P. N. R. and Grundfest, H. (1965). Peripheral inhibition in skeletal muscle of insects. *J. Neurophysiol.* **28**, 497–518.

Usherwood, P. N. R. and Machili, P. (1968). Pharmacological properties of excitatory neuromuscular synapses in the locust. *J. Exp. Biol.* **49**, 341–361.

Walker, R. J., Woodruff, G. N. and Kerkut, G. A. (1971). The effect of ibotenic acid and muscimol on single neurones of the snail, *Helix aspersa. Comp. Gen. Pharmacol.* **2**, 168–174.

Walther, C. and Usherwood, P. N. R. (1972). Characterization of glutamate receptors at the locust excitatory neuromuscular junction. *Verh. Deut. Zool. Ges. Helgoland* **65**, 309–312.

Wood, D. W. (1963). The sodium and potassium composition of some insect skeletal muscle fibres in relation to their membrane potentials. *Comp. Biochem. Physiol.* **9**, 151–159.

10

The Role of Activity in the Control of Membrane and Contractile Properties of Skeletal Muscle

T. Lömo

Institute of Neurophysiology, University of Oslo, Norway

I Introduction

Neural influences are essential for the control of membrane and contractile properties of muscle as shown by the changes in these properties following denervation and their return towards normal following subsequent reinnervation. The motoneurone may influence the properties of a muscle in two basically different ways; either by virtue of its ability to control the activity of the muscle cell or by means of some other mechanism independent of muscle activity, e.g. the release of substances from the nerve terminals often referred to as trophic factors. The effects of denervation could be explained by either of these two not mutually exclusive mechanisms and there is continuing argument about their relative importance (Drachman, 1974; Fambrough, 1974; Harris, 1974; Thesleff, 1974). Here I would like to review some recent experimental results from this laboratory which bear upon this problem.

II Membrane properties

A EXTRAJUNCTIONAL SENSITIVITY TO ACETYLCHOLINE (ACH)

Since the pioneering studies by Axelsson and Thesleff (1959) and Miledi (1960) the distribution of ACh sensitivity along individual muscle fibres has been studied in great detail and with considerable accuracy with iontophoretic application of ACh from micropipettes. Such studies show that denervation causes a dramatic and largely uniform increase in the sensitivity of the extrajunctional membrane to ACh (Fambrough, 1970; Lömo and Westgaard, 1975). In rat soleus muscle the increase (1000 fold or more) occurs mainly between the second and the third day after denervation, with little further increase after this time, when

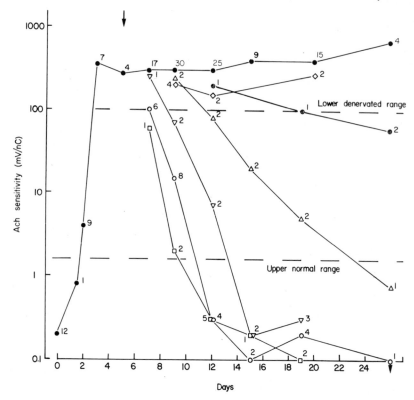

Fig. 1. Direct muscle stimulation causes denervation hypersensitivity to ACh to decline at a rate which depends on the number of stimuli. Each symbol gives the median sensitivity of many soleus fibres from a number of soleus muscles (as indicated) measured 2 mm from the Achilles tendon at different days after the denervation. ACh sensitivity in this and subsequent figures is expressed as mV depolarization per nC charge passed through the micropipette containing 3 M AChCl. ● denervated control fibres. Open symbols: Fibres stimulated from day 5 after the denervation (arrow) with identical 10 s trains of stimuli at 10 Hz. Each train was interrupted by 0 s (□), 1 min 30 s (○), 16 min 40 s (▽), 2 h 46 min 30 s (△), 5 h 33 min 10 s (⊕) or 12 h (◇) so that the average frequency of stimulation was 10, 1, 0·1, 0·01, 0·005 or 0·0023 Hz respectively. The interrupted line at 100 mV/nC indicates the lowest ACh sensitivity found in more than 3 days denervated muscles. The interrupted line at 1·6 mV/nC gives the highest sensitivity found in normal fibres. (Reprinted, with permission, from Lömo and Westgaard, 1975.)

measured in this way (Fig. 1). Direct stimulation through chronically implanted electrodes abolishes this hypersensitivity at a rate which depends critically on the amount and pattern of the stimuli (Figs 1, 2). In these experiments stimulation was started 5 days after the denervation. At this time the nerve terminals are completely degenerated and removed by phagocytosis (Miledi and Slater, 1968; Gonzenbach and Waser, 1973) and the hypersensitivity is virtually fully developed (Fig. 1). Thus, the effectiveness of different stimulation procedures could be assessed in terms of the rate at which they caused the hypersensitivity to decline and any concomitant neural effects, mediated for example by the release of trophic substances, could be ruled out.

The amount of muscle activity was varied by presenting identical trains of stimuli (10 Hz for 10 s) interrupted by pauses of up to 12 h, so that the overall average frequency ranged from 10 to 0·0023 Hz (Fig. 1). These procedures reduced the extrajunctional sensitivity to 1 mV/nC (within the normal range) within times varying from 5 days (10 Hz mean) to 20 days (0·01 Hz mean). One train (100 stimuli) every 5 h 33 min (0·005 Hz mean) gradually reduced the sensitivity to one tenth of the denervated control values after 3 weeks of stimulation, whereas one train every 12 h had no apparent effect.

The pattern of activity was varied by changing the timing of a constant number of stimuli presented to the muscle every day (Fig. 2). Stimulation at 100 Hz for 1 s every 100 s reduced the sensitivity to 1 mV/nC within 4 days, whereas 10 Hz for 10 s every 100 s and continuous 1 Hz stimulation required 6 and 10 days respectively to bring the sensitivity to the same low levels. The stimulation became much less effective when long pauses (18 h) alternated with shorter periods (6 h) of more vigorously imposed activity (10 Hz for 10 s every 25 s).

The differences in rate of decline of ACh sensitivity for each of the stimulation series, shown in Figs 1 and 2, were of a high statistical significance ($P < 0·001$, Wilcoxon Two-sample test, Lömo and Westgaard, 1975). For most stimulation procedures the sensitivity of the stimulated fibres eventually became as low as or lower than that of normally innervated fibres. The sensitivity declined at roughly the same rate along the entire length of the muscle fibres except at the end plate where the sensitivity remained high and indistinguishable from that of normally innervated fibres.

In a different series of experiments, stimulation was stopped on day 15 when the extrajunctional membrane had become insensitive to ACh.

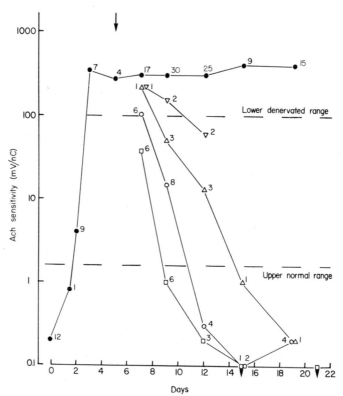

Fig. 2. Direct muscle stimulation causes denervation hypersensitivity to ACh to decline at a rate which depends on the pattern of stimuli. The average frequency of stimulation was 1 Hz in each of the four stimulation procedures. Each symbol gives the median sensitivity of many fibres from a number of soleus muscles (as indicated) measured 2 mm from the Achilles tendon at different days after the denervation. ● denervated control fibres. Open symbols: fibres stimulated from day 5 after the denervation (arrow) with 1 s stimulus trains at 100 Hz every 100 s (□), 10 s stimulus trains at 10 Hz every 100 s (○), continuous 1 Hz stimulation (△) or 10 s stimulus trains at 10 Hz every 25 s for 6 h every 24 h (▽). The interrupted horizontal lines have the same meaning as in Fig. 1. (Reprinted, with permission, from Lömo and Westgaard, 1975.)

This caused the hypersensitivity to reappear but later, after stimulation with brief 100 Hz trains than with longer 10 Hz trains (Fig. 3, average frequency 1 Hz in both cases). Relative to the onset of inactivity, ACh hypersensitivity in an acutely denervated soleus develops still faster (Fig. 1).

Some aspects of the results shown in Figs 1–3 may be emphasized and commented upon. The rate of decline of ACh hypersensitivity in-

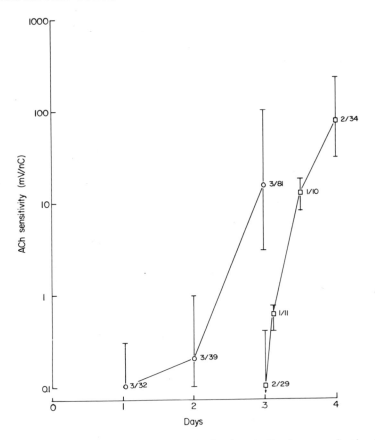

Fig. 3. Development of ACh hypersensitivity in chronically denervated, stimulated soleus muscles following cessation of stimulation. The pattern of stimulation was either "slow" (○, 10 Hz for 10 s every 100 s) or "fast" (□, 100 Hz for 1 s every 100 s). The muscles had been denervated and stimulated for the 15 days preceding day 0 (abscissa) when the stimulation was stopped. Each symbol gives the median ACh sensitivity of single fibres 2 mm from the Achilles tendon. The number of muscles and fibres is indicated.

creases in an apparently continuously graded manner with increases in the number of stimuli or, if the number of stimuli is kept constant, with the stimulus frequency within the trains. The greater effectiveness of high frequency stimulation is strikingly illustrated by the finding that 100 Hz stimulation caused the sensitivity to decline faster than 10 times as many stimuli at 10 Hz and 10 Hz stimulation, about as fast as 10 times as many stimuli at 1 Hz. High frequency stimulation was also the more effective in delaying the reappearance of hypersensitivity when stimulation was stopped. This suggests that suppression of ACh receptor formation is graded even at levels when no extrajunctional ACh sensitivity can be detected by the methods used.

The rate of decline of ACh sensitivity was approximately exponential with a half-time of about 9 h during optimal stimulation. A half-time of about 8 h has been reported for the loss of α-bungarotoxin from the extrajunctional membrane of the denervated diaphragm (Berg and Hall, 1974). Although other possibilities exist, this loss may represent degradation of the receptor-toxin complex and reflect a normal turnover of the extrajunctional receptors. It seems an attractive possibility, therefore, that optimal stimulation blocks all synthesis of extrajunctional receptors with the result that the ACh sensitivity falls at a rate determined by the degradation of the extrajunctional receptors.

We have called the 100 Hz and the 10 Hz stimulus patterns, both having an average frequency of stimulation of 1 Hz, fast and slow because they resemble somewhat the pattern of motor unit activity of fast and slow rat muscles respectively (Fischbach and Robbins, 1969) and because only the fast pattern induces contractile properties characteristic of fast muscle in the denervated slow soleus muscle (Lömo et al., 1974; and below). The fact that a fast pattern suppressed ACh receptor synthesis more effectively than a slow pattern, together with possible differences in overall amount of activity, may help to explain why denervation hypersensitivity appears more quickly in the slow soleus than in the fast (EDL) extensor digitorum longus (Albuquerque and McIsaac, 1970), and why some normal soleus fibres have a low but clearly detectable extrajunctional ACh sensitivity while all EDL fibres are insensitive (Miledi and Zelená, 1966).

Even very low levels of activity strongly suppress ACh sensitivity. As few as 100 stimuli (10 Hz for 10 s) given every 5 h 33 min or 2 h 45 min for 3 weeks caused respectively a 10 fold or 100 fold reduction in ACh sensitivity and would probably have caused a further reduction had the stimulation been continued. This may explain why immobilization by limb fixation (Solandt et al., 1943; Fischbach and Robbins, 1971) or by de-afferentation and isolation of the spinal centres (Solandt and Magladery, 1942; Johns and Thesleff, 1961) causes only a moderate or very small increase in the ACh sensitivity. In these preparations muscle activity is greatly reduced but probably not completely eliminated.

Six hours of frequent (every 25 s) stimulus trains (10 Hz for 10 s) repeated every 24 h had much less effect than the same number of similar trains given throughout the 24 h (every 100 s). Indeed, one hundredth of the number of trains, i.e. one train given regularly every 2 h 45 min, was equally effective. Evidently, long periods without externally imposed activity (18 h in this example) markedly reduce the effectiveness of the stimulation and more than offsets the effects of more vigorous activity during the active periods. This may help to explain why fibrillatory activity, which on average is relatively high (1 Hz) in mammalian muscles kept in organ culture (Purves and Sakmann, 1974) and presumably also *in vivo* (Belmar and Eyzaguirre, 1966), does not prevent denervation hypersensitivity. Fibrillatory activity, at least in organ culture, is cyclic with long inactive periods (probably 2–3 days) alternating with shorter periods (22–23 h) of relatively intense activity (Purves and Sakmann, 1974). These authors mimicked periods of spontaneous activity by stimulating directly for 24 h at 10 Hz. Subsequent spontaneous activity was suppressed for 1–3 days suggesting that fibrillatory activity is self-inhibiting, but extrajunctional ACh sensitivity was only slightly reduced. *In vivo*, direct stimulation for 24 h at 10 Hz has negligible effects on denervation hypersensitivity (Lömo and Westgaard, unpublished). It appears that fibrillatory activity is likely to be too short lasting relative to the duration of the inactive periods to cause any significant reduction in extrajunctional ACh sensitivity.

B ELECTRICAL CHARACTERISTICS

1 *Resting membrane potential*

One to two days after denervation the resting membrane potential (RMP) of soleus fibres falls rapidly by about 10–15 mV. If the muscle is stimulated directly from day 5 after denervation with the "slow" stimulus pattern (average frequency of stimulation 1 Hz, see above) the RMP returns to normal with a time course resembling that of its fall after denervation (Fig. 4). With more vigorous stimulation (average frequencies of stimulation about 8 Hz) the RMP becomes higher than normal (Lömo, 1974; Westgaard, 1975).

Albuquerque *et al.* (1971) have reported a marked fall in RMP of the EDL as early as 2–3 h after a crush to the nerve close to the muscle. This is contrary to other reports (Redfern and Thesleff, 1971). In more recent, similar experiments other workers have failed to find such an early fall in RMP (Diana Card, personal communication). In the soleus, the fall in RMP does not appear clearly to precede the onset of

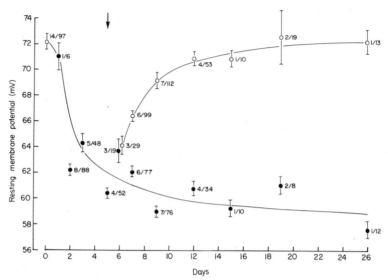

Fig. 4. Mean resting membrane potential plotted against time after denervation for merely denervated soleus fibres (●) and for denervated fibres stimulated at 10 Hz for 10 s every 100 s from day 5 after denervation (○). Numbers near each symbol indicate number of muscles and fibres, bars indicate standard errors of the mean. (Reprinted, with permission, from Lömo and Westgaard, 1975.)

other denervation changes like failing neuromuscular transmission, fibrillation (Salafsky *et al.*, 1968) and ACh hypersensitivity (Fig. 1). The fall in RMP may be related to a fall in resting potassium conductance (Klaus, *et al.*, 1960; Kernan and McCarthy, 1972) or in active transport of Na (Locke and Solomon, 1967) but its precise mechanism is not known.

2 *Specific membrane resistance* (Rm) *and capacitance* (Cm)

These properties begin to change approximately 2–3 days after denervation (Albuquerque and Thesleff, 1968; Albuquerque and McIsaac, 1970). In a study on rat soleus muscles Westgaard (1975) found the average R_i to increase from 766 to 2291 Ω cm^2 and Cm to decrease from 3·6 to 2·7 μF/cm^2 during 19 days of denervation. Direct stimulation from day 5 to day 19 restored these properties completely to normal (760 Ω cm and 3·5 μF/cm^2). The input resistance did not become completely normal (413 kΩ *v.* the normal 262 kΩ) but this could be accounted for by the finding that the stimulation only arrested further atrophy when it was started on day 5 but did not return the fibres to their original size (see also below).

C RECEPTIVITY TO INNERVATION

Mammalian skeletal muscle fibres are normally innervated by a single axon, and the presence of the original innervation prevents synapse formation by an implanted foreign nerve. If the original nerve is cut (Elsberg, 1917), poisoned with botulinum toxin (Fex *et al.*, 1966), or reversibly blocked (Jansen *et al.*, 1973), however, the foreign axons will innervate the muscle. This synapse formation is prevented in the denervated soleus by direct electrical stimulation indicating that muscle activity *per se* normally prevents hyperinnervation in adult muscles (Jansen *et al.*, 1973).

Electrical stimulation, however, does not prevent reinnervation at the original denervated end plate (Fig. 5.1). Figure 5.1 is also an illus-

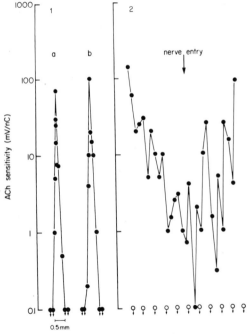

Fig. 5. Direct electrical stimulation abolishes extrajunctional ACh sensitivity but has no effect on junctional ACh sensitivity nor on reinnervation of denervated end plates. 1. Examples of ACh sensitivity at a reinnervated (a) and not yet reinnervated (b) end plate 14 days after a crush to the soleus nerve and continuous stimulation at 100 Hz for 10 s every 100 s. 2. Extrajunctional ACh sensitivity of fibres spaced at approximately equal intervals across the muscle, 2 mm from the Achilles tendon, in the stimulated (○) and in the unstimulated control muscle (●).

tration of the essentially normal sensitivity to ACh at denervated end plates in electrically stimulated muscles. One of the two illustrated end plates (a) was reinnervated 14 days after a crush to the soleus nerve. 16 days after the crush 98% and 96% of the soleus surface fibres were reinnervated in stimulated and unstimulated muscles respectively. Thus, no extrajunctional ACh sensitivity is required for reinnervation of denervated end plates at least at these relatively short times after denervation when the end plates retain their high ACh sensitivity despite electrical stimulation. We do not know if electrical stimulation would prevent reinnervation by regenerating nerves arriving much later at the denervated end plates but this seems a possibility (see below). It has been suggested that ACh receptors may be required for synapse formation in muscle (Katz and Miledi, 1964; Fex *et al.*, 1966) but there is now evidence that synapses form also when the ACh binding part of the ACh receptor is blocked by curare or α-bungarotoxin (Crain and Peterson, 1971; Cohen, 1972; Van Essen and Jansen, 1974).

Figure 2 shows that 14 days after the crush the unstimulated muscle was least hypersensitive near the nerve entry and progressively more sensitive towards both edges of the muscle where reinnervation occurs later. The first signs of contraction were seen near the nerve entry 9 days after the crush. At this time all fibres tested in the unstimulated muscle were fully hypersensitive. At 14 days, most fibres near the nerve entry had not quite returned to normal levels of sensitivity (Fig. 5.2). These findings suggest that the sensitivity begins to fall after the onset of contraction and with a time course which resembles that in directly stimulated muscles (Fig. 1).

Ectopic innervation and direct electrical stimulation have similar effects on the denervated soleus. Both abolish all extrajunctional ACh hypersensitivity but fail to affect the sensitivity and subsequent reinnervation at the old denervated end plates. Single denervated soleus fibres may therefore become doubly innervated, first by the transplanted foreign nerve and later by regeneration of the original nerve. Studies of such preparations show that once the two types of synapses have formed on the same fibre they may coexist and function for very long times (a year or more). In this preparation there was no indication that one synapse, such as the "appropriate" soleus synapse, could suppress the other "inappropriate" one (Frank *et al.*, 1975). It appears that muscle activity whether electrically or neurally evoked, effectively prevents *de novo* synapse formation but has no effect on already established synapses.

How quickly new synapses become established or resistant to activity is presently under study (Lömo and Slater, in preparation). Examination of the distribution of ACh sensitivity in doubly innervated soleus fibres 2–3 weeks after section of the foreign fibular nerve shows a high focal ACh sensitivity associated with AChE activity at the denervated new end plates (Frank et al., 1975). Evidently the new nerve, before it is cut, induces an enduring postsynaptic change in the muscle membrane comparable to that at a normal end plate.

If the tibial nerve is resected rather than crushed many fewer regenerating nerve fibres innervate muscle fibres already innervated by the foreign fibular nerve probably because they arrive much later at the denervated end plates. This suggests that the denervated end plates gradually loose their "innervatability" although they may still be quite sensitive to ACh (Frank et al., 1975). Thus it is possible, but not shown, that continued long-term electrical stimulation eventually would make the denervated muscle fibre entirely unreceptive of reinnervation.

D POSSIBLE MECHANISMS FOR THE EFFECTS OF MUSCLE ACTIVITY/INACTIVITY

All the properties of the extrajunctional membrane so far examined are restored to normal by direct stimulation of the denervated rat soleus. Stimulation also reverses the increased uptake of uridine 5–^{3}H in denervated muscles (Muchnik and Kotsias, 1975) and mimicks neural influences in decreasing the extrajunctional AChE activity and release of AChE from chick muscle fibres maintained in vitro (Walker and Wilson, 1975). The effect does not require the presence of nerve terminals. Direct stimulation also suppresses ACh sensitivity of adult denervated muscles (Purves and Sakmann, 1974) or embryonic non-innervated myotubes (Cohen and Fischbach, 1973) maintained in vitro. Therefore, the effect of stimulation, at least on extrajunctional ACh sensitivity, seems to be independent of influences carried in the nerve, in the blood or remaining in the fibre from previous innervation. Instead it seems that the fibre responds by some intrinsic mechanism to variations in muscle activity per se.

The continuously graded relation between different types and levels of activity and the rate of change of extrajunctional ACh sensitivity suggests a close coupling between activity and ACh receptor synthesis and turnover. The nature of this coupling is unknown but any substance whose free concentration in the muscle fibre varies with the type of activity might be a mediator. Calcium ions and cAMP are both possible candidates. In heart muscle the concentration of cAMP varies in synchrony with the contractions (Brooker, 1973). In skeletal muscle it increases after denervation and decreases after reinnervation (Carlsen, 1975). Tomkins (1975) has recently advanced the general hypothesis that particular environmental changes, e.g. starvation, elicit an intracellular accumulation of certain molecules, e.g. cAMP, which serve as symbols of the environmental change to bring about a complex coordinated cellular response of adaptive value to the cell. There are increasing indications that Ca^{2+} might assist in such a function (Rasmussen, 1970; McMahon, 1974; Brostrom et al., 1975). For lymphocytes and fibroblasts there are suggestions that variations in the internal concentration of calcium may control cell proliferation which involves long-term activation of numerous intracellular systems (Freedman et al., 1975; Dulbecco and Elkington, 1975).

In muscle persistent changes in the concentration of Ca^{2+} or cAMP (or some other molecule) might represent a symbol of inactivity which elicits all or most of the extrajunctional responses to denervation. The value of these responses might be facilitation of subsequent reinnervation by sprouts from neighbouring intact nerves or from regenerating nerves and better maintenance of the fibre by fibrillatory activity during nerve regeneration.

Denervation hypersensitivity is probably a result of de novo synthesis of ACh receptor protein (Fambrough, 1970; Grampp et al., 1972; Brockes and Hall, 1975) and other specific properties of the denervated membrane may have a similar basis (Grampp et al., 1972). In agreement with this the RNA content and uracil and uridine $5-^3H$ uptake increase after denervation (Manchester and Harris, 1968; Muchnik and Kotsias, 1975). Extrajunctional ACh receptors, fibrillations, fall in resting membrane potential and increase in specific membrane resistance all appear at roughly the same time after denervation; in the rat soleus around the second day. Whether on a finer timescale some changes precede others is at present not clear. However, the data are consistent with the notion that denervation causes the concentration of some molecule ("symbol") to change with the result that numerous genes become activated simultaneously (Tomkins, 1975).

The suppression of ACh sensitivity might be linked to electrical activity in the membrane, the excitation-contraction coupling, contractile activity or some other accompaniment of muscle activity. Which of these factors is the more important is not known. It has been reported that stimulation *in vivo* with stimuli too weak to cause contraction reduces somewhat the sensitivity of denervated muscles to bath applications of ACh. Stronger stimuli giving maximal contractions *in vivo* had no greater effect (Gruener *et al.*, 1974). On the other hand, in our experience only vigorous contractions *in vivo* prevent denervation hypersensitivity in all fibres. With slightly weaker stimulation or some displacement of the stimulating electrodes the contractions may still appear vigorous by inspection and palpation and yet fully hypersensitive fibres, usually in one part of the muscle, may appear next to insensitive fibres. This suggests to us that the critical difference in this preparation may be absence or presence of impulse activity. Impulse activity, however, is always followed by Ca^{2+} release, activation of myofibrils and various other effects and these may well be equally important for suppression of ACh sensitivity.

Denervation hypersensitivity has been reported to decline before transmission is reestablished by regenerating axons (Miledi, 1960; Bennet *et al.*, 1973). However, it is difficult to rule out that fibres which do not respond to nerve stimulation *in vitro* and yet have a low extrajunctional ACh sensitivity are not subject to a presynaptic block of impulse conduction since this often occurs *in vitro* at early stages of reinnervation. In the avian posterior latissimus dorsi (PLD) crossreinnervation with the nerve to the slow anterior latissimus dorsi (ALD) also reduces the extrajunctional hypersensitivity to ACh (Vyskočil and Vyklický, 1974), although stimulation of the foreign nerve *in vitro* gives small e.p.p.s with a mean quantal content of only 2·4 (Vyskočil *et al.*, 1971). However, more than six months after the operation the fibres had a low input resistance (mean value 0·38 MΩ for cross-reinnervated and 0·51 MΩ for control PLD muscles), and this suggests that some activity may have occurred *in vivo* since this muscle undergoes rapid atrophy after denervation (Jirmanová and Zelená, 1970). Such activity, however, may not be dependent on impulse activity of the muscle (Vyskočil and Vyklický, 1974). In multiply innervated fibres, for example tonic (or slow) frog muscle fibres or avian ALD muscles, extrajunctional ACh sensitivity is low and increases after denervation (Nasledov and Thesleff, 1974; Bennett *et al.*, 1973) although the frog fibres (Burke and Ginsborg, 1956) and possibly ALD fibres (see Hess, 1970) lack an action potential generating mechanism. Thus suppression of ACh sensitivity may not be coupled to action potential generation *per se* but, more likely, to some other event such as distributed membrane depolarizations, excitation-contraction coupling or contraction which occurs whether the fibres have action potentials or not.

Denervation of parasympathetic ganglion cells in the frog heart also increases the ACh sensitivity of the extrajunctional membrane (Kuffler *et al.*, 1971). Thus, suppression of ACh sensitivity may not necessarily be related to contractile activity. Instead, it seems likely that the maintenance of a normal cell membrane is related to the particular activity for which the cell is specialized. Although this is contraction in a muscle cell, impulse activity and transmitter release in a nerve cell or secretion in a glandular cell, the mediator might be the same in each cell.

E   FACTORS, UNRELATED TO ACTIVITY, WHICH INFLUENCE MEMBRANE PROPERTIES

In some situations, factors other than activity may influence the properties of the muscle membrane. For example, ACh hypersensitivity and other signs of denervation develop sooner if the nerve is cut close to rather than far from the muscle, although presumably in both cases muscle activity stops immediately (Luco and Eyzaguirre, 1955; Harris and Thesleff, 1972). ACh hypersensitivity develops around denervated end plates in frog sartorius fibres whose activity is maintained by intact end plates on the same fibres (Miledi, 1960). These effects could be due to lack of neurotrophic substances. The observation that colchicine, a drug known to interfere with the transport of material along axons, causes ACh hypersensitivity in active innervated muscles has been raised in support of this notion (Albuquerque *et al.*, 1972; Hofmann and Thesleff, 1972). However, it is possible that the various results cited above are the result of abnormal factors present only in these experimental situations. Considerable evidence to support the latter view is discussed in some detail elsewhere (Lömo and Westgaard, 1976) and reviewed only briefly here.

When soleus muscles are stimulated with a slow pattern (see above) from the time of denervation they develop a marked (> 100 mV/nC) and transient (maximal at four days, absent after 8–10 days) hypersensitivity to ACh (Fig. 6.1, filled circles). A similarly transient hypersensitivity develops in doubly innervated soleus muscles if the foreign fibular nerve is cut and activity maintained by the regenerated soleus

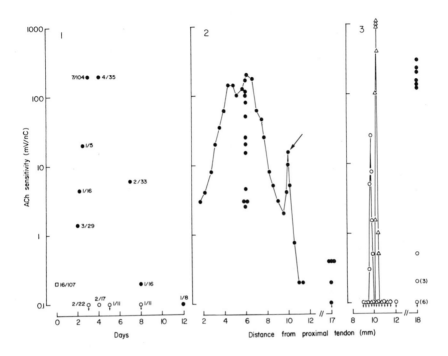

Fig. 6. Transient ACh hypersensitivity after nerve section in soleus muscles kept active by "slow" pattern stimulation or, if the muscle was dually innervated, by the intact "slow" soleus nerve. Prevention of such hypersensitivity by a "fast" stimulation pattern or by an intact "fast" nerve. 1. Median ACh sensitivity 2 mm from the Achilles tendon of normal fibres (open square) and of fibres stimulated at 10 Hz for 10 s every 100 s (●, slow pattern) or at 100 Hz for 1 s every 100 s (○, fast pattern) from the time of denervation (abscissa). Number of muscles and fibres is indicated. 2. ACh sensitivity along single soleus fibre (symbols joined by lines) innervated by both the soleus nerve ("slow") and the foreign fibular nerve ("fast") three days after cutting the fibular nerve. Arrow points to the high sensitivity at the intact solus nerve end plate where transmission was present. Other symbols at 6 mm and 17 mm (2 mm from the Achilles tendon) give the sensitivity of neighbouring fibres. A large number of denervated fibular end plates, revealed by staining for AChE, were found corresponding to the high sensitivities in the proximal half of the muscle. After 17 days the extrajunctional hypersensitivity had disappeared leaving only the junctional sensitivity at the denervated end plates (not shown). 3. Absence of extrajunctional ACh hypersensitivity around two degenerating soleus nerve fibre end plates (at 10 mm) three days after cutting the soleus nerve to a muscle with many fibres also innervated by the fibular nerve. At 18 mm (2 mm from the Achilles tendon) the sensitivity was high in fibres not innervated by the fibular nerve and thus completely denervated (●) and absent or low in fibres innervated by the fibular nerve (○).

nerve (Fig. 6.2, filled circles). That the hypersensitivity is transient and often maximal around the degenerating end plate suggests that it may be related to the degeneration of the nerve terminals. This is supported by a further two sets of results. ACh hypersensitivity appears more slowly if muscle activity is suddenly stopped by locally anaesthetizing rather than by cutting the nerve (Cangiano, personal communication; Lömo and Westgaard, unpublished). Similarly, if soleus muscles are made insensitive by stimulating for 15 days after denervation and the stimulation is then stopped, then the subsequent hypersensitivity develops more slowly than after acute denervation. In both the anaesthetized and the chronically stimulated denervated preparations, there is no active nerve terminal degeneration because the terminals are intact in the first (Lömo and Rosenthal, 1972) and already phagocytosed (Miledi and Slater, 1968; Gonzenbach and Waser, 1973) in the second preparation. Thus the development of hypersensitivity in these situations probably reflects inactivity alone while the faster development of hypersensitivity after acute denervation may reflect the combined influence of inactivity and factors associated with nerve terminal degeneration. Cutting the nerve close to the muscle is known to make the terminals degenerate faster than cutting the nerve far from the muscle and this, rather than a quicker disappearance of some essential neurotrophic substance may cause the earlier onset of postsynaptic changes with short nerve stumps. The hypersensitivity around degenerating end plates in doubly innervated frog sartorius muscles may be similarly attributed to nerve terminal degeneration (Jones and Vrbová, 1974).

If degeneration of nerve terminals can induce ACh hypersensitivity, its mechanism is not known. In normally innervated muscle, an intense hypersensitivity (< 1000 mV/nC) can be induced by placing a foreign body for example a piece of thread or nerve on the surface of the muscle (Jones and Vrbová, 1974; Jones and Vyskočil, 1975; Lömo and Slater, unpublished). Its timecourse is similar to that seen in contracting muscles with degenerating nerves, being maximal after about four days and absent after more than 10–12 days. It may be that substances released from activated phagocytes induce the hypersensitivity in both types of experiments and this possibility is now being further explored. The effects of colchicine referred to above should be interpreted with caution because colchicine has been shown to raise the ACh sensitivity of innervated soleus fibres without any demonstrable effect on axonal transport (Cangiano and Fried, 1974); to raise the ACh sensitivity of denervated stimulated soleus fibres where its action is independent of neural influences (Lömo, 1974); because, in experiments with colchicine containing cuffs around the sciatic nerve, the nerve develops early local conduction blocks more frequently than in control experiments with no colchicine in the cuffs (Cangiano, personal communication) and may finally degenerate (Albuquerque *et al.*, 1972).

F   INTERACTING FACTORS IN THE CONTROL OF MEMBRANE PROPERTIES

If the "slow" soleus nerve to a doubly innervated soleus muscle is cut and the "fast" foreign fibular nerve is left intact, then no hypersensitivity develops around the degenerating soleus end plates (Fig. 6.3, open symbols). This finding surprised us initially since cutting the fibular nerve and not the soleus nerve did induce hypersensitivity (Fig. 6.2, filled circles). We then denervated the ordinary soleus, stimulated it with a fast pattern and found that the transient hypersensitivity seen with a slow stimulus pattern failed to appear (Fig. 6.1, open circles). This showed that a fast activity pattern, whether it was imposed by a fast nerve or a fast stimulus pattern effectively suppressed the transient hypersensitivity which with a slow activity pattern appeared around the degenerating end plates. A fast pattern also causes denervation hypersensitivity to disappear more rapidly during stimulation and to reappear more slowly after the stimulation is stopped than a slow pattern (Figs 2 and 3).

It is possible, therefore, that sensitivity inducing factors on the outside of the fibre for example from degenerating nerve terminals, may interact competitively with activity linked, sensitivity suppressing influences within the fibre; the actual sensitivity in a given circumstance reflecting which of the two opposing influences dominates.

A mechanism of competitive interaction seems to be supported by the results obtained with a foreign body placed on the surface of the muscle (see above). The hypersensitivity induced in this situation falls off gradually as the distance from the foreign body and the dense infiltrate of inflammatory cells increases, suggesting that as the sensitivity inducing stimulus becomes less potent activity gradually becomes dominant. Furthermore, fast muscles respond with less hypersensitivity than slow muscles to the same foreign body (Jones and Vyskočil, 1975).

G JUNCTIONAL v. EXTRAJUNCTIONAL RECEPTORS

Several differences exist between junctional and extrajunctional parts
of the muscle membrane apart from gross morphological differences.
The ACh receptors at the end plate are more densely packed (Fambrough and Hartzell, 1972), turn over more slowly (Berg and Hall,
1974) and appear to have different pharmacological and physical
properties from the extrajunctional receptors (Beránek and Vyskočil,
1967; Brockes and Hall, 1976). These properties and a highly localized
AChE activity are induced by nerve terminals but only in a receptive
muscle membrane for example before innervation or after denervation.
Once induced, junctional properties are very stable. Junctional ACh
receptors persist for weeks or months after denervation and are resistant
to activity evoked elsewhere in the fibre in sharp contrast to extrajunctional receptors which may appear or disappear within a few days.
Evidently, junctional and extrajunctional properties are controlled in
different ways but the basis for this difference is not known. Some recent
results indicate that cholinergic transmission and muscle activity may
not be required for induction of junctional properties (Steinbach et al.,
1973; Steinbach, 1974).

H CONCLUSIONS

I would like to suggest a scheme for the overall control of ACh sensitivity which is consistent with the data presented here and which seems
to be testable with respect to its more speculative parts.

In each region along a normal, adult, mammalian muscle fibre there
is an internal influence, linked in a graded manner to activity, which
suppresses ACh receptor synthesis. If activity stops the suppressive
influence terminates and ACh receptors appear after a delay which
depends on the type or level of the previous activity. Opposing this are
external influences possibly in the form of diffusable substances released
from nearby cells, which can stimulate receptor synthesis. Such factors
are present when for example nerve terminals degenerate or the surrounding tissue becomes inflamed. In the extrajunctional membrane of
normal fibres, activity dominates and there is little or no sensitivity.
During an inflammatory response the external influence may dominate
so that local sensitivity develops even in a normally active muscle fibre.
In this context the process of nerve terminal degeneration may be seen
as a small scale inflammatory response. In normal fibres external

influences, originating from the nerve, locally overcome the suppressive influence of activity and cause the high sensitivity at the neuromuscular junction. These influences might come from contact with the nerve terminal or substances released from it and might act similarly to the influences from inflammatory cells. This analogy is far from complete, however, since the hypersensitivity induced by inflammation is transient and more diffuse while that induced by the nerve is remarkably stable and discrete.

III Contractile properties

In their pioneering cross-reinnervation experiments Buller, Eccles and Eccles (1960) pointed out that a foreign nerve might change the con-tractile properties of a muscle either by imposing a different pattern of activity on the muscle or by affecting the muscle "trophically" in a different way. A large variety of experimental procedures on inner-vated muscles have not clearly separated these mechanisms because any change in impulse pattern might also alter the amount or kind of trophic substance released from the nerve. Direct stimulation of denervated muscle might, on the other hand, provide information about the effects of muscle activity *per se* and help to distinguish between the two pos-sibilities. For this reason we have stimulated the slow rat soleus for 3–6 weeks after denervation with a "slow" (10 Hz for 10 s every 50 s) or a "fast" (100 Hz for 0·5 s every 25 s) stimulus pattern, both patterns having the same average frequency of stimulation of 2 Hz (Lömo *et al.*, 1974).

A CONTRACTION SPEED

On average, the fast pattern reduced the isometric twitch contraction time to 16 ms while the slow pattern maintained it at the essentially normal value of 33 ms (normal 35 ms, denervated 41 ms)(Figs 7.1, 8.2). In contrast to the slow pattern, the fast stimulus pattern also reduced the twitch half-relaxation time and the twitch tetanus ratio (Fig. 7.2); increased the tetanic fusion frequency and induced the development of post tetanic potentiation of the twitch (Fig. 9) and of alkali resistance of the myosine ATPase activity (Fig. 10). The results indicated that the fast pattern shortened the time course of the active state. The intrinsic shortening velocity is closely correlated with the activity of the myosin

ATPase (Bárány and Close, 1969; Buller *et al.*, 1971). It was not directly measured in these experiments but the change in the myosin ATPase made it seem likely that the shortening velocity had also increased. The

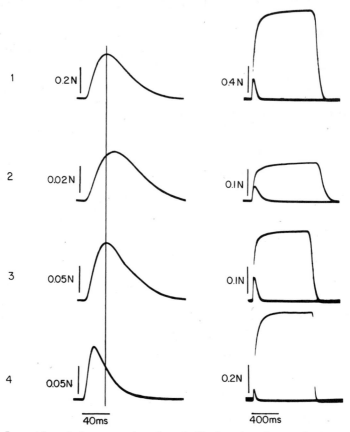

Fig. 7. Isometric twitch contraction time, half-relaxation time and twitch tetanus ratio are reduced in denervated slow soleus muscles by chronic stimulation with a fast pattern (4) but not with a slow pattern (3). To the left are isometric twitch responses and to the right tetanic responses to stimulation at 200 Hz with a single twitch superimposed to indicate twitch tetanus ratio. The soleus was curarized but otherwise normal in 1, denervated 39 days in 2, denervated and stimulated for 36 days at 10 Hz for 10 s every 50 s in 3 (slow pattern) or for 39 days at 100 Hz for 0·5 s every 25 s in 4 (fast pattern). The responses were recorded at optimal lengths. The soleus, with its main circulation intact, was in a pool of warm (34·5–35·5°C) oxygenated Ringer solution made from the skin. Stimuli of 1 ms duration were delivered between a platinum electrode encircling the muscle belly and an indifferent electrode in the bath.

properties induced by the fast pattern are similar to those of normal fast muscle or of slow muscle cross-reinnervated with a fast nerve (Close, 1969; Close and Hoh, 1968; Guth *et al.*, 1970).

Experiments with cross-reinnervation of slow muscles have given the puzzling result that reinnervation of the cat soleus with the nerve to the fast flexor digitorum longus (FDL) increased the maximum shortening velocity with only 10–15% (Buller *et al.*, 1971) while reinnervation of the rat soleus with the nerve to the fast EDL increased it with 116% to about the value of the EDL (Close, 1969). The discrepancy may now be attributed to the more recent finding that the FDL contains more slow motor units than the EDL (Close, 1967; Bagust *et al.*, 1973). If the nerve to the EDL is used also in the cat, the effect on the shortening velocity is similar to the rat (Luff, 1975).

Long-term stimulation of denervated adult fast muscles with a slow pattern has not been reported. However, if innervated fast muscles are stimulated at 10 Hz the twitch contraction time increases (Salmons and Vrbová, 1969), the shortening velocity decreases (Al-Amood *et al.*,

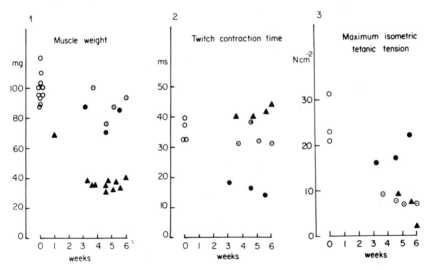

Fig. 8. Muscle weight (1), isometric twitch contraction time (2) and maximal isometric tension (3) plotted against time after denervation. The soleus muscles were either normal (○), merely denervated (▲) or stimulated since the time of denervation at 10 Hz for 10 s every 50 s (◎, slow pattern) or at 100 Hz for 0·5 s every 25 s (●, fast pattern). The tension in 3 is given per unit cross-sectional area (cm²) of the muscle estimated from the muscle weight and the average length of the muscle fibres. (Close, 1972.)

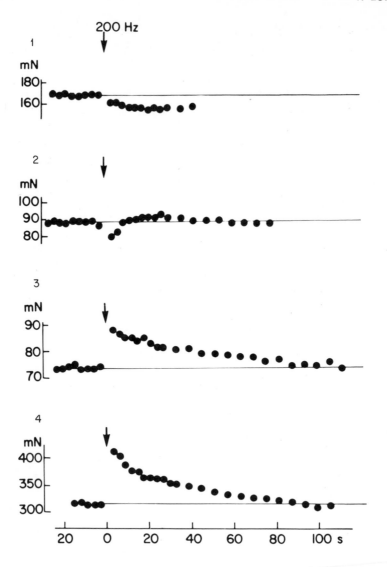

Fig. 9. Post tetanic potentiation of the twitch is absent in a normal soleus (1) and after 36 days of denervation and intermittent stimulation at 10 Hz (2, slow pattern) but present after 39 days of denervation and intermittent stimulation at 100 Hz (3, fast pattern) and in a normal EDL (4). Each symbol is the single isometric twitch tension before and after a 1 s tetanus at 200 Hz delivered at the time indicated by the arrows. The time to peak tension did not change during the potentiation which can therefore be attributed entirely to a faster rise in tension.

1973) and myosin light chains characteristic of slow muscle appear (Stréter *et al.*, 1973). Reinnervation of a fast muscle with a "slow" nerve has similar effects (Weeds, *et al.*, 1974).

From these experiments it appears that variation in muscle activity *per se*, in particular in the frequency of the muscle impulses, is a major factor in determining the contractile speed of a muscle.

B   FIBRE SIZE

The size of muscle fibres varies independently of neural influences when for example stretch causes hypertrophy of denervated muscles (Shiaffino and Hanzlíková, 1970; Gutmann *et al.*, 1971) or when denervated muscles of young rats continue to grow at a rate corresponding to the elongation of the bones (Stewart, 1968). Increases in work load by denervation of synergists (Rowe and Goldspink, 1968) or by the use of heavy weights (Goldspink, 1964) cause considerable hypertrophy in innervated muscles but in this case changes in impulse pattern probably occur together with changes in muscle tension. The hypertrophy of actively used innervated muscles and of passively stretched denervated muscles is associated with increases in contractile and sarcoplasmic proteins (Stewart, 1955; Hamosch *et al.*, 1967) and a preferential growth of small diameter fibres (Goldspink, 1964; Feng and Lu, 1965; Sola, O. M., cited in Stewart, 1972). It appears that muscle tension, whether influenced by neural or non-neural mechanisms, is an important factor in the control of fibre size.

The size of muscle fibres may be controlled independently of many other properties of the muscle. Tenotomy produces marked atrophy and yet twitch contraction time, twitch tetanus ratio (Nelson, 1969) as well as sensitivity of the membrane to ACh (Lömo and Rosenthal, 1972) remain normal. Directly stimulated denervated muscles are usually more or less atrophic, but the membrane may have normal resting potential, passive electrical characteristics (Westgaard, 1975) and content of extrajunctional ACh receptors (Frank *et al.*, 1975). The denervated stimulated soleus with a normal ACh sensitivity may become very atrophic if the stimulating electrodes are placed so that neighbouring fast muscles such as the gastrocnemius are also stimulated, presumably because the simultaneously contracting fast muscles take much of the load off the slower contracting soleus. Furthermore, in this preparation

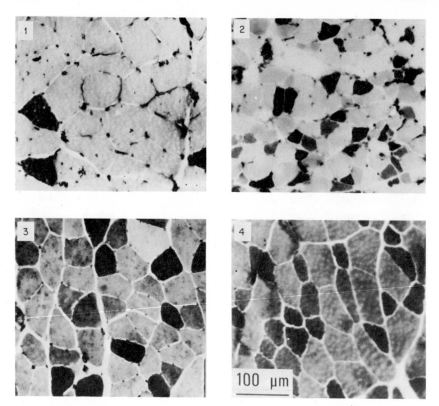

Fig. 10. Cross-sections showing relative amount of muscle fibres with alkaline resistant actomyosin ATPase activity (darkly stained) in a normal soleus (1), a soleus denervated for 22 days (2) or denervated and stimulated intermittently at 100 Hz for 22 days (3) and a normal extensor digitorum longus (4). The stimulation was 100 Hz for 1 s every 6 s (average frequency of stimulation 16·7 Hz) and all the muscles from this figure are from a different series of experiment than those of Figs 7, 8, 9 and 11. The actomyosin ATPase activity of muscles in the latter series, as revealed by this histochemical method, is shown in Lömo *et al.* (1974) which also gives more information about the technique used.

the entire leg is usually denervated and the little resistance offered by antagonists may contribute to the atrophy.

As shown in Figs 8.1 and 10, direct stimulation may also largely prevent denervation atrophy. In these experiments the electrodes were placed on the surface of the soleus and there was little co-contraction from neighbouring muscles. It would seem that denervation atrophy is best prevented by having the stimulated muscle work isometrically or against some load.

C INTRINSIC STRENGTH

The denervated slow soleus stimulated with a fast pattern produced on average 2·5 times more tetanic force per cm^2 than similar muscles stimulated with a slow pattern (Fig. 8.3). The intrinsic strength during a tetanus is also higher in normal fast muscles (1·4 times) or in slow muscle cross-reinnervated with a fast nerve (1·3 times) than in normal slow muscle (Bárány and Close, 1969). The stimulation experiments are somewhat marred by the lower than normal intrinsic strength in both groups of muscles, but this may be attributed to the abnormal loading conditions in the completely denervated leg and possibly to the damage of superficial fibres from the closely placed stimulating electrodes. Nevertheless, the different effects of the two stimulation patterns are striking and together with the results of the cross-reinnervation experiments they suggest that high frequency bursts of muscle activity characteristic of fast muscle cause a high intrinsic strength of the fibre. The effect might be mediated by the higher tension produced by the high frequency bursts of activity.

D FIBRE LENGTH

If rat soleus muscles are stretched by flexion of the foot and the position maintained for some weeks by a cast then both the length and the number sarcomeres of individual muscle fibres increase by as much as 25%. A forced, maintained shortening of the muscle has an even greater opposite effect (35%, Goldspink et al., 1974). The response is similar in innervated and denervated muscles. It is therefore independent of neural influences and may instead depend on changes in the tension of the muscle. Cross-reinnervation of the soleus with a fast nerve increases both the length and the number of sarcomeres of individual muscle fibres (Close, 1969). This explains why the slow fibres become as fast as the fast fibres without a complete conversion of the shortening velocity of individual sarcomeres. Close (1969) has suggested that the neural influence determines the speed of shortening of whole fibres rather than merely altering the intrinsic speed of shortening of the contractile material to a predetermined level. Another possibility is that episodes of high frequency impulses or of high tension during fast activity increase the number of sarcomeres while failure of the activity in the cross-reinnervated muscle to completely mimic the activity of a normal fast muscle may account for the incomplete conversion of the intrinsic shortening velocity. The similar shortening velocities for the whole muscles might therefore be coincidental. Merely cutting the nerve to antagonistic muscles causes a moderate change in shortening velocity of the soleus (Guth and Wells, 1972) probably because the impulse activity in the soleus motoneurons and the type of work done by the soleus is altered. Cross-reinnervation with a fast nerve probably does not cause the soleus to work in a manner identical to that of the fast muscle and a complete conversion would therefore not be expected.

It is of some interest that the shortening velocities of muscles correlate with the size of the animal species; muscles from small animals shortening faster than muscles from larger animals (see Close, 1972). It has been suggested that the force required to accelerate the body parts cannot exceed some upper limit set by the tensile strength of the tissue (Hill, 1950). Thus, tension may influence and set an upper limit to shortening velocity.

E ENDURANCE

Chronic stimulation of the denervated soleus with either a fast or a slow stimulus pattern did not make the muscle less resistant to fatigue (Fig. 11). Nor did it cause a "sag" in tension during a tetanus (Fig. 7.2) characteristic of fast muscles or of slow muscles cross-reinnervated with a fast nerve (Close, 1969). This is the only example of some fast muscle property which failed to develop during stimulation with the fast pattern. It shows that the muscle may become considerably faster without becoming less resistant to fatigue. From other experiments we know that particular training procedures such as running on a treadmill may considerably increase endurance without affecting contraction speed or myosin ATPase activity (Barnard et al., 1970). Thus, endurance and contraction speed may be independently controlled. It may be important that in the stimulation experiments the fast and the slow pattern were different only in impulse frequency but not in the total number of stimuli imposed on the muscle.

F CONCLUSIONS

Muscle activity per se appears to be a very important factor in the regulation of the contractile properties of a muscle. Differences in impulse frequency within bursts of activity have marked effects on contraction

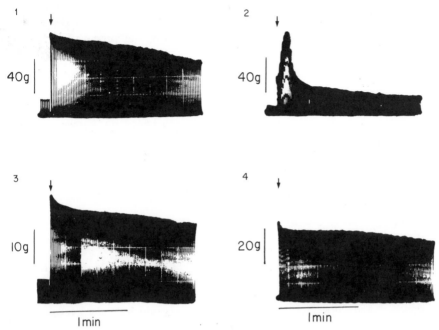

Fig. 11. Isometric tetanic responses to trains of stimuli at 80 Hz in a curarized but otherwise normal soleus (1) or normal extensor digitorum longus (2), and in a soleus denervated and stimulated intermittently at 10 Hz (slow pattern) for 42 days (3) or at 100 Hz (fast pattern) for 39 days (4). The trains started at the times indicated by arrows. Each train lasted 0·5 s and was repeated every second for 2 minutes.

speed while differences in overall activity may mainly influence endurance. Other factors, such as passive or active tension, the latter depending both on the type of the activity and the work load have important effects on the size and the length of muscle fibres. There is no direct evidence that neurotrophic substances influence contractile properties, but more detailed studies of how different types of denervated muscle fibres respond to a variety of patterns and amounts of stimuli under different loading conditions might provide important information in this connection. At present it seems most likely that the muscle continually adapts itself by intrinsic mechanisms to the type of work imposed on it.

IV   Acknowledgements

Most of the experiments from this laboratory in which the author has taken part were done in collaboration with Drs H. A. Dahl, E. Frank, J. K. S.

Jansen, K. Nicolaysen and R. H. Westgaard. I thank Dr. C. R. Slater for most valuable discussion and suggestions during the preparation of the manuscript.

V References

Al-Amood, W. S., Buller, A. J. and Pope, R. (1973). Long-term stimulation of cat fast-twitch skeletal muscle. *Nature Lond.* **244**, 225–227.

Albuquerque, E. X. and Thesleff, S. (1968). A comparative study of membrane properties of innervated and chronically denervated fast and slow skeletal muscles of the rat. *Acta Physiol. Scand.* **73**, 471–480.

Albuquerque, E. X. and McIsaac, R. J. (1970). Fast and slow mammalian muscles after denervation. *Exp. Neurol.* **26**, 183–202.

Albuquerque, E. X., Schuh, F. T. and Kauffman, F. C. (1971). Early membrane depolarization of the fast mammalian muscle after denervation. *Pflügers Arch.* **328**, 36–50.

Albuquerque, E. X., Warnick, J. E., Tasse, J. R. and Sansone, F. M. (1972). Effects of vinblastine and colchicine on neural regulation of the fast and slow skeletal muscles of the rat. *Exp. Neurol.* **37**, 607–634.

Axelsson, J. and Thesleff, S. (1959). A study of supersensitivity in denervated mammalian skeletal muscle. *J. Physiol.* **147**, 178–193.

Bagust, J., Knott, S., Lewis, D. M., Luck, J. C. and Westerman, R. A. (1973). Isometric contractions of motor units in a fast twitch muscle of the cat. *J. Physiol.* **231**, 87–104.

Bárány, M. and Close, R. I. (1971). The transformation of myosin in cross-innervated rat muscles. *J. Physiol.* **213**, 455–474.

Barnard, R. J., Edgerton, V. R. and Peter, J. B. (1970). Effect of exercise on skeletal muscle. II. Contractile properties. *J. Appl. Physiol.* **28**, 767–770.

Belmar, J. and Eyzaguirre, C. (1960). Pacemaker site of fibrillation potentials in denervated mammalian muscle. *J. Neurophysiol.* **29**, 425–441.

Bennett, M. R., Pettigrew, A. G. and Taylor, R. S. (1973). The formation of synapses in reinnervated and cross-reinnervated adult avian muscles. *J. Physiol.* **230**, 331–357.

Beránek, R. and Vyskočil, F. (1967). The action of tubocurarine and atropine on the normal and denervated rat diaphragm. *J. Physiol.* **188**, 53–66.

Berg, D. K. and Hall, Z. W. (1974). Fate of α-bungarotoxin bound to acetylcholine receptors of normal and denervated muscle. *Science* **184**, 473–475.

Brockes, J. P. and Hall Z. W. (1975). Synthesis of acetylcholine receptor by denervated rat diaphragm muscle *Proc. Nat. Acad. Sci. U.S.A.* **72**, 1368–1372.

Brockes, J. P. and Hall, Z. W. (1976). Acetylcholine receptors in normal and denervated muscle. Cold Spring Harbor Symposium on Quantitative Biology, Volume XL: The Synapse (in press).

Brooker, G. (1973). Oscillations of cyclic adenosine monophosphate concentration during the myocardial contraction cycle. *Science* **182**, 933–934.

Brostrom, C. O., Huang, Y-C., Breckenridge, B. McL. and Wolff, D. J. (1975).

Identification of a calcium-binding protein as a calcium-dependent regulator of brain adenylate cyclase. *Proc. Nat. Acad. Sci. U.S.A.* **72**, 64–68.

Buller, A. J., Eccles, J. C. and Eccles, R. M. (1960). Interactions between motoneurones and muscles in respect of the characteristic speeds of their responses. *J. Physiol.* **150**, 417–439.

Buller, A. J., Kean, C. J. C. and Ranatunga, K. W. (1971). The force-velocity characteristics of cat and fast and slow-twitch skeletal muscle following cross-innervation. *J. Physiol.* **213**, 66–67B.

Buller, A. J., Mommaerts, W. F. H. M. and Seraydarian, K. (1971). Neural control of myofibrillar ATPase activity in rat skeletal muscle. *Nature New Biol.* **233**, 31–32.

Burke, W. and Ginsborg, B. L. (1956). The electrical properties of the slow muscle fibre membrane. *J. Physiol.* **132**, 586–598.

Cangiano, A. and Fried, J. A. (1974). Neurotrophic control of skeletal muscle of the rat. *J. Physiol.* **239**, 31–33P.

Carlsen, R. C. (1975). The possible role of cyclic AMP in the neurotrophic control of skeletal muscle. *J. Physiol.* **247**, 343–361.

Close, R. (1967). Properties of motor units in fast and slow skeletal muscles of the rat. *J. Physiol.* **193**, 45–55.

Close, R. (1969). Dynamic properties of fast and slow skeletal muscles of the rat after nerve cross-union. *J. Physiol.* **204**, 331–346.

Close, R. I. (1972). Dynamic properties of mammalian skeletal muscle. *Physiol. Rev.* **52**, 129–197.

Close, R. and Hoh, J. F. Y. (1969). Post-tetanic potentiation of twitch contractions of cross-innervated rat fast and slow muscles. *Nature Lond.* **221**, 179–181.

Cohen, M. W. (1972). The development of neuromuscular connexions in the presence of D-tubocurarine. *Brain Res.* **41**, 457–463.

Cohen, S. A. and Fischbach, G. D. (1973). Regulation of muscle acetylcholine sensitivity by muscle activity in cell culture. *Science* **181**, 76–78.

Crain, S. M. and Peterson, E. R. (1971). Development of paired explants of fetal spinal cord and adult skeletal muscle during chronic exposure to curare and hemicholinium. *In Vitro* **6**, 373.

Drachman, D. B. (1974). The role of acetylcholine as a neurotrophic transmitter. *Ann. N.Y. Acad. Sci.* **228**, 160–175.

Dulbecco, R. and Elkington, J. (1975). Induction of growth in resting fibroblastic cell cultures by Ca^{2+}. *Proc. Nat. Acad. Sci. U.S.A.* **72**, 1584–1588.

Elsberg, C. A. (1917). Experiments on motor nerve regeneration and the direct neurotization of paralyzed muscles by their own and by foreign nerves. *Science* **45**, 318–320.

Fambrough, D. M. (1970). Acetylcholine sensitivity of muscle fiber membranes: Mechanism of regulation by motoneurones. *Science* **168**, 372–373.

Fambrough, D. M. (1974). Cellular and developmental biology of acetylcholine receptors in skeletal muscle. *In* "Neurochemistry of Cholinergic Receptors" (F. de Robertis and J. Schacht, eds). Raven Press, New York.

Fambrough, D. M. and Hartzell, H. C. (1972). Acetylcholine receptors: Number and distribution at neuromuscular junctions in rat diaphragm. *Science* **176**, 189–191.

Feng, T. P. and Lu, D. K. (1965). New lights on the phenomenon of transient hyper-trophy in the denervated hemidiaphragm of the rat. *Sci. Sinica* **14**, 1772–1784.

Fex, S., Sonesson, B., Thesleff, S. and Zelená, J. (1966). Nerve implants in botulinum poisoned mammalian muscle. *J. Physiol.* **184**, 872–882.

Fischbach, G. D. and Robbins, N. (1969). Changes in contractile properties of disused soleus muscles. *J. Physiol.* **201**, 305–320.

Fischbach, G. D. and Robbins, N. (1971). Effects of chronic disuse of rat soleus neuro-muscular junctions on postsynaptic membrane. *J. Neurophysiol.* **34**, 562–569.

Frank, E., Gautvik, K. and Sommerschild, H. (1975). Cholinergic receptors at denervated mammalian motor end-plates. *Acta Physiol. Scand.* **95**, 66–76.

Frank, E., Jansen, J. K. S., Lömo, T. and Westgaard, R. H. (1975). The interaction between foreign and original nerves innervating the soleus muscles of rats. *J. Physiol.* **247**, 725–743.

Freedman, M. H., Raff, M. C. and Gomperts, B. (1975). Induction of increased calcium uptake in mouse T lymphocytes by concanavalin A and its modulation by cyclic nucleotides. *Nature Lond.* **255**, 378–382.

Goldspink, G. (1964). The combined effects of exercise and reduced food intake on skeletal muscle fibers. *J. Cell. Comp. Physiol.* **63**, 209–216.

Goldspink, G., Tabary, C., Tabary, J. C., Tardieu, C. and Tardieu, G. (1974). Effect of denervation on the adaptation of sarcomere number and muscle exten-sibility to the functional length of the muscle. *J. Physiol.* **236**, 733–742.

Gonzenbach, H. R. and Waser, P. G. (1973). Electron microscopic studies of degeneration and regeneration of rat neuromuscular junctions. *Brain Res.* **63**, 167–174.

Grampp, W., Harris, J. B. and Thesleff, S. (1972). Inhibition of denervation changes in skeletal muscle by blockers of protein synthesis. *J. Physiol.* **221**, 743–754.

Gruener, R., Baumbach, N. and Coffee, D. (1974). Reduction of denervation super-sensitivity of muscle by submechanical threshold stimulation. *Nature Lond.* **248**, 68–69.

Guth, L. and Wells, J. B. (1972). Physiological and histochemical properties of the soleus muscle after denervation of its antagonists. *Exp. Neurol.* **36**, 463–471.

Guth, L., Samaha, F. J. and Albers, R. W. (1970). The neural regulation of some phenotypic differences between the fiber types of mammalian skeletal muscle. *Exp. Neurol.* **26**, 126–135.

Gutmann, E., Schiaffino, S. and Hanzliková, V. (1971). Mechanism of compensatory hypertrophy in the skeletal muscle of the rat. *Exp. Neurol.* **31**, 451–464.

Hamosch, M., Lesch, M., Baron, J. and Kaufman, S. (1967). Enhanced protein synthesis in a cell-free system from hypertrophied skeletal muscle. *Science* **157**, 935–937.

Harris, A. J. (1974). Inductive functions of the nervous system. *Ann. Rev. Physiol.* **36**, 251–305.

Harris, J. B. and Thesleff, S. (1972). Nerve stump length and membrane changes in denervated skeletal muscle. *Nature New Biol.* **236**, 60–61.

Hess, A. (1970). Vertebrate slow muscle fibers. *Physiol. Rev.* **50**, 40–62.

Hill, A. V. (1950). The dimensions of animals and their muscular dynamics. *Sci. Progr. (London)* **38**, 209–230.

Hofmann, W. W. and Thesleff, S. (1972). Studies on the trophic influence of nerve in skeletal muscle. *European J. Pharmacol.* **20**, 256–260.

Jansen, J. K. S., Lömo, T., Nicolaysen, K. and Westgaard, R. H. (1973). Hyper-innervation of skeletal muscle fibers: Dependence on muscle activity. *Science* **181**, 559–561.

Jirmanová, I. and Zelená, J. (1970). Effect of denervation and tenotomy on slow and fast muscles of the chicken. *Z. Zellforsch.* **106**, 333–347.

Johns, T. R. and Thesleff, S. (1961). Effects of motor inactivation on the chemical sensitivity of skeletal muscle. *Acta Physiol. Scand.* **51**, 136–141.

Jones, R. and Vrbová, G. (1974). Two factors responsible for the development of denervation hypersensitivity. *J. Physiol.* **236**, 517–538.

Jones, R. and Vyskočil, F. (1975). An electrophysiological examination of the changes in skeletal muscle fibers in response to degenerating nerve tissue. *Brain Res.* **88**, 309–317.

Katz, B. and Miledi, R. (1964). The development of acetylcholine sensitivity in nerve-free segments of skeletal muscle. *J. Physiol.* **170**, 389–396.

Kernan, R. P. and McCarty, I. (1972). Effects of denervation on ^{42}K influx and membrane potential of rat soleus muscles measured *in vivo*. *J. Physiol.* **226**, 62–63.

Klaus, W., Lullmann, H. and Muscholl, E. (1960). Der Kalium-Flux des normalen und denervierten Rattenzwerchfells. *Pflügers Arch.* **271**, 761–775.

Kuffler, S. W., Dennis, M. J. and Harris, A. J. (1971). The development of chemo-sensitivity in extrasynaptic areas of the neuronal surface after denervation of para-sympathetic ganglion cells in the heart of the frog. *Proc. Roy. Soc. B* **177**, 555–563.

Luco, J. V. and Fyzaguirre, C. (1955). Fibrillation and hypersensitivity to ACh in denervated muscle: effect of length of degenerating nerve fibres. *J. Neurophysiol.* **18**, 65–73.

Luff, A. R. (1975). Dynamic properties of fast and slow skeletal muscles in the cat and rat following cross-reinnervation. *J. Physiol.* **248**, 83–96.

Locke, S. and Solomon, H. C. (1967). Relation of resting potential of rat gastro-cnemius and soleus muscles to innervation, activity and the Na–K pump. *J. Exp. Zool.* 377–386.

Lömo, T. (1974). Neurotrophic control of colchicine effects on muscle? *Nature Lond.* **249**, 473–474.

Lömo, T. and Rosenthal, J. (1972). Control of ACh sensitivity by muscle activity in the rat. *J. Physiol.* **221**, 493–513.

Lömo, T. and Westgaard, R. H. (1975). Further studies on the control of ACh sensitivity by muscle activity in the rat. *J. Physiol.* (in press).

Lömo, T. and Westgaard, R. H. (1976). Control of ACh sensitivity in rat muscle fibers. Cold Spring Harbor Symposium on Quantitative Biology, *XL*: The Synapse (in press).

Lömo, T., Westgaard, R. H. and Dahl, H. A. (1974). Contractile properties of muscle: Control by pattern of muscle activity in the rat. *Proc. R. Soc. B* **187**, 99–103.

Manchester, K. L. and Harris, F. J. (1968). Effect of denervation on the synthesis of ribonucleic acid and deoxyribonucleic acid in rat diaphragm muscle. *Biochem. J.* **108**, 177–183.

McMahon, D. (1974). Chemical messengers in development: a hypothesis. *Science* **185**, 1012–1021.

Miledi, R. (1960). The acetylcholine sensitivity of frog muscle fibres after complete and partial denervation. *J. Physiol.* **151**, 1–23.

Miledi, R. and Zelená, J. (1966). Sensitivity to acetylcholine in the rat slow muscle. *Nature Lond.* **210**, 855–856.

Miledi, R. and Slater, C. R. (1968). Electrophysiology and electronmicroscopy of rat neuromuscular junctions after nerve degeneration. *Proc. R. Soc. B* **169**, 289–306.

Muchnik, S. and Kotsias, B. A. (1975). Effect of chronic stimulation of denervated muscles on the uridine-5-3H incorporation and fibrillation activity. *Life Sciences* **16**, 543–550.

Nasledov, G. A. and Thesleff, S. (1974). Denervation changes in frog skeletal muscle. *Acta Physiol. Scand.* **90**, 370–380.

Nelson, P. G. (1969). Functional consequences of tenotomy in hind limb muscles of the cat. *J. Physiol.* **201**, 321–333.

Purves, D. and Sakmann, B. (1974). The effect of contractile activity on fibrillation and extrajunctional acetylcholine-sensitivity in rat muscle maintained in organ culture. *J. Physiol.* **237**, 157–182.

Rasmussen, H. (1970). Cell communication, calcium ion and cyclic adenosine monophosphate. *Science* **170**, 404–412.

Redfern, P. and Thesleff, S. (1971). Action potential generation in denervated rat skeletal muscle. **I.** Quantitative aspects. *Acta Physiol. Scand.* **81**, 557–564.

Rowe, R. W. D. and Goldspink, G. (1968). Surgically induced hypertrophy in skeletal muscles of the laboratory mouse. *Anat. Rec.* **161**, 69–75.

Salafsky, B., Bell, J. and Prewitt, M. (1968). Development of fibrillation potentials in denervated fast and slow skeletal muscle. *Amer. J. Physiol.* **215**, 637–643.

Salmons, S. and Vrbová, G. (1969). The influence of activity on some contractile characteristics of mammalian fast and slow muscles. *J. Physiol.* **201**, 535–549.

Schiaffino, S. and Hanlíková, V. (1970). On the mechanism of compensatory hypertrophy in skeletal muscles. *Experientia* **26**, 152–153.

Solandt, D. Y. and Magladery, J. W. (1942). A comparison of effects of upper and lower motor neurone lesions on skeletal muscle. *J. Neurophysiol.* **5**, 373–380.

Solandt, D. Y., Partridge, R. C., and Hunter, J. (1943). The effect of skeletal fixation on skeletal muscle. *J. Neurophysiol.* **6**, 17–22.

Steinbach, J. H. (1974). Role of muscle activity in nerve-muscle interaction *in vitro*. *Nature Lond.* **248**, 70–71.

Steinbach, J. H., Harris, A. J., Patrick, J., Schubert, D. and Heinemann, S. (1973). Nerve-muscle interaction in vitro. Role of acetylcholine. *J. Gen. Physiol.* **62**, 255–270.

Stewart, D. M. (1955). Changes in the protein composition of muscles of the rat in hypertrophy and atrophy. *Biochem. J.* **59**, 553–558.

Stewart, D. M. (1968). Effect of age on the response of four muscles of the rat to denervation. *Amer. J. Physiol.* **214**, 1139–1146.

Stewart, D. M. (1972). The role of tension in muscle growth. *In* "Regulation of Organ and Tissue Growth" (R. J. Goss, ed.), pp. 77–100. Academic Press, New York and London.

Stréter, F. A., Gergely, J., Salmons, S. and Romanul, F. (1973). Synthesis of fast muscle of myosin light chains characteristic of slow muscle in response to long-term stimulation. *Nature New Biol.* **241,** 17–19.

Thesleff, S. (1974). Physiological effects of denervation of muscle. *Ann. N.Y. Acad. Sci.* **228,** 89–103.

Tomkins, G. M. (1975). The metabolic code. *Science* **189,** 760–763.

Van Essen, D. and Jansen, J. K. S. (1974). Reinnervation of the rat diaphragm during perfusion with α-bungarotoxin. *Acta Physiol. Scand.* **91,** 571–573.

Vyskočil, F., Vyklický, L. and Huston, R. (1971). Quantum content at the neuro-muscular junction of fast muscle after cross-union with the nerve of slow muscle in the chick. *Brain Res.* **26,** 443–445.

Vyskočil, F. and Vyklický, L. Acetylcholine sensitivity of the chick fast muscle after cross-union with the slow muscle nerve. *Brain Res.* **72,** 758–161.

Walker, C. R. and Wilson, B. W. (1975). Control of acetylcholinesterase by contractile activity of cultured muscle cells. *Nature Lond.* **256,** 215–216.

Weeds, A. G., Trentham, D. R., Kean, C. J. C. and Buller, A. J. (1974). Myosin from cross-reinnervated cat muscles. *Nature Lond.* **247,** 135–139.

Westgaard, R. H. (1975). Influence of activity on the passive electrical properties of soleus muscle fibres in the rat. *J. Physiol.* **251,** 683–697.

11

Problems in Differentiating Trophic Relationships between Nerve and Muscle Cells

E. Gutmann

Institute of Physiology, Czechoslovak Academy of Sciences, Prague, Czechoslovakia

I The problem

The dependence of muscle on nerve has been primarily demonstrated by denervation and reinnervation experiments (Gutmann, 1964; Guth, 1968). However, the mechanisms by which this dependence is mediated and maintained are apparently complex and still not elucidated.

The manifold properties of muscle cells are primarily regulated by neuronal mechanisms, and this was clearly revealed by the transformation of the different properties of fast and slow muscles brought about by cross union of nerves. This was first demonstrated with respect to contractile behaviour, which changed according to the new, foreign nerve supply (Buller *et al.*, 1960). The basic condition for transformation of muscle type appears to be the formation of a "foreign" type of the neuro-muscular junction (NMJ) (Hník *et al.*, 1967; Koenig, 1970) which will mediate the characteristics of the "foreign" neuron. The first studies (Buller *et al.*, 1960) attributed this transformation of muscle type to "neurotrophic" effects. However, chronic stimulation of denervated muscle resulted in a shortening of contraction time (Lömo *et al.*, 1974; Melichna and Gutmann, 1974) and long-term stimulation of muscles through implanted electrodes to the nerve resulted in a change of speed of contraction (Salmons and Vrbová, 1969). Non-neural factors also affected different properties of muscle (see Gutmann, 1975). It would appear that it is necessary to delimit and define more exactly and consistently the term "neurotrophic relations and functions", to use adequate indicators of the effects of neurotrophic functions and to differentiate neurotrophic functions from other neural and non-neural mechanisms operating in the multiple control of muscle properties (see Gutmann, 1973).

"Neurotrophic relations and functions" are increasingly defined as "long-term maintenance regulations not mediated by nerve impulses" (Gutmann and Hník, 1962; Miledi, 1963; Harris, 1974). It appears useful to discuss "maintenance of synaptic connections" (Harris, 1974) and the "interactions between nerves and other cells which initiate or control molecular modifications in the other cell" (Guth, 1969) with respect to this restriction. The NMJ has been successfully studied as a chemical synapse operating by the release of ACh from the presynaptic nerve terminal, acting on a component of the postsynaptic membrane, the AChR, activation of which alters membrane permeability. The possibility to study molecular events elicited in the endplate membrane by ACh (see Katz and Miledi, 1972), the distribution, turnover and biochemical nature of AChR, have focussed the main interest in the study of neural control of muscle to events related to spontaneous and evoked release of ACh. The characteristic denervation changes in membrane properties (see Thesleff, 1974), acetylcholine sensitivity (Ginetzinski and Shamarina, 1942; Axelsson and Thesleff, 1959), distribution of AChR (see Fambrough et al., 1974) and experiments reproducing denervation changes by agents blocking the release of ACh suggested that the neurotransmitter ACh might be the sole agent mediating neurotrophic effects (Drachman, 1967). This could be due to the continuous spontaneous secretion of ACh even in the absence of nerve impulses (Katz, 1966) or to impulse evoked release of ACh. Finally, direct electrical stimulation of denervated muscle was found to reduce or prevent the onset of development of extrajunctional hypersensitivity in vivo (Jones and Vrbová, 1970; Drachman and Witzke, 1972; Lömo and Rosenthal, 1972). This has led to the conclusion that muscle activity per se controls most if not all properties of the extrajunctional membrane and that there is no need to postulate any neurotrophic (non-impulse) control mechanism (see Chapter XI; Lömo and Westgaard, 1975).

The concept that dependence of muscle on nerve is also affected by neuronal activity not related to neurotransmitter (ACh) release cannot as yet muster direct evidence gained at a molecular level. However, the evidence for neurotrophic (not impulse-related) agents has been progressively strengthened, especially by observations on: (a) trophic effects of neurons in tissue culture without synaptic transmission and even without contact between nerve and muscle cell (e.g. Harris *et al.*, 1971; Lentz, 1972; Crain and Peterson, 1974); (b) release from nerve of both proteins and neurotransmitters (Musick and Hubbard, 1972), axoplasmic transport of proteins moving with slow and fast rates and involved apparently in the synthesis and renewal of a wide population of synaptic proteins serving different functions (see Lubinska, 1975); (c) neurotrophic effects mediated by afferent nerves (Olmsted, 1920; Zelená, 1964); (d) growth-promoting influences of nerves in amphibian limb regeneration independent of impulse conduction (see Singer, 1952); (e) metabolic effects of the distal nerve stump after nerve section, dependent on the length of the peripheral nerve stump (Gutmann *et al.*, 1955; Luco and Eyzaguirre, 1955); and (f) some specific changes in muscle after nerve section affecting stability of the membrane and intracellular constituents (see Gutmann, 1964; Guth, 1968; Thesleff, 1974). The evidence for neurotrophic mechanisms has been summarized in more detail in recent reviews (Harris, 1974; Gutmann, 1975).

Neurochemical communications can evidently be accomplished by different mechanisms. Hormonal actions, especially neurosecretory ones, have helped us to envisage the route and the mechanisms of neurotrophic communications. The mechanisms conveyed by the "peptidergic" neurons in the hypothalamo-neurohypophysal system, the neural secretion of peptides, the transport of bound hormones along the related pathways and their release by exocytosis have served as a model for secretory neurons which act on remote target cells (Bargmann, 1966; Scharrer, 1965). Mechanisms conveyed by "peptidergic" neurons and a number of aminergic neurons are very similar indeed, and further implications defining the neuron generally as a secretory cell are gaining general confirmation (Smith, 1971). Thus it appears plausible to envisage the motor nerve cell also as a neurosecretory cell which mediates long term information to the muscle cell, pertinent for maintenance of structure and metabolism of the post-synaptic cell not directly connected with transmitter release and related impulse activity (see Gutmann, 1964). The concept ascribing a neurosecretory action to all neurons is an old one (see Smith, 1971) and is increasingly detailed, especially from a morphological point of view (see Andres, 1975).

There is no doubt that some trophic effects depend on activation of impulse activity and muscle contraction respectively. Since the biochemical nature of neurotrophic agents has not been established, an indirect approach and evidence are unavoidable. Such approaches also show a growing necessity to differentiate neurotrophic mechanisms from impulse-related ones. The experiments reported here do not give any direct evidence of neurotrophic mechanisms. They attempt to outline examples of the multiple regulation of muscle properties, the study of which should in time allow a more direct approach to the definition of neurotrophic agents.

II Supporting evidence for neurotrophic functions

Two types of experiments are reported: (a) effects of the distal nerve stump after nerve section dependent on the length of the peripheral stump; (b) "dissociation" of impulse activity and sensitivity to ACh. Figure 1 shows the effect of nerve section made at various distances from the muscle on some properties of muscle. It can be seen that degeneration of terminal axons, decrease of glyocgen synthesis (Gutmann et al., 1955) and increase of protein degradation (as evidenced by increased proteolytic activity) appear earlier in the muscle with a short nerve stump. Such a temporary dependence of early denervation changes on length of sectioned axon has also been observed with respect to other properties of muscle (see Gutmann, 1975) and suggests a progressive depletion of trophic substances occurring earlier in a muscle with a short nerve stump. Axonal transport of trophic factors thus apparently persists for some time before depletion in the peripheral segment of the sectioned nerve. Its neurotrophic maintenance function can thus be detected and its velocity calculated on the basis of such experiments (see Lubinska, 1975). Calculations indicate a velocity of 30 mm/day for decrease of synthesis of glycogen and increase of proteolytic activity. A similar dependence of onset of changes in contraction time (which shows progressive prolongation in the fast muscle after denervation) and of myosin ATP-ase activity on the length of the sectioned nerve stump was, however, not observed. At no time after nerve section could a difference in contraction properties dependent on the length of the nerve stump be observed. This indicated that contractile properties are not primarily regulated or affected by trophic substances migrating in the axon, but by nerve-impulse activity which is lost simultaneously

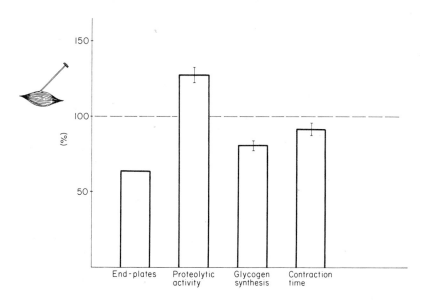

Fig. 1. Number of intact end plates (terminal axons), proteolytic activity, glycogen content (after injection of glucose) and contraction time, 24 h after section of the peroneal nerve near entry into the muscle. The values in the muscle with the "short" nerve stump are expressed as percentage of the corresponding values of the muscle of the contralateral side, where the nerve was cut high up in thigh. (Data from Gutmann *et al.*, 1955; Gutmann *et al.*, 1975.)

in the muscle with short and long nerve stump and triggers off the denervation—disturbance of the contractile proteins (Gutmann *et al.*, 1975).

Figure 2 shows the results of experiments in which the diaphragm muscle of the rat was crushed at one point distant from the NMJ, thus producing three segments, i.e. the intact one with the NMJ (1) the crushed one in which degeneration of the muscle fibres is induced (3) and the "decentralized" one, temporarily disconnected from the intact segment (2). Neuro-muscular transmission to segment 2 is abolished three days after the crush lesion and extrajunctional sensitivity to ACh is recorded. Fast recovery of neuro-muscular transmission is observed (7 and 10 days after the crush lesion). Extrajunctional sensitivity to ACh is, however, maintained (Vyskočil and Gutmann, 1976). This dissociation between impulse activity and ACh sensitivity is interesting with respect to the normal function of the NMJ maintained by respiration. It suggests that extrajunctional ACh sensitivity is controlled not

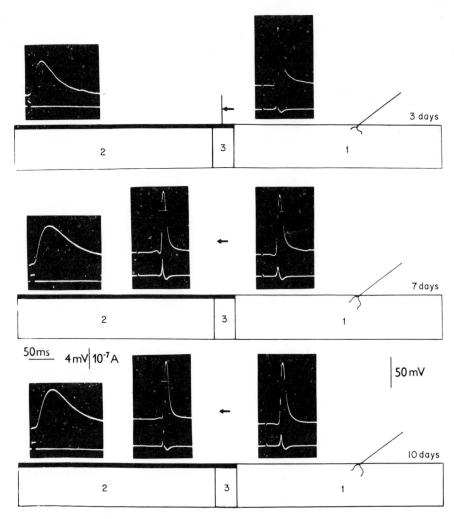

Fig. 2. Evoked action potentials after stimulation of the phrenic nerve and extra-junctional sensitivity to iontophoretic application of ACh in a strip of the diaphragm muscle which was crushed at a distinct point distant from nerve entry, thus producing three segments, i.e. 1 (with the intact nerve), 2 (intact nerve free segment) and 3 (crushed segment with regenerating muscle fibres). Recording 3, 7 and 10 days after operation. (From Vyskočil and Gutmann, 1976.)

only by impulse activity. In this connection other "dissociation" experiments may be cited. The application of colchicine or vinblastine to motor nerve axons in doses that block fast axonal transport induces denervation changes in the muscle without affecting nerve muscle transmission (Hofmann and Thesleff, 1972; Albuquerque *et al.*, 1972). However, long-lasting reversible block of impulse conduction did not result in axonal degeneration at end plates not in denervation-like changes in RNA and DNA metabolism (Gutmann and Žák, 1961) though it did result in denervation-like spread of the ACh-sensitive area (Lömo and Rosenthal, 1972). Botulinum toxin applied to a mammalian skeletal muscle in a dose sufficient to block neuro-muscular transmission alters chemical and electrical excitability of muscle fibres in the same way as denervation (Thesleff, 1960). However, muscle fibres, partially poisoned by botulinum toxin, could still contract in response to nerve stimulation but be supersensitive to ACh (Bray and Harris, 1975). Trophic substances, carried by axoplasmic transport, may thus exert effects on muscle independently of nerve-muscle transmission and activation of contraction. On the other hand control of extrajunctional sensitivity to ACh was reduced or abolished by stimulation of denervated muscle (Lömo and Rosenthal, 1972; Drachman and Witzke, 1972; Lömo and Westgaard, 1975), which mimics impulse activity. It is not yet possible to reconcile these different sets of observations, but possibly, release of trophic factors is coupled to the release of ACh (Thesleff, 1974). The most plausible explanation is, that both neurotrophic and impulse activity affect properties of the muscle.

III Indicators of neurotrophic functions

Adequate indicators of neurotrophic functions should relate clearly to (a) non-impulse activities of the neuron connected with axoplasmic transport mediating such functions and (b) a stimulation of synthesis of specific proteins in muscle. The latter might include a mechanism of transfer of neurotrophic agents from presynaptic terminals to post-synaptic structures, implying direct transfer of macromolecules, diffusion of substances or complex changes at the synapse including synthesis and release of synaptic macromolecules participating apparently in synthesis of neurotransmitter, storage, vesicle release, exocytosis and pinocystic activity of the muscle membrane (see Barondes, 1974; Bloom *et al.*, 1970). Such mechanisms, although not proven, are clearly indicated by tissue culture work (see Harris, 1974). They are, however, difficult to reproduce or to study in the organism. Seen under the restrictions mentioned above, none of the indicators used appear to be convincing. This is due to the lack of chemical definition of the neurotrophic agents and to the circumstance that many properties of muscle may be regulated by both neurotrophic and impulse activity. Considerable interest was directed to regulation of ACh sensitivity and distribution and synthesis of AChR respectively. On the basis of denervation experiments, it appeared that these properties are regulated by neurotrophic agents (Miledi, 1963), spontaneous release of ACh producing the miniature endplate potentials (e.p.p.) being a likely candidate. However, no direct effect of ACh on protein synthesis is known and many denervation changes such as the decrease of resting membrane potential (RMP) (Thesleff, 1963; Albuquerque et al., 1971) and the increase in absolute RNA and DNA content (Gutmann and Žák, 1961) occur prior to disappearance of m.e.p.p. An increase of absolute RNA content by $12 \pm 3.6\%$ is observed in the denervated extensor digitorum longus (EDL) muscle 6 h after denervation (unpublished results). Moreover, spread of ACh sensitivity (Fambrough, 1970) and other alterations of membrane excitability (Grampp et al., 1972) can be prevented by inhibition of RNA and protein synthesis. Very marked atrophy of muscle, e.g. in the androgen-sensitive levator ani muscle of the rat, is not accompanied by any decrease of spontaneous mediator release (Vyskočil and Gutmann, 1969), whereas its extreme decrease in muscles of senescent animals does not result in an increase of ACh sensitivity (Gutmann et al., 1971). A regulation of AChR by activation of muscle contraction has been demonstrated (Jones and Vrbová, 1970; Lömo and Rosenthal, 1972; Purves and Sackmann, 1974), however, it concerns only extrajunctional receptors. Junctional receptors are apparently not affected by muscle stimulation (Lömo and Westgaard, 1975), moreover other factors, such as release of agents from dividing cells, apparently participate (Blunt et al., 1975). Thus, the use of ACh sensitivity and synthesis of AChR as indicators of neurotrophic functions, though suggestive, remains ambiguous.

Global changes, such as atrophy and hypertrophy of muscle, can of course not serve as adequate indicators being affected by neural and non-neural influences. Basic properties of the myosin molecule (Samaha et al., 1970), protein capacity of RNA (Burešová et al., 1975) and the relation of glycolytic to oxidative enzymes (see Close, 1972) are apparently neurally regulated, as evidenced by the reversal of these properties in fast and slow muscles after heteroinnervation. Implantation of the fast peroneal nerve into the denervated or self-reinnervated soleus muscle, a procedure which produces hyperinnervation (Gutmann and Hanzlíková, 1967) results in a reversal of protein synthetic capacity of proteins (i.e. increase corresponding to levels found in fast muscle) and a differential change in histochemical fibre pattern (increase of glycolytic, decrease of oxidative enzymes) (Burešová et al., 1975). If this transformation is affected by neurotrophic influences it would support the contention that neurotrophic regulation may concern the regulation of gene expression, or of synthesis of specific proteins in the muscle cell (see Guth, 1971).

The regulatory influence of the neuron is revealed particularly by observations on contractile properties and histochemical muscle fibre pattern during development and after heteroinnervation achieved by cross-union of nerves or cross-transplantation of muscles (see Close, 1972; Gutmann, 1973). In slow and fast mammalian motor units, properties of the neuron are closely matched to properties of the muscle fibres. The motor unit is apparently homogeneous, as revealed by the same histochemical pattern of muscle fibres innervated by one axon (Edström and Kugelberg, 1968). The reversal of contractile properties and histochemical muscle fibre pattern suggested therefore the operation of neuronal mechanisms. However, the primary contention that contraction properties are regulated by a trophic factor released from the nerve (Buller et al., 1960) can no longer be maintained. Chronic stimulation of a fast denervated muscle results in shortening of contraction time (CT) in vivo (Brown, 1973; Melichna and Gutmann, 1974) and in vitro (Gutmann, 1969). The frequency pattern was apparently a decisive factor in the transformation of the muscle with respect to CT in stimulation experiments of both, denervated (Lömo et al., 1972) and innervated tissue (Salmons and Vrbová, 1969). Moreover, there was no early, temporary change of CT after denervation dependent on length of the sectioned nerve stump (see Fig. 1) which would suggest operation of axoplasmic flow and neurotrophic influences respectively.

A developmental study suggests the mechanisms by which neuronal impulse activity changes contractile properties of muscle. Slow and fast muscle of the guinea pig exhibit a differential postnatal behaviour— slow muscles show a prolongation, fast muscles a progressive shortening of CT. The slow soleus muscle is fast at birth and slow in the adult muscle, and this is apparently related to the increasing response of the "tonic" muscle to antigravity forces. Correspondingly, the muscle fibre pattern changes according to ATP-ase activity from a mixed (fibres with high and low enzyme activity) to an homogeneous one (fibres with low activity only) and this is accompanied by related changes in myosin-ATP-ase activity and the molecular light chain pattern of myosin (Gutmann et al., 1974). The number of muscle fibres does not change and the developmental transformation is apparently due to a decrease of ATP-ase activity in originally fast muscle fibres. Guinea pigs are born with a relatively mature motor system. Transplantation of a muscle results in successive degeneration and regeneration of the muscle cells and does in fact produce a developmental

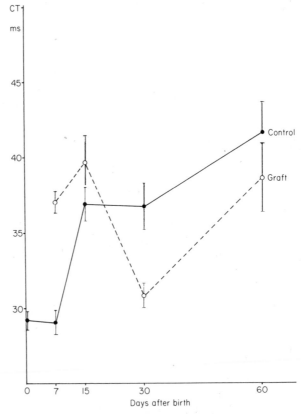

Fig. 3. Changes in isometric twitch contraction time (measured *in vitro* at 37°C) in the soleus muscle of the guinea pig at birth (full line) and in the soleus muscle graft 7, 30 and 60 days after orthotopic autotransplantation (interrupted line). (Data from Gutmann et al., 1976.)

recapitulation of properties in the grafted muscle (Carlson and Gutmann, 1972). The grafted soleus muscle of the guinea pig (see Fig. 3)
shows a temporary shortening of CT with later prolongation (Gutmann
et al., 1976) and a related homogeneous pattern of fibres with low ATP-
ase activity. It appears that a prenatal phase has been produced by the
transplantation, implying a temporary developmental shortening of
CT of the slow muscle which has also been observed postnatally in the
cat (Buller *et al.*, 1960) and rat (see Close, 1972) and in transplantation
of this muscle (Carlson and Gutmann, 1974). These animals are born
with a relatively immature motor system. The biphasic change in
muscle properties may be a general developmental feature and may
suggest myogenic mechanisms. Contractile properties and muscle fibre
pattern are apparently subject to multiple regulation. For instance,
increased motor activity in adult rats resulted in shortening of CT and
in a relative increase of muscle fibres with high ATP-ase activity.
Impulse activity is probably the decisive factor regulating contractile
properties. However, one can conclude that contractile properties and
the muscle fibre pattern do not afford adequate indicators of neurotrophic function.

The search for adequate indicators leads to a recognition of a
multiple regulation of many properties of muscle. The difficulties
involved may be discouraging but the differentiation of the components
of regulation is unavoidable. Only a few examples of this will be given.

IV Multiple regulation of muscle properties

A MYOGENIC INFLUENCES

The first stages of muscle development proceed independently of
neural influences (see Fishman, 1972). This is well documented by
transplantation experiments, in which a Marcaine-treated† EDL
muscle is grafted into the bed of the muscle of the contralateral leg,
left either denervated or innervated. This procedure results in complete
degeneration and regeneration of all the muscle fibres (Carlson and
Gutmann, 1976). In this type of experiment the changes in twitch CT
and maximal tetanus were followed. CT showed a shortening from
41.37 ± 6.43 to 26.50 ± 1.13 ms and tetanic tension output increased
from 0.47 ± 0.27 to 9.91 ± 0.61 gm. This gradient is very similar

---

† A local anaesthetic (Bupivacaine) with selective myotoxicity.

to that found in the graft on the intact leg before it becomes reinnervated. Thus, considerable synthesis of contractile proteins, together
with a temporary shortening of CT, occurs also in the denervated
graft. At later stages, CT in the innervated graft shortens to values
found in the control EDL muscle and tetanic tension output regains
values of about 75% of that of the control muscle. In the denervated
graft, however, CT remains slow and tetanic tension output shows a
progressive reduction. It remains to be seen, whether the biphasic
development is a general "myogenic" feature. This was indicated by
the biphasic development of histochemical muscle fibre types in fetal
muscles (Ashmore *et al.*, 1972) and in grafted slow muscles of the
guinea pig (see Fig. 3). Neural influences are certainly necessary for
the maintenance of differentation and postnatal growth (see Fishman,
1972). However, some myogenic characteristics are apparently
primarily coded. For examples, the electrophoretic pattern of light
chains of embryonic myosin was found to be the same as that of adult
fast, white muscle (Sreter *et al.*, 1972). The early coordinated development of ACh sensitivity and elaboration of contractile proteins suggests
such a myogenic control at the level of genetic information (Fambrough
et al., 1974). The evidence for primarily encoded "myogenic" properties, which might be reflected in the emergence of different muscle
fibre types is, however, so far not consistent. Myogenic properties are
shown in experiments in which a slow muscle is transplanted into the
site of the androgen-sensitive (fast) levator ani muscle of the rat.
According to the "fast" nerve supply of the pudendal nerve, now
reinnervating the slow muscle, the CT of the slow soleus muscle is
changed to a fast one. However, the grafted skeletal muscle does not
acquire hormone sensitivity to androgens, i.e. no change of CT or
weight is observed after castration or testosterone administration, as is
the case in the androgen-sensitive ("target") muscle (Hanzlíková and
Gutmann, 1974). Thus the neuronal impulse activity transforms the
muscle with respect to CT and muscle fibre pattern, but hormone
sensitivity is not changed, being primarily of myogenic origin.

B HORMONAL INFLUENCES

Many hormones have a regulatory action on some phases of development and maintenance of muscles, affecting synthesis and breakdown of
proteins. The effects are marked in target cells, having specific receptors.

Though such muscles are rare, they afford models for a study of differentiation of neuronal and hormonal mechanisms. We will restrict ourselves to a description of this interaction in the highly androgen-sensitive levator ani muscle (LA) of the rat. This muscle undergoes complete postnatal involution in females and was therefore thought not to exist in female animals (Čihák et al., 1970). The muscle persists, however, and shows normal contractile behaviour and ChE activity, if testosterone (T) is continuously applied from birth (Gutmann et al., 1976). It can be maintained temporarily following T treatment even after perinatal denervation (Burešová et al., 1972), the hormone having primarily a relative nerve independent effect during early stages of development. As shown previously by transplantation experiments of the soleus muscle into the bed of the LA muscle, the specific receptors to androgens are primary myogenic ones, as also shown by biochemical studies (Jung and Baulieu, 1972). In the adult animal interaction of hormonal and neural influences in this highly hormone-sensitive muscle is indicated by a lack of activation of protein (Burešová et al., 1972) and glycogen synthesis (Pellagrino et al., 1970) after denervation. The differentiation of hormonal and neural influences can be demonstrated in cross-union experiments of nerves to the LA muscle after castration and T administration, especially with respect to the histochemical muscle fibre pattern which is homogeneous in the LA muscle. The myotropic hormonal influence on the target muscle results in an increase of muscle fibre size, but not in reversal of muscle fibre pattern which is affected by "foreign" innervation (Hanzlíková and Gutmann, 1972). Figure 4 shows the homogenous histochemical fibre pattern of the normal LA muscle (1), the decrease of fibre size and reversal (heterogenity) of the muscle fibre pattern with castration performed two months after cross union of the tibial (which originally innervated a mixed fibre population) with the pudendal nerve (2) and the increase in muscle fibre size in the case in which T was applied after cross-union and castration (3). The interaction of hormonal and neuronal influences is, however, a complex one. After castration the LA muscle, ChAc and even more ChE, activities decrease considerably with concomitant changes of the end-plate structure (Gutmann et al., 1969). Even one year after castration,when the weight of the muscle is decreased to 15%, ChAC activity by 42% and ChE activity by 79% of the control values, treatment with T rapidly increased the enzyme activities near control values (Tuček et al., 1976). It is interesting that

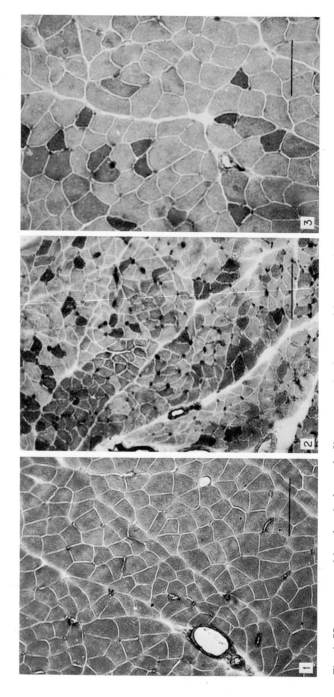

Fig. 4. Homogeneous histochemical muscle fibre pattern in the normal levator ani (LA) muscle with respect to ATPase activity (1), in the cross-reinnervated LA muscle from male rats, in which castration was performed two months after cross- uniting the tibial with the pudendal nerve (2) and in the LA muscle of a rat which received testosterone propionate daily for a period of one month after castration (3). The bar represents 100 μ. (From Hanzlíková and Gutmann, 1972.)

even two years after castration, when CT of the LA muscle showed a prolongation to 19.5 ± 1.05 ms, T application resulted in a fast return to normal values, i.e. 12.0 ± 2.07 ms (unpublished results). As T acts primarily on the LA muscle which contains specific T receptors, all the effects on presynaptic (ChAc) and postsynaptic (ChE) enzymes and on contractile properties are apparently affected by a retrograde trans-synaptic regulatory (trophic) influence on the motoneuron. There is a correlation between muscle fibre size and e.g. release of ACh quanta (Kuno *et al.*, 1971). A feedback system, including activation of specific receptors in the target LA muscle, increase of protein-synthesis of the muscle and an increase of the presynaptic enzyme ChAc, inducing in turn increased synthesis of ChE, is suggested.

C OTHER INFLUENCES

The neurotrophic regulation should be considered as a component in the framework of intercellular regulations (see Gutmann, 1975). This indeed complicates the search and analysis of neurotrophic influences. A few examples of humoral and peripheral regulations will be cited, which may affect nerve-muscle cell interactions. Developmental changes in blood flow and the resulting changes in substrate supply may be an important additional factor in differentiating metabolic muscle types (see Hudlická, 1973). The differentiation of muscle fibre types with immunofluorescent methods (Gröschel-Stewart *et al.*, 1973) indicates participation of the immune response. A changed immuno-chemical behaviour of the denervated muscle has also been described (Hollán *et al.*, 1965). Finally, the effects of peripheral factors may be cited. Compensatory hypertrophy of a muscle with concomittant prolongation of CT occurs also in denervated muscle, apparently due to the stretch induced in such experiments (Gutmann *et al.*, 1971). A denervated muscle shows a different change in muscle fibre pattern and CT according to whether it is immobilized in a flexed or a stretched position (Melichna and Gutmann, 1974).

V Conclusions

In spite of the extensive evidence for the existence of neurotrophic, long-term interactions between nerve and muscle cells, progress has been slow because often less than adequate indicators of neurotrophic functions have been used, because only a single neurotrophic agent was often assumed. The best evidence comes from tissue culture work (see Harris, 1974) and indicates that such neurotrophic effects are independent of nerve transmission and may be mediated by diffusable chemical substances. The evidence for neurotrophic long-term influences regulating the muscle cell in terms of synthesis and degradation of proteins has been strengthened by the increasing knowledge of mechanisms of axoplasmic transport (see Lubinská, 1975) by which apparently these effects are mediated. The experiments on different changes in denervated muscle dependent on the length of nerve stump, and the "dissociation" experiments blocking axoplasmic transport are also interpreted in view of these mechanisms. However, the evidence is still indirect and cannot substitute for direct approaches, which would have to define the neurotrophic agents biochemically. Research on nerve and muscle with respect to conduction, neuromuscular transmission and contraction of muscle fibres has of course been much more successful, the unit response in each case being a short-lasting event of a physical character. The complex interaction of neural (impulse and non-impulse) and non-neural mechanisms in long-term regulations will need new, more adequate models and application of methods such as those used or attempted in isolation of mediators and receptors. This report is not intended as a detailed analysis of neurotrophic mechanisms, which would have to include a discussion, e.g. of collateral regeneration, maintenance of synaptic contacts and questions of transfers of molecules at the synapse (see Harris, 1974; Gutmann, 1975). Its task was primarily to show that the multiple regulation of muscle properties makes it necessary to differentiate neurotrophic from other mechanisms operating in intercellular regulations. In the search for neurotrophic mechanisms, this has proved to be a complicating but unavoidable approach.

VI References

Andres, K. H. (1975). Morphological criteria for the differentiation of synapses in vertebrates. *Neural Transmission*, Suppl. **XII**, 1–37.

Albuquerque, E. X., Schuch, F. T. and Kauffman, F. C. (1971). Early membrane depolarization of the fast mammalian muscle after denervation. *Pflügers Arch.* **328**, 36–50.

Albuquerque, E. X., Warnich, J. E., Tasse, J. R. and Sansone, F. M. (1972). Effects of vinblastine and colchicine on neural regulation of fast and slow skeletal muscles of the rat. *Exp. Neurol.* **37**, 607–634.

Ashmore, C. R., Robinson, D. W., Ratrray, P. and Doerr, L. (1972). Biphasic development of muscle fibres in the fetal lamb. *Exp. Neurol.* **37**, 241–255.

Axelsson, J. and Thesleff, S. (1959). A study of supersensitivity in denervated mammalian skeletal muscle. *J. Physiol.* **149**, 178–193.

Bargman, W. (1966). Neurosecretion. *Int. Rev. Cytol.* **19**, 183–201.

Barondes, S. H. (1974). Synaptic macromolecules: identification and metabolism. *Ann. Rev. Biochem.* **43**, 147–168.

Bloom, F. E., Iversen, L. L. and Schmitt, F. (1970). Macromolecules in synaptic function. *Neurosc. Res. Progr. Bull.* **8**, 329–449.

Blunt, R. J., Jones, R. and Vrbová, G. (1975). The use of local anaesthetics to produce prolonged motor nerve blocks in the study of denervation hypersensitivity. *Pflügers Archiv.* **355**, 189–204.

Bray, J. J. and Harris, A. J. (1975). Dissociation between nerve-muscle transmission and nerve trophic effects on rat diaphragm using type D Botulinum toxin. *J. Physiol.* **253**, 53–77.

Brown, M. D. (1973). Role of activity in the differentiation of slow and fast muscles. *Nature Lond.* **244**, 178–179.

Buller, A. J., Eccles, J. C. and Eccles, R. M. (1960). Interactions between motoneurons and muscles in respect to the characteristic speeds of their responses. *J. Physiol.* **150**, 417–439.

Burešová, M., Gutmann, E. and Hanzlíková, V. (1972). Differential effects of castration and denervation on protein synthesis in the levator ani muscle of the rat. *J. Endocrin.* **54**, 3–14.

Burešová, M., Hanzlíková, V. and Gutmann, E. (1975). Changes of ribosomal capacity for protein synthesis, contraction and histochemical properties of muscle after implantation of fast nerve into denervated and into self reinnervated slow soleus muscle of the rat. *Pflügers Arch.* **360**, 95–108.

Carlson, B. M. and Gutmann, E. (1972). Development of contractile properties of minced muscle regenerated in the rat. *Exp. Neurol.* **36**, 239–249.

Carlson, B. M. and Gutmann, E. (1974). Transplantation and "cross-transplantation" of free muscle grafts in the rat. *Experientia* **30**, 1292–1294.

Carlson, B. M. and Gutmann E. (1976). The free grafting of the extensor digitorium longus muscle of the rat after Morcaine pre-treatment. *Exp. Neurol.* (in press).

Crain, S. M. and Peterson, E. R. (1974). Development of neural connections in culture. *Ann. N. Y. Acad. Sci.* **228**, 6–34.

Čihák, R., Gutmann, E. and Hanzlíková, V. (1970). Involution and hormone-induced persistence of the M. sphincter (levator) ani in female rats. *J. Anat.* **106**, 93–110.

Close, R. I. (1972). Dynamic properties of mammalian skeletal muscles. *Physiol. Rev.* **52**, 129–197.

Drachman, D. B. (1967). Is acetylcholine the trophic neuromuscular transmitter? *Arch. Neurol.* **17**, 206–218.

Drachman, D. B. and Witzke, F. (1972). Trophic regulation of acetylcholine sensitivity of muscle: Effect of electrical stimulation. *Science* **176**, 514–516.

Edström, L. and Kugelberg, E. (1968). Histochemical composition, distribution of fibers and fatiguability of single motor units. *J. Neurol. Neurosurg. Psychiat.* **31**, 424–433.

Fambrough, D. M. (1970). Acetylcholine sensitivity of muscle fiber membrane: mechanisms of regulation by motoneurons. *Science* **168**, 372–378.

Fambrough, D. M., Hartzell, H. C., Powell, J. A., Rash, J. E. and Joseph, N. (1974). On the differentiation and organization of the surface membrane of a postsynaptic cell—the skeletal muscle fibre. *In* "Synaptic Transmission and Neuronal Interaction" (M. V. L. Bennet, ed.), pp. 285–313. Raven Press, New York.

Fischman, D. A. (1972). Development of striated muscle. *In* "The Structure and Function of Muscle" (G. H. Bourne, ed.), I/1, pp. 75–148. Academic Press, London and New York.

Ginetzinski, A. G. and Shamarina, N. M. (1942). The tonomotor phenomenon in denervated muscles. *Usp. Sovrem. Biol.* (In Russian), **15**, 283–294.

Grampp, W., Harris, J. B. and Thesleff, S. (1972). Inhibition of denervation changes on skeletal muscle by blockers of protein synthesis. *J. Physiol.* **221**, 743–754.

Gröschel-Stewart, U., Meschede K. and Lahr, I. (1973). Histochemical and immunochemical studies on mammalian striated muscle fibres. *Histochemie* **33**, 79–85.

Guth, L. (1968). "Trophic" influences of nerve on muscle. *Physiol. Rev.* **48**, 645–687.

Guth, L. (1969). "Trophic" effects of vertebrate neurons. *Neurosc. Res. Progr. Bull.* **7**, 1–73.

Guth, L. (1971). A review of the evidence for the neural regulation of gene expression in muscle. *In* "Contractility of Muscle Cells and Related Processes" (R. J. Podolsky, ed.), pp. 189–201.

Gutmann, E., Vodička, Z. and Zelená, J. (1955). Changes in cross striated muscle after nerve interruption depending on the length of the nerve stump. *Physiol. Bohemoslov.* (In Russian), **4**, 200–204.

Gutmann, E. and Žák, R. (1961). Nervous regulation of nucleic acid level in crossstriated muscle. Changes in denervated muscle. *Physiol. Bohemoslov.* **10**, 493–500.

Gutmann, E. and Hník, P. (1962). Denervation studies in research of neurotrophic relationships. *In* "The Denervated Muscle" (E. Gutmann, ed.), pp. 13–56. Czechosl. Acad. Sci., Prague.

Gutmann, E. (1964). Neurotrophic relations in the regeneration process. *Progr. Brain Res.* **13**, 72–112.

Gutmann, E. (1969). The trophic function of the nerve cell. *Scientia* **104**, 1–20.

Gutmann, E. and Hanzlíková, V. (1967). Effects of accessory nerve supply to muscle achieved by implantation into muscle during regeneration of its nerve. *Physiol. Bohemoslov.* **16**, 244–250.

Gutmann, E., Tuček, S. and Hanzlíková, V. (1969). Changes in cholinacetyltransferase and cholinesterase activities in the levator ani muscle of rats following castration. *Physiol. Bohemoslov.* **18**, 195–203.

Gutmann, E., Hanzlíková, V. and Vyskočil, F. (1971). Age changes in cross striated muscle of the rat. *J. Physiol.* **219**, 331–343.

Gutmann, E., Schiaffino, S. and Hanzlíková, V. (1971). Mechanism of compensatory hypertrophy in skeletal muscle of the rat. *Exp. Neurol.* **31**, 451–464.

Gutmann, E. (1973). The multiple regulation of muscle fibre pattern in cross striated muscle. *Nova Acta Leopoldina* **38**, 193–218.

Gutmann, E., Melichna, J. and Syrový, I. (1974). Developmental changes in contraction time, myosin properties and fibre pattern of fast and slow skeletal muscles. *Physiol. Bohemoslov.* **23**, 19–27.

Gutmann, E., Melichna, J., Herbrychová, A. and Štichová, J. (1976). Different changes in contractile and histochemical properties of reinnervated slow soleus muscles of the guinea pig. *Pflügers Arch.* **364**, 191–194.

Gutmann, E., Hanzlíková, V. and Niemerko, S. (1975). Persistence of the levator ani muscle of female rats after testosterone—administration (in preparation).

Gutmann, E. (1975). Neurotrophic relations. *Ann. Rev. Physiol.* **38**, 177–216.

Gutmann, E., Melichna, J. and Syrový, I. (1976). Independence of changes in contraction properties and myofibrillar ATP-ase activity of denervated mammalian muscle on length of nerve stump. *Physiol. Bohemoslov.* **25**, 43–50.

Hanzlíková, V. and Gutmann, E. (1972). Effects of foreign innervation on the androgen-sensitive levator ani muscle of the rat. *Z. Zellforsch.* **135**, 165–174.

Hanzlíková, V. and Gutmann, E. (1974). The absence of androgen-sensitivity in the grafted soleus muscle innervated by the pudental nerve. *Cell. Tiss. Res.* **145**, 121–129.

Harris, A. J. (1974). Inductive functions of the nervous system. *Ann. Rev. Physiol.* **36**, 251–305.

Harris, A. J., Heinemann, S., Schubert, D. and Tarakis, H. (1971). Trophic interaction between cloned tissue culture lines of nerve and muscle. *Nature Lond.* **231**, 296–301.

Hník, P., Jirmanová, I., Vyklický and Zelená, J. (1967). Fast and slow muscles of the chick after nerve cross-union. *J. Physiol. Lond.* **193**, 309–325.

Hofmann, W. W. and Thesleff, S. (1972). Studies on the trophic influence of nerve in skeletal muscle. *Eur. J. Pharmacol.* **20**, 256–260.

Hollán, S. R., Novák, E., Koszegy, S. and Stark, E. (1965). Immunochemical study of denervated muscle proteins. *Life Sciences* **4**, 1779–1783.

Hudlická, O. (1973). "Muscle Blood Flow," p. 219. Swets and Zeitlinger, Amsterdam.

Jones, R. and Vrbová, G. (1970). Effect of muscle activity on denervation hypersensitivity. *J. Physiol.* **210**, 144–145.

Jung, I. and Balieu, E. E. (1972). Testosterone cytosol "receptors" in the rat levator ani muscle. *Nature New Biol.* **237**, 24–26.

Katz, B. (1966). "Nerve, Muscle and Synapse," p. 193. McGraw-Hill, New York.

Katz, B. and Miledi, R. (1972). The statistical nature of the acetylcholine potential and its molecular components. *J. Physiol.* **224**, 665–699.

Koenig, J. (1970). Contribution à l'étude de la morphologie des plaques matrices des grands dorsaux antérieur et postérieur du poulet après innervation croisée. *Arch. Anat. Micr. Morph. Exp.* **59**, 403–426.

Kuno, M., Turkanis, S. A. and Weakly, J. N. (1971). Correlation between nerve terminal size and transmitter release at the neuromuscular junction of the frog. *J. Physiol.* **213**, 546–556.

Lentz, C. T. L. (1974). Neurotrophic regulation at the neuromuscular junction. *Ann. N. Y. Acad. Sci.* **228**, 323–337.

choline acetyltransferase and cholinesterase activities in rat levator ani muscle. in the rat. *J. Physiol.* **221**, 493–513.

Lömo, T., Westgaard, R. H. and Dahl, H. A. (1974). Contractile properties of muscle: control by pattern of muscle activity in the rat. *Proc. R. Soc. Biol.* **187**, 99–103.

Lömo, T. and Westgaard, R. H. (1975). Further studies on the control of ACh sensitivity by muscle activity in the rat. *J. Physiol.* **252**, 603–626.

Lubinska, L. (1975). On axoplasmic flow. *Int. Rev. Neurobiol.* **17**, 241–296.

Luco, J. V. and Eyzaguirre, C. (1955). Fibrillation and hypersensitivity to ACh in denervated muscle. Effect of length of degenerating nerve fibres. *J. Neurophysiol.* **18**, 65–73.

Melichna, J. and Gutmann, E. (1974). Stimulation and immobilization effects on contractile and histochemical properties of denervated muscle. *Pflügers Arch.* **352**, 165–178.

Miledi, R. (1963). An influence of nerve not mediated by impulses. *In* "The Effect of Use and Disuse on Neuromuscular Functions" (E. Gutmann and P. Hník, eds.), pp. 35–40. Czechosl. Acad. Sci., Prague.

Musick, J. and Hubbard, J. I. (1972). Release of protein from mouse motor nerve terminals. *Nature Lond.* **237**, 279–281.

Olmsted, J. M. D. (1920). The nerve as a formative influence in the development of taste buds. *J. Comp. Neurol.* **31**, 465–468.

Pellegrino, C., Bergamini, E. and Pagni, R. (1970). Interference of denervation on testosterone effects on glycogen synthesis in levator ani muscle. *Exc. Med. Int. Congr. Ser.* **210**, 163.

Purves, D. and Sakmann, B. (1974). Membrane properties underlying spontaneous activity of denervated muscle fibres. *J. Physiol.* **239**, 125–153.

Salmons, S. and Vrbová, G. (1969). The influence of activity on some contractile characteristics of mammalian fast and slow muscle. *J. Physiol.* **201**, 535–549.

Samaha, F. J., Guth, L. and Albers, R. W. (1970). The neural regulation of gene expression in the muscle cell. *Exptl. Neurol.* **27**, 276–282.

Scharrer, E. (1965). The final common path in neuroendocrine integration. *Arch. Anat. Micr.* **54**, 359–370.

Singer, M. (1952). The influence of the nerve in regeneration of the amphibian extremity. *Quart. Rev. Biol.* **27**, 169–200.

Smith, A. D. (1971). Summing up: some implications of the neuron as a secreting cell. *Phil. Trans. R. Soc. Lond. B* **261**, 423–437.

Sreter, F. A., Holtzer, S., Gergely, J. and Holtzer, H. (1972). Some properties of embryonic myosin. *J. Cell. Biol.* **55**, 586–594.

Thesleff, S. (1960). Supersensitivity of skeletal muscle produced by botulinum toxin. *J. Physiol.* **151**, 598–607.

Thesleff, S. (1963). Spontaneous electrical activity in denervated rat skeletal muscle. *In* "The Effect of Use and Disuse of Neuromuscular Functions" (E. Gutmann and P. Hník, eds), pp. 41–51. Czechosl. Acad. Sci., Prague.

Thesleff, S. (1974). Physiological effects of denervation of muscle. *Ann. N. Y. Acad. Sci.* **228**, 89–103.

Tuček, S., Koštířová, D. and Gutmann, E. (1975). Testosterone-induced changes of

Lömo, T. and Rosenthal, J. (1972). Control of ACh sensitivity by muscle activity *Neurol. Sci. J.* **27**, 353–362.

Vyskočil, F. and Gutmann, E. (1969). Spontaneous transmitter release from motor nerve endings in muscle fibres of castrated and old animals. *Experientia* **25**, 945–946.

Vyskočil, F. and Gutmann, E. (1976). Control of ACh sensitivity in temporarily unconnected ("decentralized") segments of diaphragm-muscle fibres of the rat (in press).

Zelená, J. (1964). Development, degeneration and regeneration of receptor organs. *Progr. Brain. Res.* **13**, 175–213.

Subject Index